T0296414

Cellular Materials in Nature and Medicine

Bringing to life the fascinating structures and unique mechanics of natural and biomedical cellular materials, this book is an expert guide to the subject for graduates and researchers. Arranged in three parts, the book begins with a review of the mechanical properties of nature's building blocks (structural proteins, polysaccharides, and minerals) and the mechanics of cellular materials. Part II then describes a wide range of cellular materials in nature: honeycomb-like materials such as wood and cork and foam-like materials including trabecular bone, plant parenchyma, adipose tissue, coral, and sponge. Natural cellular materials often combine with fully dense ones to give mechanically efficient structures. The efficiencies of natural sandwich structures (iris leaves, bird skulls), circular sections with radial density gradients (palm, bamboo) and cylindrical shells with compliant cores (animal quills and plant stems) are all discussed. Images convey the structural similarities of very different materials, whilst color property charts provide the reader with mechanical data. Part III discusses biomedical applications of cellular materials: metal foams for orthopedic applications and porous tissue engineering scaffolds for regenerating tissue. It includes the effect of scaffold properties on cell behavior (e.g. attachment, morphology, migration, contraction).

Lorna J. Gibson is the Matoula S. Salapatas Professor of Materials Science and Engineering at the Massachusetts Institute of Technology (MIT), where she has been a faculty member since 1984. Her research interests focus on the mechanics of materials with a cellular structure such as honeycombs and foams and she is co-author, with Michael F. Ashby, of *Cellular Solids: Structure and Properties* (2nd edition, Cambridge University Press, 1997).

Michael F. Ashby is Emeritus Professor in the Department of Engineering at the University of Cambridge, where he has been a faculty member since 1973. He is a member of the Royal Society, the Royal Academy of Engineering, and the US National Academy of Engineering. He has authored a number of books on materials, design, and the environment and he has a life-long interest in natural and cellular materials.

Brendan A. Harley is an Assistant Professor in the Department of Chemical and Biomolecular Engineering at the University of Illinois at Urbana-Champaign, as well as a core faculty member of the Institute for Genomic Biology (Regenerative Biology and Tissue Engineering Theme). His research interests focus on fabricating homogeneous and spatially patterned cellular biomaterials for tissue engineering applications.

Cellular Materials in Nature and Medicine

LORNA J. GIBSON
Massachusetts Institute of Technology (MIT)

MICHAEL F. ASHBY
University of Cambridge

BRENDAN A. HARLEY
University of Illinois at Urbana-Champaign

CAMBRIDGE
UNIVERSITY PRESS

University Printing House, Cambridge CB2 8BS, United Kingdom

One Liberty Plaza, 20th Floor, New York, NY 10006, USA

477 Williamstown Road, Port Melbourne, VIC 3207, Australia

314-321, 3rd Floor, Plot 3, Splendor Forum, Jasola District Centre, New Delhi - 110025, India

103 Penang Road, #05-06/07, Visioncrest Commercial, Singapore 238467

Cambridge University Press is part of the University of Cambridge.

It furthers the University's mission by disseminating knowledge in the pursuit of education, learning and research at the highest international levels of excellence.

www.cambridge.org
Information on this title: www.cambridge.org/9780521195447

First published 2010
3rd printing 2020

A catalogue record for this publication is available from the British Library

Library of Congress Cataloging in Publication data
Gibson, Lorna J.
Cellular materials in nature and medicine / Lorna J. Gibson, Michael F. Ashby, Brendan A. Harley.
p. ; cm.
Includes bibliographical references and index.
ISBN 978-0-521-19544-7 (hardback)
1. Cytology. 2. Biomedical materials. 3. Foamed materials.
I. Ashby, M. F. II. Harley, Brendan A. III. Title.
[DNLM: 1. Cellular Structures. 2. Biocompatible Materials. QU 350]
QH581.2.G53 2010
660.6–dc22 2010027380

ISBN 978-0-521-19544-7 Hardback

Additional resources for this publication at www.cambridge.org/9780521195447

Contents

The color plates appear between pages 246 and 247

Preface

Cellular materials are widespread in nature. Examples are found in wood and palm, plant leaves and stems, trabecular bone and coral, pith, cork and sponges. They are also increasingly used as biomaterials in medical applications: metal foams are now used as coatings for orthopedic implants and foam-like polymer, ceramic, glass and composite scaffolds are used for regenerating tissues in the body. Their mechanical behavior can be described using suitably adapted models for engineering honeycombs and foams. We were led to write this book in an attempt to bring together, in one text, a broad survey of the structure and mechanics of natural and biomedical cellular materials.

This book has grown out of a number of collaborations with insightful colleagues over many years, for which we are grateful. We would like to particularly acknowledge our collaborations with Professor Ken Easterling, formerly of the University of Lulea, Sweden (cork, wood, iris leaves); Professors Tom McMahon and Toby Hayes, formerly of Harvard University; Professor Tony Keaveny of the University of California at Berkeley; Professor Ed Guo of Columbia University; Professor Matt Silva of Washington University; Dr. Steve Bowman, Dr. Deb Cheng, Dr. Andy Kraynik, Dr. Tara Moore and Ms. Surekha Vajjhala (trabecular bone); Ms. Phoebe Cheng, Dr. Gebran Karam and Dr. Matt Dawson (plant stems and animal quills); Professor Ioannis Yannas and Dr. Myron Spector of MIT; Professors Bill Bonfield, Ruth Cameron and Serena Best of Cambridge University, Professor Fergal O'Brien of the Royal College of Surgeons of Ireland; Dr. Toby Freyman, Dr. Biraja Kanungo, Dr. Andrew Lynn and Ms. Shona Pek (scaffolds for tissue engineering and cell interactions with scaffolds); Dr. Michelle Oyen of Cambridge University (biomaterials and their molecular structure); and Professor Ulrike Wegst of Drexel University (property charts for natural materials).

We have also greatly benefited from the advice, comments and critical readings of parts of the text by other colleagues, including Professor Jacques Dumais and Alex Cobb of Harvard University, Professor David Dunand of Northwestern University, Professor George Engelmayr Jr. of Pennsylvania State University, Dr. Lisa Freed of MIT and Dr. Louis-Philippe Lefebvre of the National Research Council of Canada. Professor Sharon Swartz of Brown University and Dr. Regina Campbell-Malone provided data on the length of trabecular bone across species. Numerous colleagues generously provided original images for the figures. Professor Ralph Müller of ETH Zurich kindly provided the cover micro-computed tomography image of trabecular bone.

We would also like to thank Dr. Michelle Carey for first suggesting that we write a book on this topic and Ms. Sarah Matthews for help with editorial questions as we worked on it; both are with Cambridge University Press.

The generous sabbatical funding for Lorna Gibson provided by Provost Rafael Reif of MIT is greatly appreciated. The figures for Chapters 1, 3–6 and 8 were expertly prepared for publication by Ms. Beth Beighlie, with the assistance of Ms. Diane Rose of MIT. Assistance with scanning electron microscopy was provided by Mr. Don Galler of MIT. Assistance with data assembly was provided by Mr. Justin Breucop of MIT and Mr. Bhushan Mahadik, Ms. Ji Sun (Sunny) Choi, Mr. Manuel Ramirez, Mr. Tyler Leonard and Mr. Daniel Weisgerber of the University of Illinois at Urbana Champaign.

Lorna Gibson and Brendan Harley are deeply grateful for the support and patience of their wives, Jeannie Hess and Kathryn Clancy, throughout the writing of this book. Brendan Harley also wishes to express his gratitude for the joy that his daughter, Joan Clancy-Harley, brings.

Part I

Background

1 Introduction

1.1 Introduction

Scientists and engineers have long been fascinated by cellular materials in nature. Some, such as wood and cork, have prismatic, honeycomb-like cells, while others, such as trabecular bone and marine sponges have polyhedral cells. When Robert Hooke perfected his microscope, around 1660, one of the first materials he examined was cork (Fig. 1.1a). What he saw led him to identify the basic unit of plant structure, which he termed "the cell": the word "cell" derives from the Latin *cella*, a small compartment, an enclosed space. Hooke's book Micrographia (1665) modestly records his thoughts:

> I no sooner descern'd these (which were indeed the first microscopical pores I ever saw, and perhaps, that were ever seen, for I had not met with any Writer or Person that had made any mention of them before this) but me thought I had with the discovery of them, presently hinted to me the true and intelligible reason for all the Phenomena of Cork.

Hooke's drawing shows the box-like shape of the cells in one section and their roughly hexagonal shape in the other, with their thin walls "*as those thin films of wax in a Honey-comb.*" The early microscopist Antonie van Leeuwenhoek (1705) also examined cork and drew similar sections (Fig. 1.1b). In his classic book, *On Growth and Form*, D'Arcy Thompson (1961) discussed the relationship between the structure of trabecular bone in a human femur and in a bird wing and the internal forces acting within them (Fig. 1.2).

People have used natural cellular materials for millennia. The oldest known wooden boat is Cheops' 4600-year-old barge, found dismantled in a pit next to the Great Pyramid in Egypt. Wooden furniture and coffins have also been found inside the pyramids (Bramwell, 1982). In the 1600s, the British supply of tall, straight trees for ship masts for the Royal Navy was so limited that Eastern white pines were imported from North America for this purpose, becoming a strategic military resource. Cork has been used as bungs in bottles and for the soles of shoes since Roman times.

Today, the structure of natural cellular materials is mimicked in engineering honeycombs and foams, used in everything from lightweight structural panels to energy-absorbing padding to thermal insulation. Recently, scientists and engineers have developed cellular materials for medical applications. Metal foams are now being produced as coatings for orthopedic implants, allowing bone ingrowth into the implant, and as potential bone substitute materials. Tissue engineering scaffolds, with their highly porous, foam-like structure, are designed to provide an environment for cells to

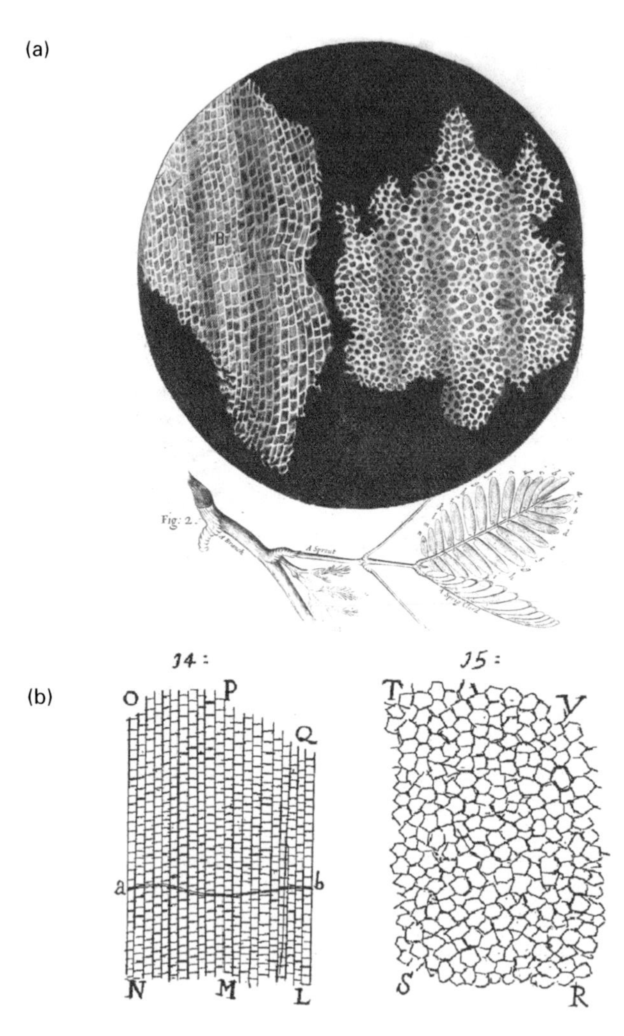

Fig. 1.1 (a) Hooke's (1665) drawing of cork, showing the box-like cells in one section and the roughly hexagonal cells in the other. (b) Leeuwenhoek's (1705) drawing of cork, showing the same two sections.

regenerate damaged or diseased tissue in the body. Collagen scaffolds, for instance, are currently being used to regenerate skin in burn patients.

In this book, we bring together the understanding of the structure and mechanical behavior of biological cellular materials in nature and in medicine. In the next two sections, we show some examples of the structure of natural and biomedical cellular materials and describe their function, mechanical performance and use in engineering applications.

1.2 Cellular materials in nature

1.2.1 Microstructure

The microstructures of several natural cellular materials are shown in Fig. 1.3. Wood and cork are *honeycomb-like cellular materials*, with prismatic cells like the hexagonal

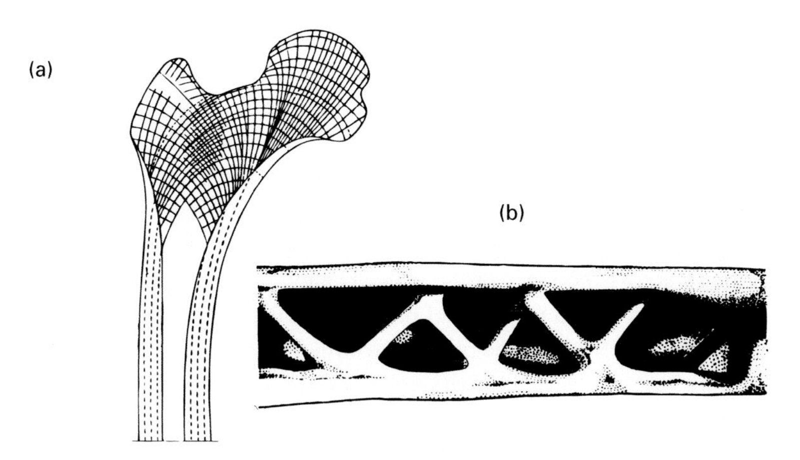

Fig. 1.2 (a) Drawing of the trabecular structure in the human femur. (b) Drawing of the trabecular structure in a bird wing. (Reproduced from D'Arcy Thompson, 1961.)

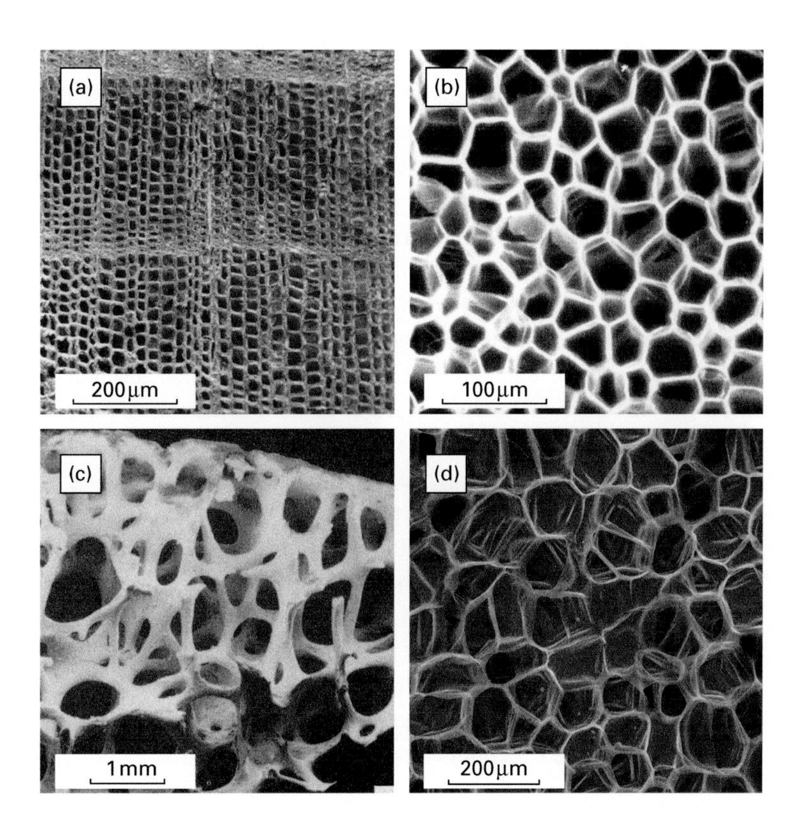

Fig. 1.3 Scanning electron micrographs showing cross-sections of (a) cedar, (b) cork, (c) trabecular bone and (d) carrot parenchyma. (a, Reprinted from Gibson and Ashby, 1997, with permission; b, reprinted from Gibson *et al.*, 1981; c, reprinted from Gibson, 1985, with permission from Elsevier.)

cells in a bee's honeycomb. Trabecular bone and plant parenchyma are *foam-like cellular materials*, with polyhedral cells. The volume fraction of solid is equivalent to the density of the cellular material normalized by that of the solid cell wall; this ratio is the *relative density*. Low-density trabecular bone is *open-celled*, with openings at the faces of the cells. As the density of the trabecular bone increases, the openings become smaller, but never fully close. In vivo, the voids in trabecular bone are filled with bone marrow, containing mesenchymal stem cells. Plant parenchyma are *closed-celled*, with membranes across the faces of the cells. The cytoplasm within parenchyma is pressurized; the mechanical response of parenchyma depends both on the response of the solid cell wall material and the pressure within the cell.

Natural structures can have both solid (or nearly so) and cellular components. The iris leaf and human skull are both sandwich structures, with two outer faces of nearly fully dense material separated by a lightweight core of a foam-like material (Fig. 1.4a,b). In the iris leaf, nearly solid longitudinal fibers on the outer surface of the leaf are separated by a core of low-density parenchyma cells. The separation of the faces by the core increases the moment of inertia (or second moment of area) of the sandwich, with little increase in weight, giving excellent resistance to bending and buckling. In the skull, the faces are nearly fully dense cortical bone, while the core is trabecular bone. Animal quills and plant stems are cylindrical shells with a foam-like core, as illustrated by the porcupine quill and milkweed stem in Fig. 1.4c,d. For a given mass, the moment of inertia (and bending resistance) for a circular tube increases with the ratio of the tube radius to the thickness, r/t; but at high r/t, the tube fails by kinking like a bent drinking straw. The foam-like core in quills and plant stems increases their resistance to kinking, or local buckling. Other natural structures have a gradient in density, rather than two distinct solid and cellular components. Bamboo epitomizes this (Fig. 1.4e): the volume fraction of dense fibers increases radially towards the periphery of the stem. Bamboo is also tubular, again increasing the bending stiffness of its cross-section.

Wood, trabecular bone and bamboo are all *anisotropic*: their mechanical properties depend on the direction of loading. Natural cellular materials exploit anisotropy to increase their mechanical efficiency, placing material where it is most needed to resist the applied loads. In trees, for instance, the highest stresses, resulting from bending in the wind, act along the length of the trunk and branches. Wood is much stiffer and stronger in this direction, along the grain, than across it, as a result of its honeycomb-like cellular microstructure as well as the composite nature of the solid cell wall material. Bone grows in response to applied loads: the trabeculae in human vertebrae, for instance, which are subjected primarily to compressive loading from the weight of the body, align in the vertical direction, increasing the stiffness and strength in that direction. Throughout this book we shall see the ways in which natural cellular materials exploit anisotropy to give exceptional mechanical performance.

1.2.2 Function and performance

Natural cellular materials perform a number of functions. Some are mechanical – a plant stem must support itself and, possibly, a flower or fruit in the wind without bending too much or breaking, for instance. Natural cellular materials are often mechanically

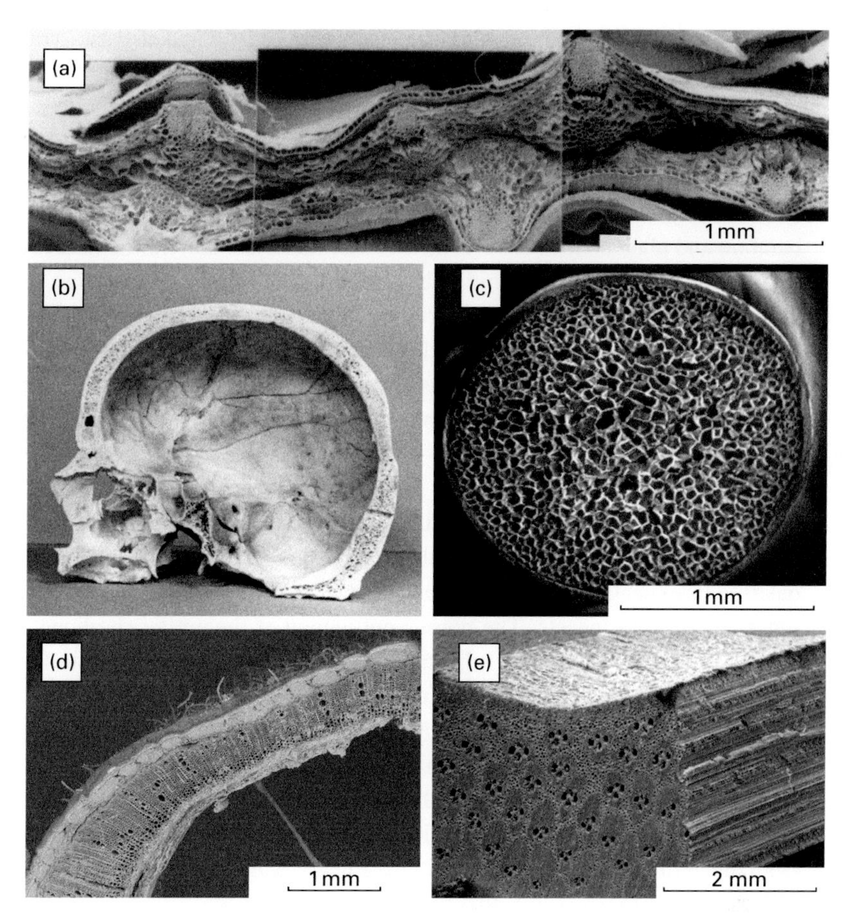

Fig. 1.4 Scanning electron micrographs showing cross-sections of (a) iris leaf, (b) skull, (c) porcupine quill, (d) milkweed stem and (e) bamboo. (a, Reprinted from Gibson *et al.*, 1988, Fig. 2b, with kind permission of Springer Science and Business Media; b, reprinted from Hodgson, 1973, Fig. 1b, with kind permission of Springer Science and Business Media; c, reprinted from Karam and Gibson, 1995, Fig. 1c, with permission from Elsevier; e, reprinted from Gibson *et al.*, 1995.)

efficient: wood has a stiffness to weight ratio equal to that of steel, for instance. The properties at four levels of microstructure, at different length scales, are of interest here. First, at the smallest scale, there are the basic "building blocks" – polysaccharides, proteins and biominerals – that make up the cell walls; these building blocks and their mechanical properties are reviewed in Chapter 2. At the next larger scale, the cell wall itself is typically a composite. In the wood cell wall, for instance, cellulose fibrils reinforce a matrix of lignin and hemicellulose; the mechanical properties of wood cell walls are well described by composites theory. At the third level, the cell walls are assembled into honeycomb-like or foam-like cellular materials (Fig. 1.3): their properties depend principally on those of the cell wall and on the fraction of space that they occupy, as described in Chapters 4 and 5. The last level is that of natural structures combining nearly fully dense and cellular components (sandwiches, filled tubes, gradient structures – see Fig. 1.4): the mechanical advantage of these arrangements is the topic of Chapter 6. In Chapter 7, the properties at these different levels are assembled into material property charts as a way of presenting a summary.

Engineers are interested in exploiting the microstructural features of natural cellular materials that make them so efficient. Biomimicking of natural cellular materials is described at the end of the sections on wood (Chapter 4), monocotyledon leaves, palm, bamboo and plant stems (Chapter 6). Cellular materials that mimic tissues in the body, such as metal foams that mimic trabecular bone and tissue engineering scaffolds that mimic the body's extracellular matrix, are the topic of Chapter 8.

Fluid flow through natural cellular materials can also be important: for instance, water has to flow through plant tissues. The cellular microstructure must also allow growth of tissues in both plants and animals. Plants store the energy of the sun in starches through photosynthesis: to do this, leaves must present a substantial surface area to the sun.

In specialized cases, natural cellular materials perform other functions. The small closed cells of cork make it an excellent thermal insulator, possibly protecting the cork oak tree from heat. The suberin in the cell walls also tastes unpleasant, perhaps providing a deterrent to insects. The porous cuttlebone of the cuttlefish allows the mollusk control of buoyancy through adjustment of the ratio of air to water in the bone. The cuttlebone must also resist, without collapse, water pressure up to the 200-meter depth to which the cuttlefish dives. The *Euplectella* sponge, also known as Venus' flower basket, lives symbiotically with a mating pair of shrimp that live within it: the shrimp clean the sponge and the lattice of the sponge protects them while allowing their food supply to enter the sponge. The spicules of the sponge, which attach it to the ocean floor, have fiber optic properties that are able to focus light from neighboring bioluminescent organisms, possibly attracting organisms that the shrimp feed on.

1.2.3 Applications for the mechanics of natural cellular materials

Wood is still one of the most widely used structural materials: 300 million tons of wood are used structurally each year. In addition to conventional sawn lumber, it is also processed into engineered wood products such as glue-laminated wood members that can be up to 50 meters in length and over a meter in depth (Fig. 1.5a). Sections of wood are bonded together in either straight or curved forms, to allow the span of longer distances than is possible with sawn lumber. Plywood and oriented strand board are examples of sheet products fabricated from either wood sheets or flakes.

Cork is still commonly used for bungs for bottles, as a result of its elasticity and inertness. Its use in the soles of shoes and for flooring exploits its elasticity as well as its frictional properties. As we will see in Chapter 4, when loaded along the prism axis, it does not expand or contract laterally, making it exceptionally good for gaskets in everything from automobile engines to musical instruments such as clarinets.

Bamboo is widely used for scaffolding in Asian countries, even on skyscrapers (Fig. 1.5b). Buckling of tubes was mentioned earlier – the example given there was that of the buckling of a drinking straw. The straw buckles when bent because it ovalizes – the circular shape begins to distort into an oval and finally pops into a totally flattened

Fig. 1.5 (a) Curved glu-lam wood members, (b) bamboo scaffolding, (c) wheat field flattened by wind and rain storm, and (d,e) normal and osteoporotic vertebral trabecular bone. (a, Reproduced courtesy of the Art Gallery of Ontario; b, reprinted from Ramanathan, 2008, with permission; c, reprinted from Blackthorn Arable, Ltd., with permission; d,e, reprinted from Vajjhala *et al.*, 2000, with permission from the American Society of Mechanical Engineers.)

kink from which it does not recover. Anisotropy in engineered tubes can be adjusted to make ovalization more difficult by increasing the elastic modulus of the tube in the circumferential direction – in Chapter 6 we see how bamboo, too, achieves this.

Agricultural crops such as grains can be flattened in windstorms, reducing yields and increasing costs (Fig. 1.5c). This phenomenon, known as stem lodging, results from kinking failure of the base of the stem. An appreciation of the mechanics of plant stems may assist in alleviating this problem.

The mechanics of trabecular bone are of interest in understanding the reduction in bone strength associated with bone loss in osteoporosis (Fig. 1.5d,e). The two most common fractures in patients with osteoporosis are fractures of the proximal femur and vertebrae: in both sites, the bulk of the bone is trabecular bone, which carries most of the loading. The mechanical properties of trabecular bone are also of interest in designing bone substitute materials: it is desirable to match the stiffness of the implant with that of the bone it is replacing to avoid stress shielding and resorption of bone around the implant.

Finally, an understanding of the microstructural features of natural cellular materials that lead to their remarkable mechanical performance gives guidance to engineers seeking to mimic that performance in engineering materials and structures.

1.3 Cellular materials in medicine

Engineers are now learning how to make materials that mimic tissues in the body for medical applications. The development of a wide range of processes for making metal foams has led to the first uses of titanium and tantalum foams as coatings for implants in orthopedic applications (Fig. 1.6a). The mechanical properties of the metal foams more closely match those of the bone they are replacing, avoiding stress shielding and bone resorption, and the open-cell structure allows bone ingrowth into the implant. The surfaces of titanium and tantalum can be treated to allow bone to bond to them. Future applications may include the replacement of bone in, for example, cases of osteonecrosis or in vertebral cages for spinal fusion procedures. The processing, microstructure and mechanical properties of metal foams for orthopedic applications are described in Chapter 8.

In tissue engineering, the goal is to provide a scaffold in which biological cells can function while they regenerate their own natural extracellular matrix and tissue. Tissue engineering scaffolds used for this purpose typically have a cellular, foam-like structure, similar to that of the natural extracellular matrix. The scaffold must satisfy a number of requirements: the solid of which it is made must be biocompatible and promote the attachment, proliferation, and function of biological cells. Since the scaffolds are designed to resorb into the body as biological cells produce their own extracellular matrix, the solid must degrade into non-toxic components that can be eliminated by the body. The scaffold must have a large volume fraction of interconnected pores to facilitate the migration of biological cells and the transport of nutrients and regulatory agents such as growth factors. The pore size must be within a critical range that depends on the particular tissue to be regenerated. The pore geometry should be conducive to the cell morphology (e.g. elongated pores for the regeneration of nerve cells) and the overall scaffold has to be able to be shaped to match irregular tissue defects. The scaffold has to have sufficient mechanical integrity to be handled during surgery and to withstand initial loads in vivo; it also has to provide an appropriate mechanical environment for cell differentiation.

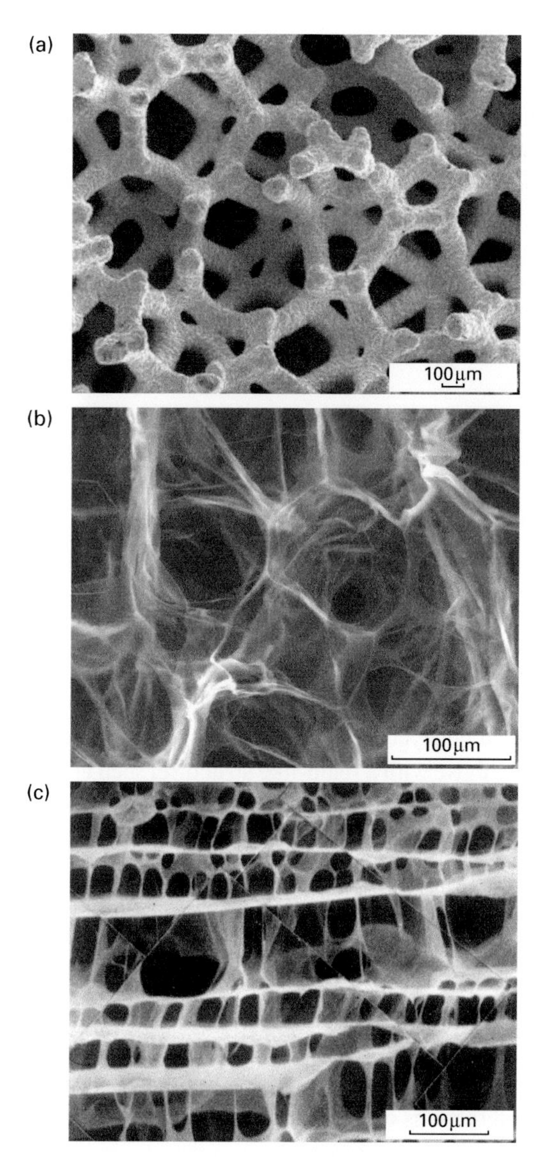

Fig. 1.6 Scanning electron micrographs of (a) a tantalum foam for a coating on a metal implant, (b) a collagen-based scaffold for skin regeneration, and (c) a collagen-based scaffold for nerve regeneration. (a, Reproduced with permission and copyright © of the British Editorial Society of Bone and Joint Surgery, from Bobyn, 1999 ; b, reprinted from Pek *et al.*, 2004, with permission from Elsevier; c, reproduced from Yannas, 2004, with permission from Wiley-VCH Verlag GmbH and Co. KGaA.)

Tissue engineering scaffolds are being developed for the regeneration of a wide range of tissues, including bone, cartilage, tendon, heart, peripheral nerve, liver and bladder. A collagen-based scaffold, designed for the regeneration of skin in burn patients, is shown in Fig. 1.6b. The scaffold is made by a freeze-drying process. It has a volume fraction of solid of 0.6% and a pore size of about 100 μm. This scaffold was one of the first to receive FDA approval for clinical use in 1996. Similar collagen-based scaffolds, but with elongated, rather than equiaxed, pores are being developed for nerve regeneration (Fig. 1.6c). The processing, microstructure and mechanical properties of a range of tissue engineering scaffolds are described in Chapter 8.

The chemical composition, microstructure and mechanical properties of scaffolds can affect the function of biological cells. For instance, cell attachment depends on the availability of binding sites on the surface of the scaffold, cell migration speed depends on the scaffold pore size and cell contractility depends on the scaffold stiffness. In Chapter 9, we review the way in which scaffold composition, microstructure and mechanical properties affect the behavior of biological cells.

1.4 Overview of the book

The book is divided into three parts. In Part I, we summarize background information on the structure and properties of the biological materials making up the solid cell walls in natural cellular materials (Chapter 2) as well as the mechanics of cellular solids and the structural components, such as sandwich panels, that use them (Chapter 3). In Part II, we describe the structure and mechanical behavior of a wide range of natural cellular materials and give examples of engineering materials mimicking them. Natural materials with a honeycomb-like structure, such as wood and cork, are described in Chapter 4, while those with a foam-like structure, such as trabecular bone and plant parenchyma, are described in Chapter 5. Some natural structures have both solid and cellular components (for instance, the sandwich structure of the iris and the filled tubes of porcupine quills) or have a gradient in the density of the cellular material (e.g. bamboo). The mechanical behavior of natural structures with a cellular component (sandwich structures, filled tubes, gradient structures), and their performance relative to the equivalent solid structures, are described in Chapter 6. The mechanical performance of natural cellular materials is summarized on material property charts and compared with that of engineering materials in Chapter 7. Finally in Part III, we describe the structure and mechanical properties of cellular materials used in medicine (Chapter 8) as well as the interactions of biological cells with tissue engineering scaffolds (Chapter 9).

References

Bobyn JD, Stackpool GJ, Hacking SA, Tanzer M and Krygier JJ (1999) Characteristics of bone ingrowth and interface mechanics of a new porous tantalum biomaterial. *J. Bone Joint Surgery* **81-B**, 907–14.

Bramwell M (1982) (editor) *The International Book of Wood*. London: AH Artists House.

Gibson LJ, Easterling KE and Ashby MF (1981) The structure and mechanics of cork. *Proc. Roy. Soc. Lond.* **A377**, 99–117.

Gibson LJ (1985) The mechanical behaviour of cancellous bone. *J. Biomech.* **18**, 317–28.

Gibson LJ, Ashby MF and Easterling KE (1988) Structure and mechanics of the iris leaf. *J. Mat. Sci.* **23**, 3041–8.

Gibson LJ, Ashby MF, Karam GN, Wegst U and Shercliff HR (1995) The mechanical properties of natural materials II: Microstructures for mechanical efficiency. *Proc. Roy. Soc. Lond.* **A450**, 141–62.

Gibson LJ and Ashby MF (1997) *Cellular Solids: Structure and Properties*, 2nd edn. Cambridge: Cambridge University Press.

Hodgson VR (1973) Head model for impact tolerance. In *Human Impact Response: Measurement and Simulation*, ed. King WF and Mertz HJ. New York: Plenum Press, pp. 113–28.

Hooke R (1665) *Micrographia,* reproduced by Dover Phoenix Editions.

Karam GN and Gibson LJ (1995) Elastic buckling of cylindrical shells with elastic cores I: Analysis. *Int. J. Solids Struct.* **32**, 1259–83.

Pek YS, Spector M, Yannas IV and Gibson LJ (2004) Degradation of a collagen-chondroitin-6-sulfate matrix by collagenase and chondroitinase. *Biomaterials* **25**, 473–82.

Ramanathan M (2008) Hong Kong – bastion of bamboo scaffolding. *Proc. Inst. Civil Eng. Civil Eng.* **161**, 177–83.

Thompson DW (1961) *On Growth and Form*, abridged edition, edited by JT Bonner. Cambridge: Cambridge University Press.

Vajjhala S, Kraynik AM and Gibson LJ (2000) A cellular solid model for modulus reduction due to resorption of trabeculae in bone. *J. Biomech. Eng.* **122**, 511–15.

van Leeuwenhoek A (1705) A letter from Mr. Anthony van Leeuwenhoek concerning the barks of trees. *Phil. Trans.* **24**, 1843–55.

Yannas IV (2004) Synthesis of tissues and organs. *Chem. Bio. Chem.* **5**, 26–39.

2 The materials of nature

2.1 Introduction

The cellular solids that are the subject of this book are structured from materials that are themselves composites. These composites are built up from a relatively small number of polymeric and ceramic components or *structural building blocks*. Woods, bamboos and palms are built from cellulose fibers in a lignin/hemicellulose matrix, shaped as hollow prismatic cells. Collagen is the basic structural element of soft tissues like tendon, cartilage, ligament, skin, blood vessels and lung. Hair, quills, hooves, horn, wool and reptilian scales are built with keratin as the main structural component. Insect cuticle relies on chitin in a matrix of protein. Mineralized tissues – antler, bone, dentine and enamel for instance – depend on hydroxyapatite with varying degrees of residual collagen for their structural integrity; shell and coral are largely calcite or aragonite. This chapter introduces these structural building blocks and some of the materials that nature creates by assembling them.

2.2 Basic building blocks: proteins, polysaccharides and minerals

2.2.1 Structural proteins

Proteins are large, complex biological molecules that perform many functions. One of these is structural. Proteins form part of the body of an organism (about 18% of the human body is protein). Examples are the collagen that makes up tendons and ligaments and is a component of bone, the keratin that is the major component of hair, wool and fur, and the silk of spiders' webs.

Protein chains in nature are all synthesized from just 20 amino acids. Nineteen of these contain the same two functional groups: an amino group, $-NH_2$, and a carboxylic acid group, -COOH. In all 20 the two functional groups are attached to the same carbon atom. This carbon is also attached to a hydrogen and a side group, -R, as in Fig. 2.1 below. The 20 amino acids differ in the nature of the side group, ranging from a single H atom in glycine or single methyl (CH_3) group in alanine to complex aromatic groups (Table 2.1). The amino acids polymerize, releasing water (a "condensation reaction") to form long-chain proteins. The figure shows part of a chain.

Proteins are made up of chains of amino acids called polypeptide chains. They have several levels of structure. The primary structure is the sequence of amino acids that make up the chain. For any protein to perform its function it must contain the correct amino

Table 2.1 The amino acids

AMINO ACID		R-GROUP	AMINO ACID		R-GROUP
Glycine	G	H	Aspartic acid	D	CH_2COOH
Alanine	A	CH_3	Asparagine	N	CH_2CONH_2
Valine	V	$CH(CH_3)_2$	Glutamic acid	E	$(CH_2)_2COOH$
Leucine	L	$CH_2CH(CH_3)_2$	Gluatmine	Q	$(CH_2)_2CONH_2$
Isoleucine	I	$CH(CH_3)CH_2CH_3$	Proline	P	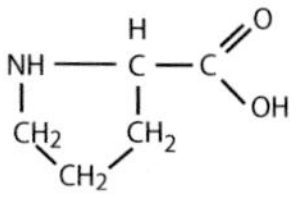 [structure]
Serine	S	CH_2OH	Tryptophan	W	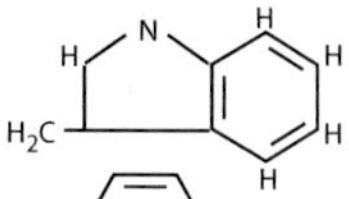 [structure]
Threonine	T	$CH(OH)CH_3$	Phenylalanine	F	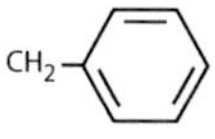 [structure]
Lysine	K	$(CH_2)_4NH_2$	Tyrosine	Y	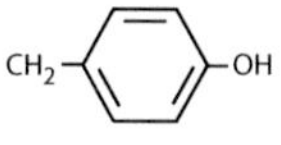 [structure]
Arginine	R	$(CH_2)_3NHCNHNH_2$	Methionine	M	$(CH_2)_2SCH_3$
Histidine	H	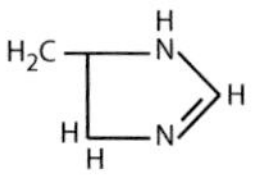 [structure]	Cysteine	C	CH_2SH

An amino acid

A protein chain

An α helix

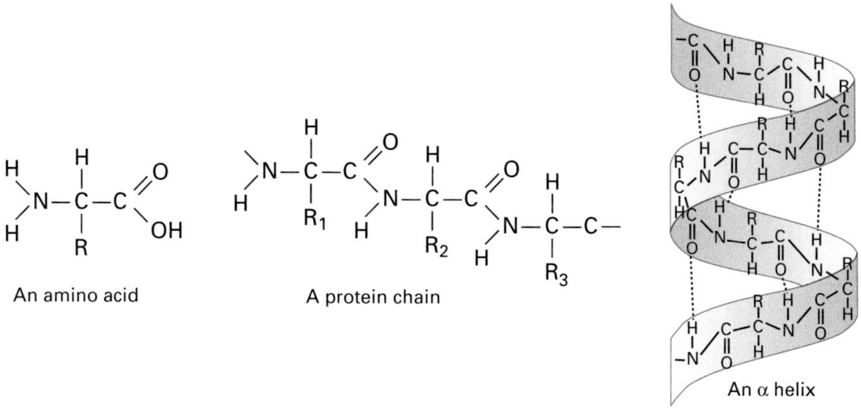

Fig. 2.1 An amino acid and a protein chain made by polymerization of amino acids. The chain wraps into an alpha helix, shown on the right (broken lines are hydrogen bonds).

acids arranged in a precise order – its function is compromised if one is out of place. When the amino acids link to form a polypeptide chain, hydrogen bonds form between the -N-H group of one amino acid and the –C=O group of another, causing the chain to coil into an *alpha helix* (Fig. 2.1) or fold into a *beta-pleated sheet*, or to form yet more complex structures, such as the triple helix of the collagen molecule, described below.

Collagens

Collagens form a family of closely related fibrous proteins. They are found throughout the animal kingdom and in virtually every tissue, ranging from shock-absorbing cartilage to light-transmitting corneas, and from tendons to blood vessels and skin. Collagens provide the stiffness and strength of soft tissues and also play a major role determining properties of mineralized tissues like bone. Collagen has a triple-helix structure formed by the assembly of three protein chains. There are at least 20 different sub-types of collagen, depending on the amino-acid sequence in the protein chains. Type I collagen has two chains with the same sequence (called alpha-1) entwined with one of a different sequence (called alpha-2). In type II collagen the three chains are the same (alpha-1). The basic motif in the amino acid sequence of collagens is Gly-X-Y, where X or Y are commonly proline or hydroxyproline; a common motif is Gly-Pro-Hyp. The glycine molecule that appears as every third residue assists in triple helix formation. Individual tropocollagen triple helices 280 nm long self-assemble into a quarter-staggered array with a 67 nm characteristic spacing as shown in Fig. 2.2.

Glycoproteins and proteoglycans

Glycoproteins and *proteoglycans* are covalently linked protein-polysaccharide complexes in which sugars are added as post-translational modifications to a core protein. The sugars in glycoproteins include galactose, glucose, mannose, xylose, acetylglucosamine and acetylgalactosamine. The sugars assist in protein folding and can improve the protein stability (resistance to enzymatic degradation); they frequently bind water and control local

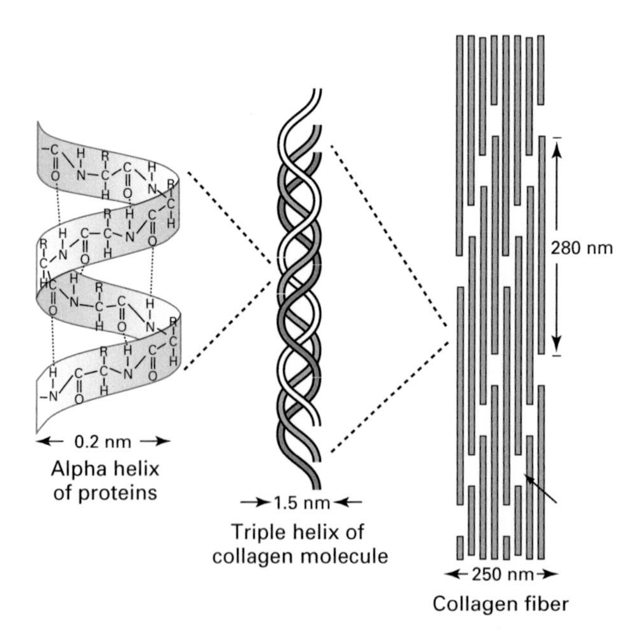

Fig. 2.2 The collagen alpha helix, the triple helical molecular structure and the formation of collagen fibers from staggered sequences of the helices.

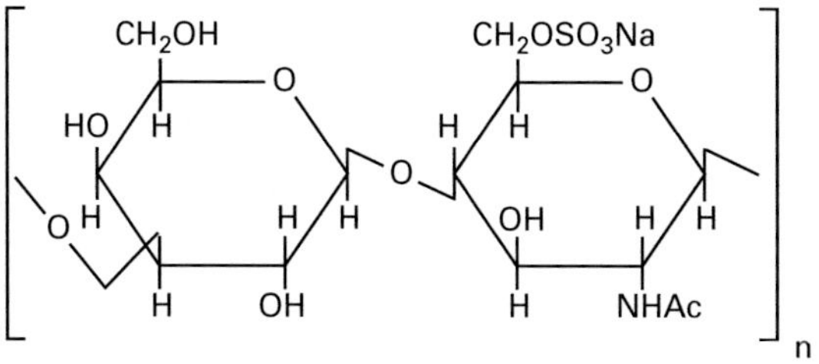

Fig. 2.3 The chemical structure of keratin.

water concentration. Proteoglycans are a special class of sulfated and negatively charged glycoproteins with critical extracellular matrix functions. The polysaccharide in proteoglycans is typically chondroitin sulfate, dermatan sulfate, heparan sulfate, or keratan sulfate.

Keratins

Keratins are fibrous structural proteins with the chemical structure shown in Fig. 2.3. Fibrous keratins form hair and wool; harder, mineralized keratins form animal quills, horn and the beaks of birds. The outermost layer of skin contains keratin. It is this that makes skin partly waterproof. When rubbed the keratin thickens, making protective calluses.

Elastomeric proteins

Elastomeric proteins, including elastin, resilin and abductin, have randomly coiled protein chains crosslinked by tyrosine links. The amino-acid chains that form the protein include proline, an angled molecule that winds the chain into an irregular tangle with high entropy. Stretching it straightens the chains, reducing the entropy and raising the internal energy. It is this that gives the rubber-like behavior. We examine the three most common elastomeric proteins in turn.

Elastin

Elastin is a natural elastomer that can be extended to nearly three times its resting length without damage. It appears in animals as connective tissue in parts that must stretch and return to their original form when unloaded: skin, lungs, arteries, the bladder and elastic cartilage. Elastin helps to keep skin smooth and tight.

Resilin

Resilin is an elastomeric protein found in the cuticle of insects (arthropods) such as fleas and locusts. It is resilin that allows fleas to jump, locusts to sing and dragonflies to hover. It has unique resilience with an elastic "efficiency" of 97% and fatigue lifetime of hundreds of millions of cycles – properties that have prompted chemists to attempt to synthesize it. Synthetic resilin is of interest as a bio-material for medical research with potential for vascular implants to restore suppleness to arteries and as prostheses to replace defective discs between vertebrae.

Abductin

Abductin does for mollusks what elastin does for mammals and resilin does for insects. Scallops, for example, swim when threatened, using a sort of jet propulsion. The two halves of the shell – the valves – quickly clamp together when the muscle connecting them contracts and the water trapped inside is squirted out. Bringing the two shells together compresses a pad of rubbery abductin. When the muscle relaxes, the rubber make the valves spring open. The protein chain of abductin has high concentrations of the amino acids glycine and methionine.

2.2.2 Polysaccharides

Polysaccharides perform two essential functions. Some, such as starch and glycogen, store energy in a way that can be retrieved when the organism needs it. Others, such as cellulose, lignin and chitin, are structural. It is these that are of interest here. The first two are the structural materials of the plant world, and because of this are the most prolific polymers on earth. The third is the structural material of the exoskeleton (the exterior shell) of insects and other arthropods – crustacea, arachnids (spiders), centipedes and other bugs – and it, too, is widespread.

Polysaccharides are made by polymerizing monosaccharides – sugars – typified by glucose (Fig. 2.4a). Glucose, a six-carbon saccharide, exists in two structural forms: as a linear molecule or as a ring. In the ring-like form it has the ability to polymerize by a condensation reaction (one releasing water) to give stiff, straight polysaccharide chains (Fig. 2.4b). The process is reversed by hydrolyzation. In the cellulose molecule, which can have many thousand monosaccharide units, the units are arranged such that the bonding oxygens are staggered (Fig. 2.4c, in which only the oxygen atoms are identified). Assemblies of these stiff, straight chains give crystalline cellulose (Fig. 2.4d). Chitin is chemically related to cellulose, but the monomer from which it is built is a glucosamine unit – glucose, with one -OH unit replaced by $-NHCOCH_3$. The molecules of

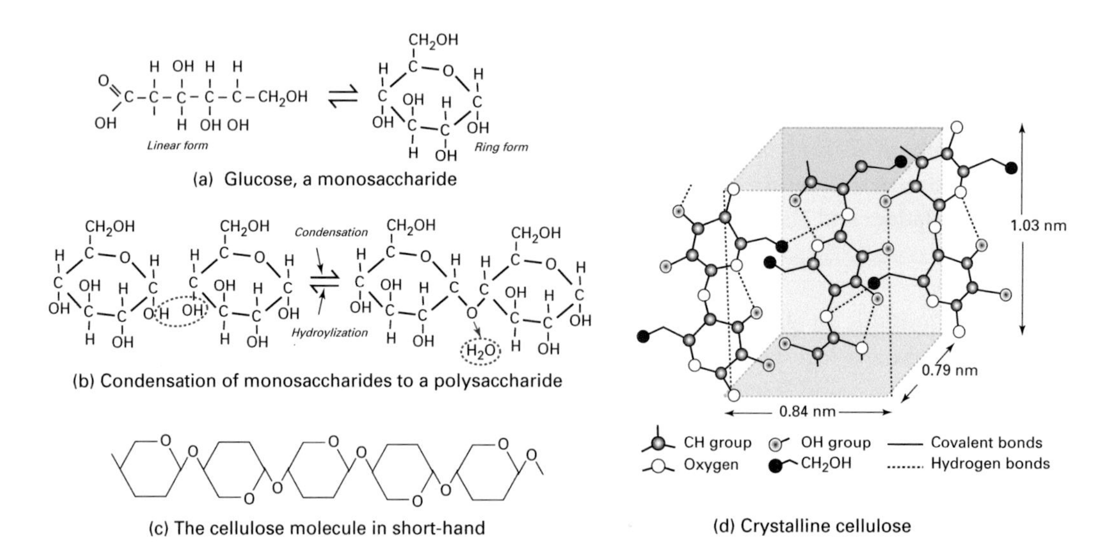

Fig. 2.4 (a) The two forms of glucose, a monosaccharide. (b) The polymerization of monosaccharides to make a polysaccharide. (c) The configuration of the monosaccharide rings in cellulose. (d) The arrangement of molecules in crystalline cellulose.

both cellulose and chitin are straight and stiff. The molecules hydrogen-bond together to form microfibrils which hydrogen-bond in turn to form fibers of great strength, comparable with that of steel.

Cellulose

Cellulose is a long-chain polysaccharide made up of 7000–15 000 units of sugar monomers. It is the primary structural material of the plant world, appearing as microfibrils 2–20 nm diameter and 100–40 000 nm long, which form the strong framework of the cell walls. Cellulose is the main constituent of cotton, flax and wood, and is the feedstock for the man-made materials rayon, cellophane, viscose, cotton wool and many other essentials of modern life. Cellulose fibers from cotton, flax or wood pulp are the basis of paper, card, newsprint and cotton wool.

Lignin

Lignin, a complex phenolic, is the second most abundant renewable polymer, after cellulose. Phenolics are a diverse group of aromatic compounds (containing benzene rings) usually with hydroxyl groups. Phenol itself, C_6H_5OH, is the simplest member of the class. Unlike cellulose, lignin is amorphous. It is the glue that holds the cellulose fibers together in woods and other plants.

Hemicellulose

Hemicellulose, a polysaccharide built up of a range of sugar monomers, is found (like cellulose and lignin) in almost all plant cell walls. But unlike cellulose, it has short chains (500 to 3000 monomer units), an amorphous structure, little strength and is easily hydrolyzed by acids or by enzymes. Hemicellulose, like lignin, acts as a binder for cellulose fibrils.

Suberin

Suberin, a waxy polysaccharide found in many plants, is so named because of the high suberin content of cork, the outer bark of the oak, *Quercus suber*. Its polymeric chains contain both aromatic and alphatic units (units involving benzene rings and units made up of open chains), the ratio varying with plant species. Suberin is hydrophobic, preventing water entering or leaving plant tissue. It appears to offer some protection to the plant against pathogen invasion and possibly protection in fire: it burns only slowly and is self-extinguishing.

Chitin

Chitin ranks with cellulose and lignin as one of the most abundant polysaccharides in nature. It forms the major structural component of the exoskeleton of invertebrates such as shrimp, lobster, crab, beetle and ant, and the cell walls of fungi. In its unmodified form, chitin is translucent, pliable and resilient, and quite tough. In arthropods, however, it is frequently modified by being embedded in a hardened proteinaceous matrix, which forms much of the exoskeleton. The difference between the unmodified and modified forms can be seen by comparing the body wall of a caterpillar (unmodified) to a beetle (modified).

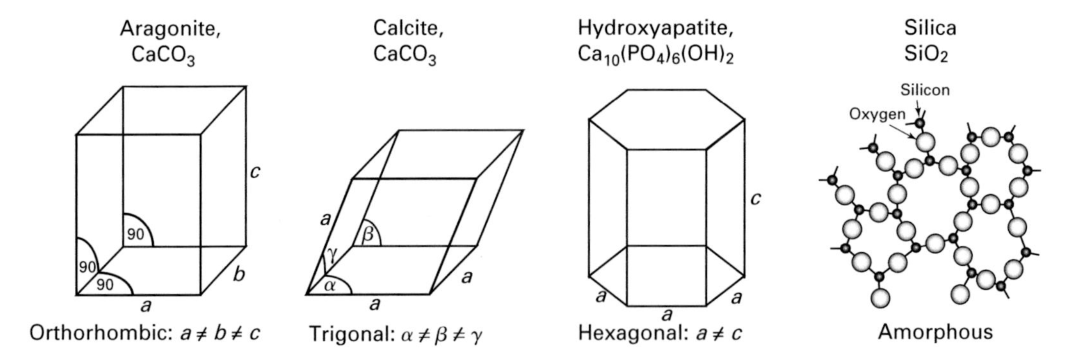

Fig. 2.5 Three of the principal structural minerals of biological systems. The fourth, bio-silica, is amorphous.

2.2.3 Minerals

Minerals, in materials-science terms, are ceramics. Natural materials use few structural ceramics, but they use them in ingenious ways. Those of principal importance are calcite, aragonite, hydroxyapatite and bio-silica. Mineralized tissue is tissue that incorporates these, but in widely varying degrees (from as little as 1% to as much as 99.9%). It is these minerals that give stiffness, hardness and strength to bone, teeth and shell. In the plant world it is bio-silica that makes some so abrasive that they can be used to clean metals, and provides others with needle or dagger-like protective armor. All four are chemically simple. Figure 2.5 shows the formulae and, since the first three are crystalline, the shape of the unit cell that describes the positions of groups of atoms in the structure. Calcite and aragonite are polymorphs – minerals with the same composition but different crystal structure. Bio-silica, SiO_2, is amorphous – its structure is like that of glass.

Calcite

Calcite, $CaCO_3$, is the most common polymorph of calcium carbonate. It has a relatively low hardness (3 on the Mohs scale) and high reactivity with even weak acids such as vinegar. It cleaves easily. Calcite is the primary material of birds' egg shells and the shells of snails and mollusks.

Aragonite

Aragonite is the second most common polymorph of natural calcium carbonate. It is found as stalactites in limestone caves and as deposits in geysers. Most bivalve animals and corals secrete aragonite for their shells; pearls are composed mostly of aragonite. The pearly and iridescent colors of the nacre in sea shells such as abalone arise from diffraction from thin, overlapping layers of aragonite.

Hydroxyapatite

Hydroxyapatite is hydrated calcium phosphate. It is the mineral component of bone, teeth and antler. It has important medical applications and is the main component of the porcelain-like ceramic called "bone china." Plasma sprayed hydroxyapatite coatings

Table 2.2 Typical property ranges for the basic materials of nature

Material	Density, kg/m^3	Young's modulus, GPa	Tensile strength, MPa	Hardness Hv	Elongation %	Thermal conductivity W/m.K	Specific heat J/kg.C
Proteins							
Collagen	1300–1450	0.2–0.8	49–95	14–23	9–14	0.4–0.45	1380–1540
Keratin	1280–1340	1–4	130–210	36–65	30–40	0.19–0.22	1500–1560
Abductin	1250–1320	0.0027–0.0045	0.7–1.0	–	150–190	0.45–0.5	3600–3680
Elastin	1260–1330	0.0018–0.0022	1.7–2.2	–	120–160	0.45–0.5	3600–3680
Resilin	1280–1320	0.0006–0.0015	1.5–2.0	–	160–200	0.45–0.5	3600–3680
Polysaccharides							
Cellulose	1450–1590	110–165	750–1010	–	20–40	0.5–0.65	1400–1500
Chitin	1500–1900	30–90	95–110	–	4–5	0.38–0.5	870–1000
Lignin	1200–1250	1.9–3.0	50–75	10–20	3–4	0.13–0.18	1650–1850
Minerals							
Aragonite	2930–2950	95–100	100–200	300–380	0.11–0.20*	4.6–5.5	860–870
Bio-silica	2050–2150	42–50	45–100	350–550	0.10–0.20*	1.4–1.5	680–730
Calcite	2710–2830	77–79	100–200	200–280	0.13–0.25*	4.6–5.5	860–870
Hydroxyapatite	3050–3150	110–130	140–300	95–140	0.13–0.25*	1.24–3	870–890

* For minerals elongation is equated to the elastic strain at the tensile strength.
Source: Data from *CES Education Database of Natural and Bio Materials* (2008). Cambridge: Granta Design (www.grantadesign.com).

are applied to orthopedic and dental implants to stimulate the development of a strong bond with the surrounding bone. Particulate hydroxyapatite is used as a filler in polymer matrices (polyethylene, for example) to create a material with bone-like properties.

Bio-silica

Bio-silica is a hydrated form of silica, $SiO_2.(H_2O)_n$, occurring widely in the plant and animal world but in much smaller quantities than calcium based minerals. It is amorphous. The water content, typically, is 3 to 10%. Bio-silica provides the hard component of seaweeds, sponges, spicules like those of the sea-urchin and the teeth of limpets. It is common in bamboos and other grasses, some of which contain up to 10% bio-silica. Its presence in certain grasses is a problem for sheep and horses that feed on them, grinding down their teeth.

2.2.4 Data for the basic building blocks

Natural materials are much more variable in their properties than those that are man-made, a consequence of species-related differences in chemistry, polymer chain length and packing, and water content. It is, none the less, useful to think of typical *ranges* within which their physical, mechanical, thermal and other properties lie.

Table 2.2 is an assembly of these ranges for seven properties of the five structural proteins (collagen and keratin, and the elastomers abductin, elastin and resilin), of the three structural polysaccharides (cellulose, lignin and chitin) and of the four minerals (aragonite, calcite, hydroxyapatite and bio-silica) that play central roles in the structural and thermal behavior of porous animal and plant tissue. The ranges listed here were assembled (or in some cases inferred) from the sources listed at the end of this chapter, listed under headings corresponding to those in the table.

2.3 The materials of nature

Virtually all the materials of nature are composites, built up from the molecules described in Section 2.2 in the form of fibers, networks and platelets that are embedded, often in a highly organized way, in matrices composed of others of these molecules. In cartilage, ligaments, skin and arteries, for example, the strength derives from collagen fibrils, and the elasticity from proteoglycans, elastin and other similar proteins (Fig. 2.6). The strength and stiffness of wood derives from the complex arrangement of cellulose fibrils in a matrix of hemicelluloses and lignin. In bone and shell the stiffness and hardness derive from the ceramic micro-platelets of hydroxyapatite, calcite or aragonite, and the toughness from the collagen, chitin or other proteins that bonds them together. Here, we examine the hierarchical structure of wood and bone in more depth.

The hierarchical structure of wood

The stiffness, strength and toughness of wood derive largely from those of the cellulose molecule, shown on the left of Fig. 2.7. Crystalline microfibrils are built up of aligned molecules some 30–60 nm in length. These form the reinforcing fibers of lamellae, the matrix of which is amorphous lignin and hemicellulose. Stacks of lamellae in the four-layer pattern and fiber orientation shown on the right become the structural material of the cell wall. The fibrils of the primary cell wall interweave randomly, as in cotton wool. In subsequent layers the fibrils are parallel and closely packed. The outer S_1 layer

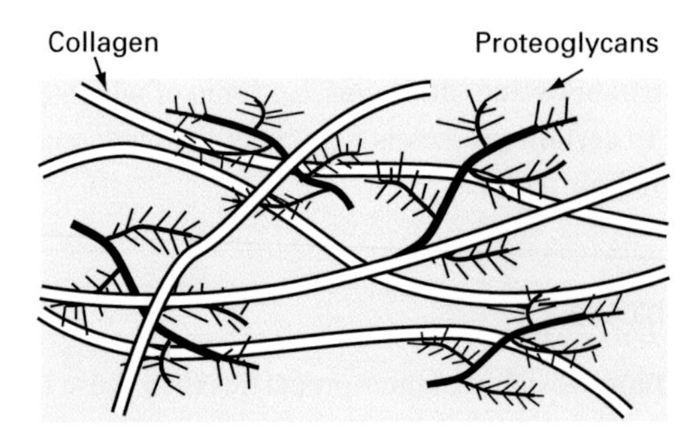

Fig. 2.6 The structure of cartilage, typically 70% water, 20% collagen and 10% proteoglycan.

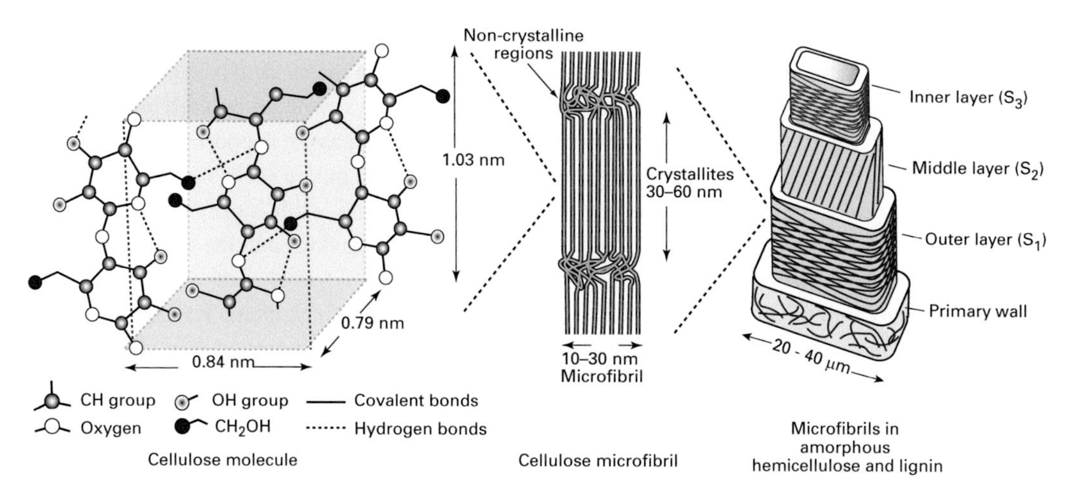

Fig. 2.7 The hierarchical structure of wood (after Dinwoodie, 1979).

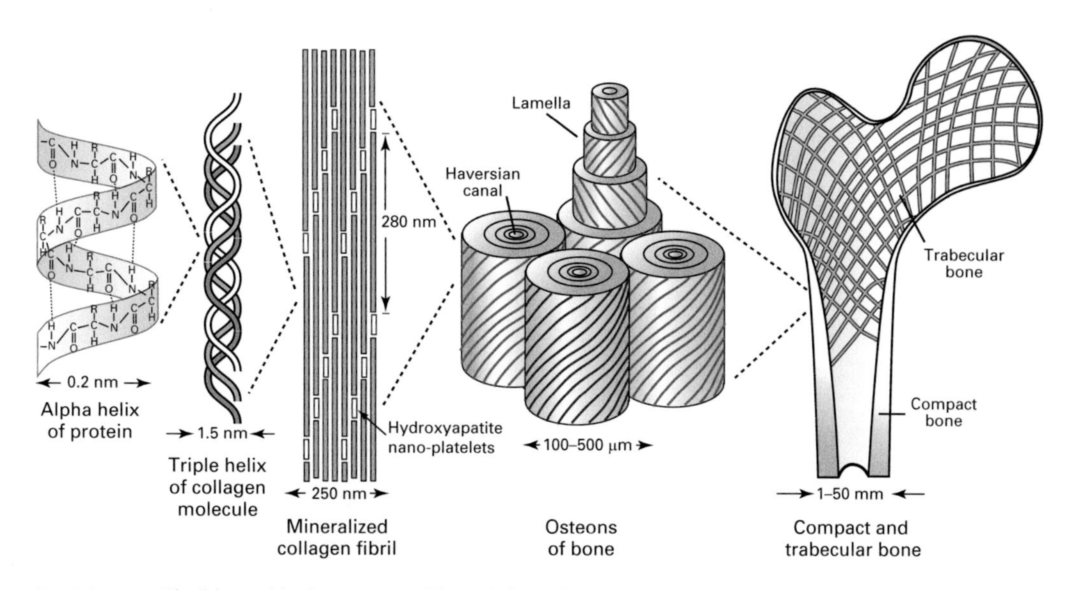

Fig. 2.8 The hierarchical structure of bone (after Rho *et al.*, 1998).

has lamellae with alternate right and left-handed spiral winding of fibrils. Beneath it, the thicker S_2 layer has fibrils that are more nearly oriented along the axis of the cell. The innermost S_3 layer has an arrangement like that of S_1. The cell as a whole is bonded to its neighbors by the middle lamella (not shown), a lignin–pectin complex devoid of cellulose.

The hierarchical structure of bone

There is an interesting parallel between the hierarchical structure of bone and that of wood, despite the great differences in their molecular chemistry (Fig. 2.8). The starting point here is the triple-helical structure of the collagen molecule, shown on the left.

But, unlike wood, this becomes the matrix, not the reinforcement of the mineralized tissue of bone. Hydroxyapatite nano-platelets deposit in the nascent tissue, and increase in volume fraction over time to give the mature osteons, with an ordered arrangement of highly mineralized fibers with the strength and stiffness to support structural loads that bone must carry in a mature organism. At the most macro scale, nearly fully dense compact bone provides the outer structure of whole bone, while highly porous trabecular bone fills the vertebrae, shell-like bones such as the skull and the ends of the long bones such as the femur.

We will return to these materials in greater detail in later chapters. For now it is helpful to have an overview of the range of properties these combinations provide – a range far greater than that of the basic building blocks of which they are made.

2.4 Overview of properties

Modulus and strength

Figure 2.9 shows the tensile strength, σ_{ts}, of natural materials plotted against the Young's modulus, E. Each small bubble encloses the range of σ_{ts} and E for one material. Large envelopes enclose material classes: soft tissue, mineralized tissue, fibers and wood-like materials, described more fully below. For presentational reasons no enve-

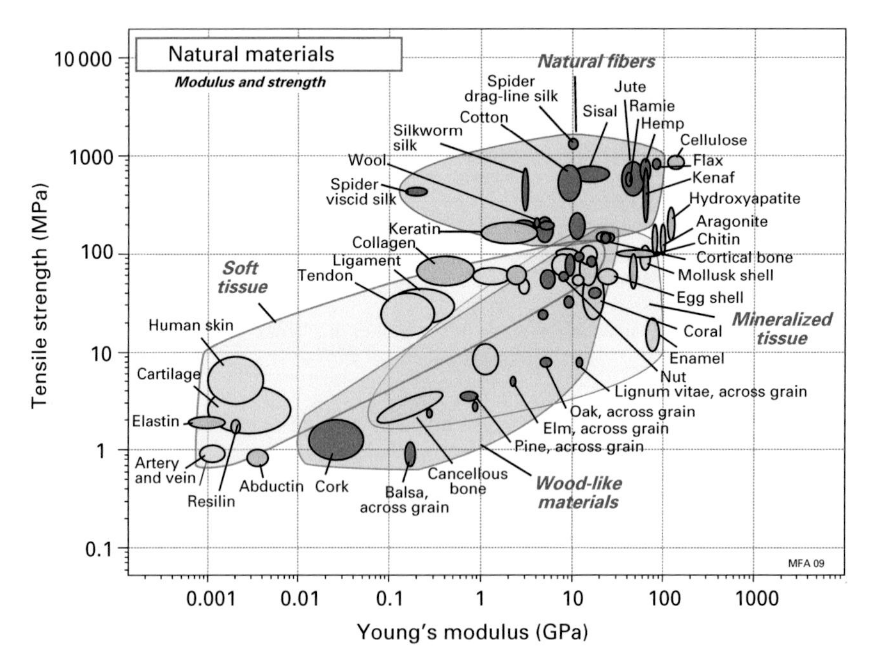

Fig. 2.9 Young's modulus and tensile strength of natural materials. For clarity the basic structural building blocks (yellow bubbles) are not enclosed in an envelope. See plate section for color version.

lope has been created for the basic structural building blocks (pink bubbles). Here we describe briefly, in turn, the groups enclosed in larger envelopes.

Soft tissue is tissue that has not been mineralized. Many have mechanical functions. There are two broad groups, distinguished by type of mechanical function: active (muscle) and passive (connective tissues). Muscle fibers are very large specialized cells containing protein fibers for voluntary contraction (the biceps, for example) or involuntary contraction (the heart) depending on the muscle type. Connective tissues (ligament, tendon, cartilage) are predominantly composite materials assembled from proteins, polysaccharides and glycoproteins, with a substantial fraction of water. The molecules are generated, organized and remodeled by living cells typically of a fibroblast morphology. Differences between tissues arise from compositional variations and variations in component organization and alignment. Keratinized tissue (hair, horn, hoof, the shells of turtles and tortoises) is much stiffer and stronger than those based only on collagen and elastin.

Mineralized tissue starts as soft tissue that is progressively mineralized as the organism develops. Bio-mineralization is the process by which minerals are deposited within living tissue, creating composites of organic and inorganic materials. The key minerals involved are those described in Section 2.2: calcite, aragonite, hydroxyapatite and bio-silica. Bio-mineralization gives organisms competitive advantage. It allows them to become mobile, supporting their own weight. It gives a rigid framework to which muscles and tendons can be attached. As exo-skeletons and shells, it gives protection from predators. Bio-mineralization also contributes to the survival of plants, allowing them to

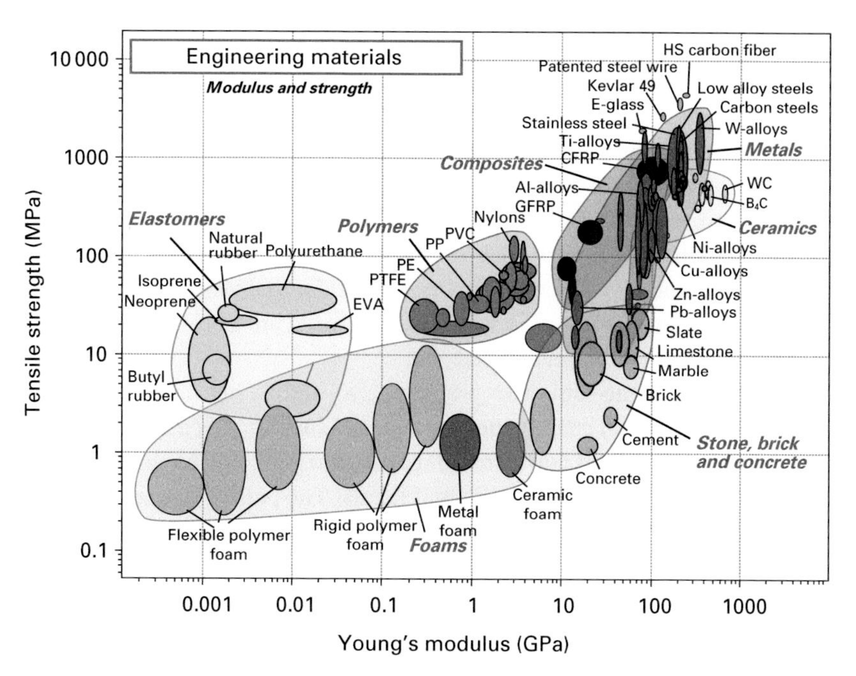

Fig. 2.10 Young's modulus and tensile strength of engineering materials plotted on exactly the same axes as Fig. 2.9. See plate section for color version.

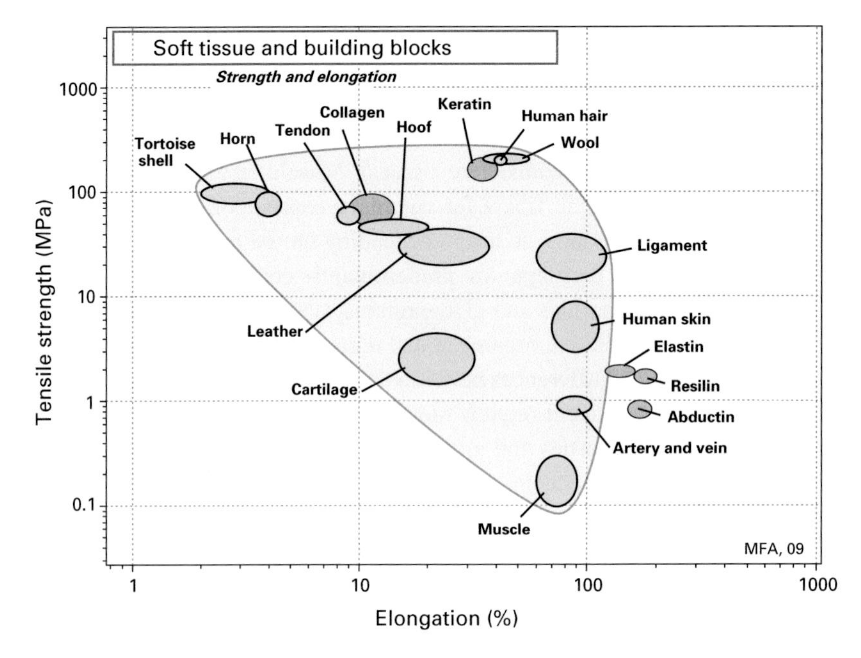

Fig. 2.11 Strength and elongation of soft tissues compared with those of the basic structural building blocks that make them up. The greater the elastin content, the greater is the elongation. See plate section for color version.

grow taller (capturing more light) and to survive aggression. It is bio-silica that gives the nettle its sting and the cactus its spines.

Woods and wood-like materials rely on cellulose for their mechanical stiffness and strength. They offer a remarkable combination of properties. They are light, and, parallel to the grain, they are stiff, strong and tough – as good, per unit weight, as all but the best man-made materials.

Many *natural fibers* have evolved to carry loads (though they perform other functions too), with the result that they are optimized for strength and stiffness at minimum weight. Others have evolved to insulate and provide warmth, giving them exceptional thermal properties. Fibers of the mammal and insect world are based on proteins such as collagen and keratin. Those of the plant world are based on cellulose and are found in the stem, the leaf and the seed pod.

In Fig. 2.10 the same two properties, σ_{ts} and E, but now for engineering materials, are plotted on exactly the same axes. As in Fig. 2.9, a small bubble encloses the range of σ_{ts} and E for one material. Large envelopes enclose material classes: metals, polymers, elastomers, ceramics and foams. The pair of figures allows a direct comparison of the properties of natural and man-made materials. They reveal that nature is able to synthesize materials with properties that cover a range almost as large as those of the engineering materials we use today.

Soft tissue and its structural building blocks

Figure 2.11 shows data for the tensile strength and elongation of soft tissues and for the basic structural materials (collagen, keratin, elastin) from which they are made. Tissues

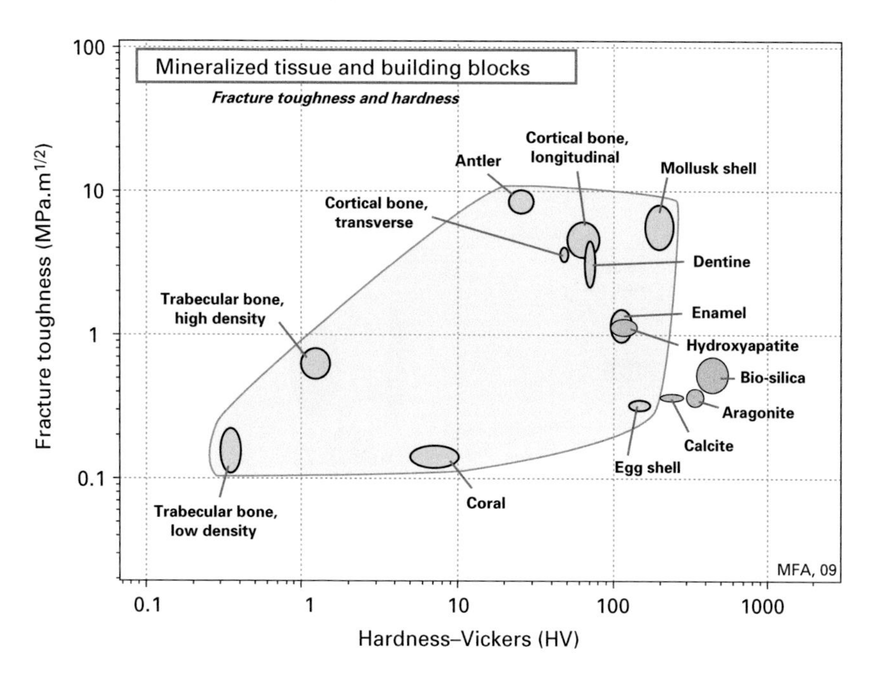

Fig. 2.12 A comparison of the fracture toughness and hardness of mineralized tissue with those of the minerals from which they are made. See plate section for color version.

that are predominantly based on keratin (horn, hoof, hair) all have strengths around 100 MPa and an elongation between 5 and 50%. Those built from collagen, elastin and proteoglycans (cartilage, ligament, skin, artery, muscle) have lower strengths, but generally higher elongation (up to 110%).

Comparison of the hardness and toughness of mineralized tissue with that of its mineral

The minerals of mineralized tissue are, when pure, hard and brittle. Mineralized tissue is less hard. No surprise there – all have, to varying degrees, a component of protein binding the mineral together. But many are much less brittle, indeed they are relatively tough. Toughness – resistance to cracking – is measured, in engineering, by the *fracture toughness*. The load needed to make a tooth chip or bone break is proportional to its fracture toughness. The toughness has a protective role as well – a tough shell, for instance, is harder for a predator to penetrate than a shell that is brittle.

Figure 2.12 presents the evidence. Hardness is no surprise: the minerals are the hardest; dilute them in any way and the hardness falls. The remarkable thing is the toughness. Antlers are used for fighting; a brittle antler is not a good idea. Antler, dense bone and dentine are between three and ten times more resistant to the propagation of a crack than the hydroxyapatite with which they are mineralized. Mollusk shell, mineralized with calcite or aragonite, too, is remarkably tough. In all four examples, the gain in toughness derives from the way the mineral and protein (e.g. collagen) are configured in the tissue, with toughness increasing with increasing protein content.

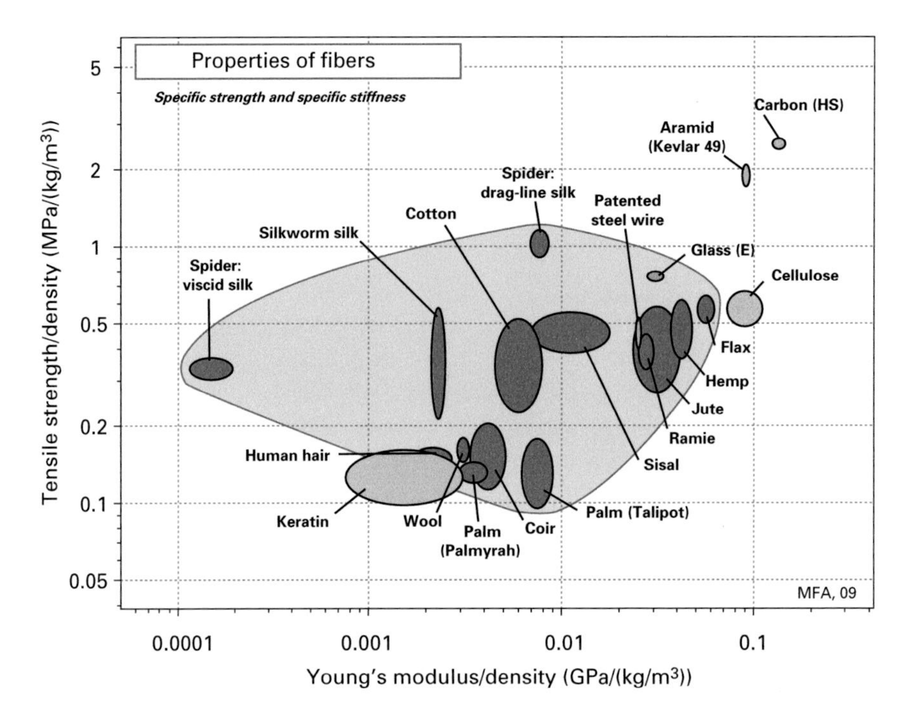

Fig. 2.13 The specific strength and specific stiffness of natural fibers compared to those of the strongest man-made fibers. Several natural fibers are as good as, or better than, steel. See plate section for color version.

The strength and stiffness of fibers

Fibers are, from a mechanical point of view, one dimensional: they are useful only when loaded in tension. Then it is the axial stiffness and strength that determines performance. When the fibers are polymeric (as are all natural and many man-made fibers) these two properties are maximized by aligning the C-C-C- chains of the polymer molecules parallel to the axis of the fiber. If, in addition, some ability to stretch is required, it can be achieved by crimping the molecules slightly so that, under tension, they progressively straighten out.

Nature makes use of all these strategies. The results are shown in Fig. 2.13, which compares the specific strength and specific stiffness (the values per unit mass) of fibers with those of man-made materials: carbon, aramid, glass and heavily drawn ("patented") steel wire. A number of natural fibers perform as well or better than steel, though none quite approach aramid or carbon.

2.5 Summary

This chapter has introduced the principal structural materials of nature. It is these that form the networks and cell walls of natural cellular solids. The basic building blocks – proteins, polysaccharides and minerals – are shaped and combined in complex ways to give materials with a wide range of mechanical properties – a range almost as great as that of the man-made materials of engineering. One way to get a feel for this is to plot the properties as charts, of which Figs. 2.9 to 2.13 are examples. They illustrate the ways in which properties are achieved by a combination of component properties and hierarchical configuration, ranging in scale from nanometers to millimeters.

References

Proteins (collagen, keratin, chitin, elastin, resilin, abductin)

Baxter S (1946) The thermal conductivity of textiles. *Proc. Phys. Soc.* **58,** 105–18.

Bertram JEA and Gosline JM (1986) Fracture toughness design in horse hoof keratin. *J. Exp. Biol.* **125**, 29–47.

Bochicchio B, Jimenez-Oronoz F, Pepe A, Blanco M, Sandberg LB and Tamburro AM (2005) Synthesis of and structural studies on repeating sequences of abductin. *Macromolecular Biosci.* **5**, 6, 502–11.

Clare S (2006) Cold scallops rely on rubber. *J. Exp. Biol.* **209**, p. i.

Denny M and Miller L (2006) Jet propulsion in the cold: mechanics of swimming in the Antarctic scallop Adamussium colbecki. *J. Exp. Biol.* **209**, 4503–14.

Elvin CM, Carr AG, Huson MG et al. (2005) Synthesis and properties of cross-linked recombinant pro-resilin. *Nature* **437**, 999–1002.

Fraser R and Macrae T (1980) Molecular structure and mechanical properties of keratins. In *The Mechanical Properties of Biological Materials (Proceedings of the Symposia of the Society for Experimental Biology; no. 34)*, ed. Vincent JFV and Currey JD. Cambridge: Cambridge University Press, pp. 211–46.

Fung YC (1993) *Biomechanics: Mechanical Properties of Living Tissues*. Berlin: Springer.

Geddes LA and Baker LE (1967) The specific resistance of biological material – a compendium of data for the biomedical engineer and physiologist. *Med. & Biol. Eng.* **5**, 271–93.

Gentleman E, Lay AN, Dickerson DA, Nauman EA, Livesay GA and Dee KC (2003) Mechanical characterization of collagen fibers and scaffolds for tissue engineering. *Biomaterials* **24**, 3805–13.

Gosline J (1980) The elastic properties of rubber-like proteins and highly extensible tissues. In *The Mechanical Properties of Biological Materials (Proceedings of the Symposia of the Society for Experimental Biology; no. 34)*, ed. Vincent JFV and Currey JD. Cambridge: Cambridge University Press, pp. 331–57.

Gosline JM and French CJ (1979) Dynamic properties of elastin. *Biopolymers* **18**, 2091–103.

Gosline J, Lillie M, Carrington E, Guerrette P, Ortlepp C and Savage K (2002) Elastic proteins: biological roles and mechanical properties. *Phil. Trans. R. Soc. Lond. B* **357**, 121–32.

Kato YP, Christiansen DL, Hahn RA, Shieh SJ, Goldstein JD and Silver FH (1989) Mechanical properties of collagen fibres: a comparison of reconstituted and rat tendon fibres. *Biomaterials* **10**, 38–41.

Mason P (1963) Density and structure of alpha-keratin. *Nature* **197**, 179–80.

Oxland H, Manschot J and Viidik A (1988) The role of elastin in the mechanical properties of skin. *J. Biomech.* **21**, 213–18.

Park JB (1979) *Biomaterials, an Introduction*. New York and London: Plenum Press.

Tao XM and Postle R (1989) A viscoeleastic analysis of keratin. *Textile Res. J.* **59**, 5, 300–6.

Vincent JFV and Currey JD (editors) (1980) *The Mechanical Properties of Biological Materials (Proceedings of the Symposia of the Society for Experimental Biology; no. 34)*. Cambridge: Cambridge University Press for Society for Experimental Biology.

Vincent JFV (1990) *Structural Biomaterials*, revised edn. Princeton, NJ: Princeton University Press.

Vogel S (1988) *Life's Devices: The Physical World of Animals and Plants*, illustrated by Calvert

RA. Princeton, NJ: Princeton University Press.

Vogel S (2003) *Comparative Biomechanics: Life's Physical World.* Princeton, NJ: Princeton University Press.

Wainwright S (1980) Adaptive materials: a view from the organism. In *The Mechanical Properties of Biological Materials (Proceedings of the Symposia of the Society for Experimental Biology; no. 34)*, ed. Vincent JFV and Currey JD. Cambridge: Cambridge University Press, pp. 438–53.

Wainwright SA, Biggs WD, Currey JD and Gosline JM (1976) *Mechanical Design in Organisms.* London: Edward Arnold.

Yamada H (1970) *Strength of Biological Materials.* Baltimore, OH: Williams & Wilkins.

Polysaccharides (cellulose, lignin)

Cousins WJ, Armstrong RW and Robinson WH (1975) Young's modulus of lignin from a continuous indentation test. *J. Mat. Sci.* **10**, 1655–8.

Dinwoodie JM (1981) *Timber, its Nature and Behaviour.* New York: Van Nostrand Reinhold.

Ferranti L Jr., Armstrong RW and Thadhani NN (2004) *Mat. Sci. Eng. A* **371**, 251–5.

Mwaikambo LY and Ansell MP (2001) The determination of porosity and cellulose content of plant fibers by density methods. *J. Mat. Sci. Lett.* **20**, 23, 2095–6.

Vincent JFV and Currey JD (editors) (1980) *The Mechanical Properties of Biological Materials (Proceedings of the Symposia of the Society for Experimental Biology; no. 34).* Cambridge: Cambridge University Press for Society for Experimental Biology.

Vogel S (2003) *Comparative Biomechanics – Life's Physical World.* Princeton, NJ: Princeton University Press.

Minerals (calcite, aragonite, bio-silica, hydroxyapatite)

Broz ME, Cook RF and Whitney DL (2006) Micro-hardness, toughness and modulus of Mohs scale minerals. *American Mineralogist* **91**, 135–42.

Evis Z and Doremus RH (2005) Coatings of hydroxyapatite. *Mat. Lett.*, **59**, 3824–928.

Khalil KA, Kim SW, Dharmaraj N, Kim KW and Kim HY (2006) Novel mechanisms to improve toughness on hydroxyapatite bioceramics. *J. Mat. Process. Tech.* **187–188**, 417–20.

Park JB (1979) *Biomaterials, an Introduction.* New York and London: Plenum Press.

Rho JY, Kuhn-Spearing L and Zioupos P (1998) Mechanical properties and the hierarchical structure of bone. *Med. Eng. Phys.* **20**, 92–102.

Skinner HCW and Jahren AH (2004) Biomineralization. In *Treatise on Geochemistry*, vol. 8, section 8.04. Oxford, UK and Burlington, MA: Elsevier, pp. 117–184.

Vincent JFV and Currey JD (editors) (1980) *The Mechanical Properties of Biological Materials (Proceedings of the Symposia of the Society for Experimental Biology; no. 34).* Cambridge: Cambridge University Press for Society for Experimental Biology.

University of Michigan (2007) Dental tables. Available at: www.lib.umich.edu/dentlib/Dental_tables/toc.html.

3 Structure and mechanics of cellular materials

3.1 Introduction

The microstructural features of cellular solids affecting their mechanical response are most easily observed in engineering honeycombs and foams (Fig. 3.1). Honeycombs, with their prismatic cells, are *two-dimensional cellular solids*, while foams, with their polyhedral cells, are *three-dimensional cellular solids*. Honeycombs with periodic hexagonal cells are straightforward to analyze and we begin by describing their behavior. Insight gained from honeycombs can be applied to the analysis of foams, as both deform and fail by the same mechanisms. The *relative density* is the density of the cellular solid divided by that of the solid from which it is made and is equivalent to the volume fraction of solid. *Open-cell foams* are solid only at the edges of the polyhedra while *closed-cell foams* have solid membranes over the faces of the polyhedra. The properties of the foam depend on those of the solid from which the honeycomb or foam is made; in many natural materials, such as wood and trabecular bone, the cell wall itself is a composite, as we saw in Chapter 2.

The stress–strain curve for a cellular solid in compression is characterized by three regimes (Fig. 3.2a,b): a *linear elastic regime*, corresponding to cell edge bending or face stretching; a *stress plateau*, corresponding to progressive cell collapse by elastic buckling, plastic yielding or brittle crushing, depending on the nature of the solid from which the honeycomb or foam is made; and a final regime of *densification*, corresponding to the collapse of the cells throughout the material and subsequent loading of the cell edges and faces against one another. Cellular solids typically have low relative densities (~ 10–20%) allowing large strains (~70–80%) before densification occurs. In tension, at small strains, the linear elastic response is the same as in compression (Fig. 3.2c,d). As the strain increases, the cells become more oriented with the loading direction, increasing the stiffness until tensile failure occurs.

The mechanical response of cellular solids has been modeled by representing the cellular structure in several ways. Initial models analyzed a *unit cell* such as a hexagon in two dimensions or a dodecahedron (a 12-faced polyhedron) or tetrakaidecahedron (a 14-faced polyhedron) in three (Fig. 3.3) (Ko, 1965; Patel and Finnie, 1970; Gibson *et al.*, 1982a; Warren and Kraynik, 1997; Zhu *et al.*, 1997a,b). The geometry of the unit cell makes the analysis tractable but may not give an exact representation of the real material (e.g. foams). An even simpler approach is to use dimensional analysis to model the mechanisms of deformation and failure observed in the cellular material without specifying the exact cell geometry (Gibson and Ashby, 1982b, 1997). This

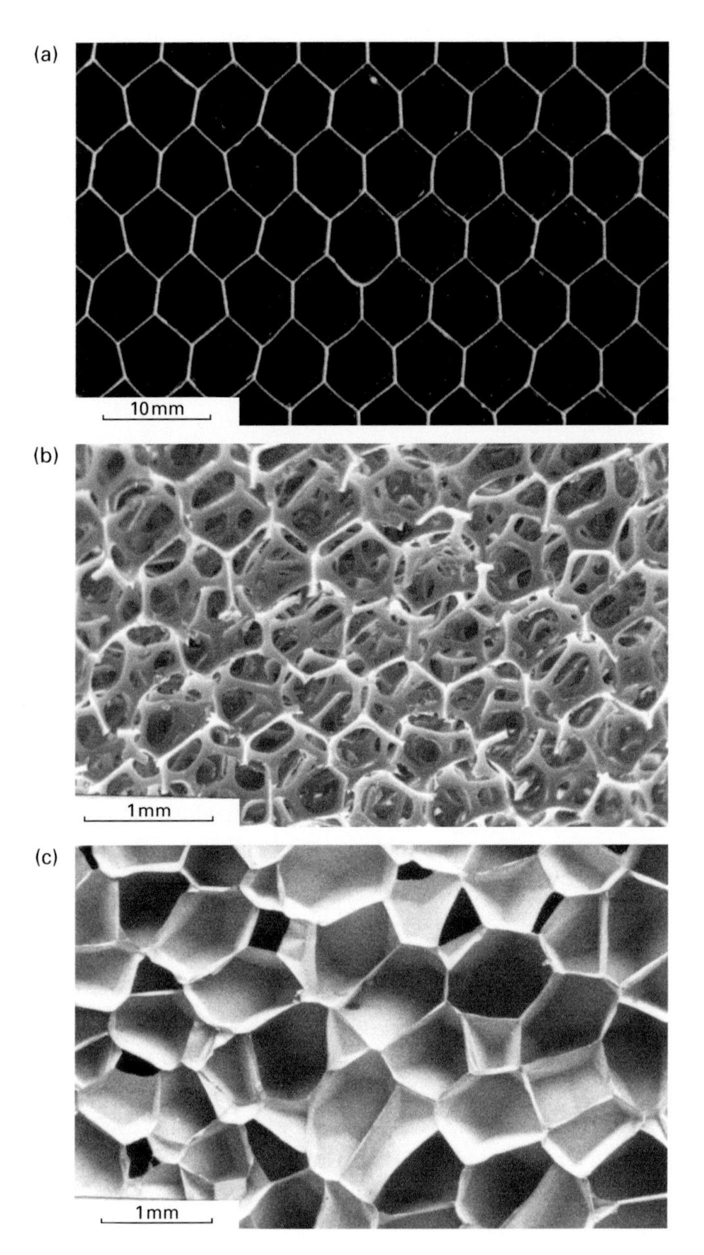

Fig. 3.1 Examples of engineering cellular solids: (a) aluminum honeycomb; (b) open-cell polyurethane foam; (c) closed-cell polyethylene foam.

approach assumes that the cell geometry is similar in foams of different relative densities. It gives the dependence of the foam properties on the relative density and the solid properties, but requires experiments to determine the constants related to the cell geometry. A third approach is to use finite element analysis of either regular or random cellular structures (Silva *et al.*, 1995, van der Burg *et al.*, 1997). Finite element analysis allows local effects, such as imperfections, to be studied (Silva and Gibson, 1997; Chen *et al.*, 2001). It can also be used in conjunction with imaging techniques such

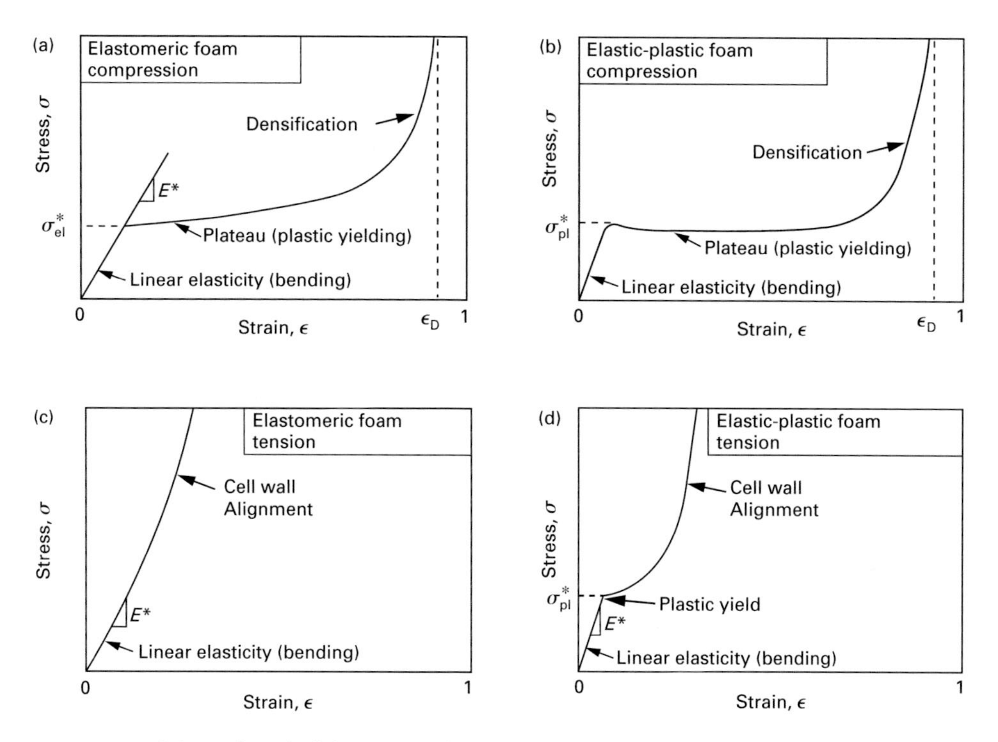

Fig. 3.2 Schematic uniaxial stress–strain curves for (a) elastomeric foam in compression, (b) elastic-plastic foam in compression, (c) elastomeric foam in tension and (d) elastic-plastic foam in tension.

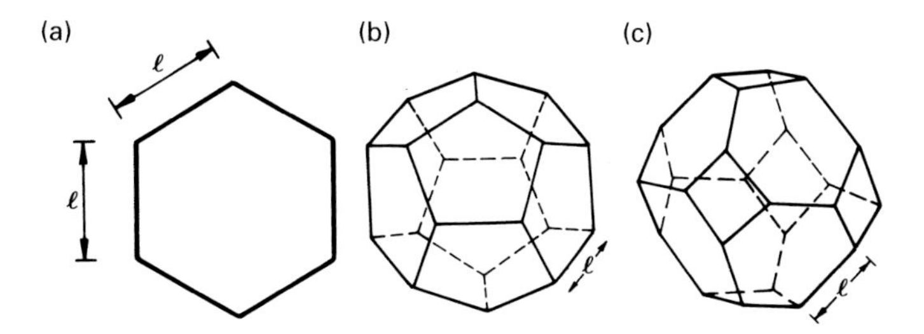

Fig. 3.3 Unit cells: (a) hexagon; (b) pentagonal dodecahedron; and (c) tetrakaidecahedron.

as micro-computed tomography to model the exact geometry of a particular sample (e.g. trabecular bone), although this is computationally intensive (van Reitbergen *et al.*, 1995; Ulrich *et al.* 1999).

Efficient structural members sometimes combine a fully dense solid material with a cellular material. In structural sandwich panels, the separation of two dense, stiff faces by a lightweight, usually cellular, core increases the moment of inertia of the panels, making them especially attractive for resisting bending and buckling. Sandwich structures appear in nature in the leaves of monocotyledon plants, in curved bones such as the skull and pelvis, and in the shells of some arthropods such as the horseshoe crab. In structural members with a circular cross-section, the flexural rigidity can be increased by placing denser, stiffer material towards the periphery of the section; palm and bamboo stems both exploit this strategy by having a radial density gradient. In palm, the cell wall thickness increases, while in bamboo, the volume fraction of dense fibers increases towards the periphery of the stem. Engineers make thin-walled cylindrical shells with internal longitudinal and circumferential stiffeners to resist local buckling; natural cylindrical shell structures, such as animal quills and plant stems, typically have either a honeycomb- or foam-like core that plays a similar role.

In engineering, designers select materials based on the function that they must perform subject to constraints: for instance, a component must carry a given load without failure. Often the designer seeks to optimize the material selection by minimizing some objective, such as cost or mass. The standard methods for selection of engineering materials can also be applied to natural materials.

In this chapter, we first summarize the results of structural analysis of unit cell models for honeycombs. We next review the results of dimensional analysis of foams and compare them to selected results from the analysis of unit cells and finite elements. Many of the results for the mechanical properties of honeycombs and foams, along with extensive data, are described in more detail in Gibson and Ashby (1997). We then examine the mechanics of structural members that have both solid and cellular components, such as structural sandwich panels. Finally, the standard methods for materials selection are described.

3.2 Two-dimensional cellular solids (honeycombs)

3.2.1 In-plane loading

When loaded uniaxially in the plane of the hexagonal cells, the cell walls of a honeycomb initially deform linearly elastically by bending (so long as the wall thickness, t, is small compared with the wall length, l) (Fig. 3.4). The Young's modulus, E^*, can be related to t/l, the modulus of the solid, E_s, and the cell geometry (h/l, θ) using structural mechanics. The relative density of the honeycomb (the density of the honeycomb, ρ^*, normalized by that of the solid, ρ_s) is related to the wall thickness to length ratio, t/l, and the cell geometry (h/l, θ) by

$$\frac{\rho^*}{\rho_s} = \frac{t}{l}\frac{(h/l+2)}{2\cos\theta(h/l+\sin\theta)} \tag{3.1}$$

A stress σ_1 acting in the x_1 direction induces a load P on the end of the inclined cell wall,

$$P = \sigma_1 (h + l\sin\theta) b \tag{3.2}$$

where b is the wall thickness. The wall deflects by

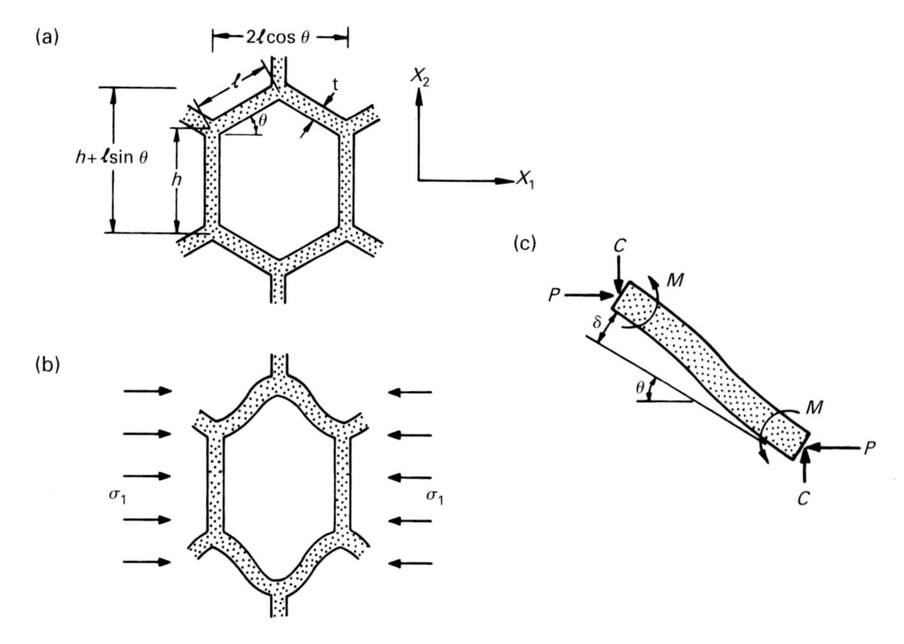

Fig. 3.4 Hexagonal cell: (a) undeformed; (b) linear elastic bending; and (c) the bending caused by the loads in the X_1 direction.

$$\delta = \frac{Pl^3 \sin\theta}{12E_s I} \tag{3.3}$$

where E_s is the Young's modulus of the solid cell wall material and I is the moment of inertia of the wall cross-section ($I = bt^3/12$). The strain in the x_1 direction is then

$$\varepsilon_1 = \frac{\delta \sin\theta}{l\cos\theta} = \frac{\sigma_1 (h + l\sin\theta) bl^2 \sin^2\theta}{12E_S I \cos\theta} \tag{3.4}$$

The Young's modulus parallel to x_1 is then

$$\frac{E_1^*}{E_S} = \left(\frac{t}{l}\right)^3 \frac{\cos\theta}{(h/l + \sin\theta)\sin^2\theta} \tag{3.5}$$

The Young's modulus for loading in the x_2 direction and the shear modulus for loading in the (x_1–x_2) plane, can be found in a similar way (Gibson and Ashby, 1997); they are listed in Table 3.1 (for regular hexagonal honeycombs see Table 3.2). All depend on the Young's modulus of the solid from which the honeycomb is made, the cube of the relative density (since ρ^*/ρ_s is proportional to t/l (3.1)) and a factor related to the cell geometry (through h/l and θ). The Poisson's ratios are found by taking the negative ratio of the strains normal to and parallel to the loading direction, $\nu_{ij}^* = -\varepsilon_j / \varepsilon_i$ (see Table 3.1). To a first approximation, ignoring shear and axial deformations of the cell wall, we find that they depend only on the cell geometry

Table 3.1 Properties of two-dimensional cellular solids (honeycombs)

In-plane properties	Out-of-plane properties
$\frac{\rho^*}{\rho_s} = \frac{t}{l}\frac{h/l+2}{2\cos\theta(h/l+\sin\theta)}$	
Elastic moduli	
$\frac{E_1^*}{E_s} = \left(\frac{t}{l}\right)^3 \frac{\cos\theta}{(h/l+\sin\theta)\sin^2\theta}$	$\frac{E_3^*}{E_s} = \frac{\rho^*}{\rho_s}$
$\frac{E_2^*}{E_s} = \left(\frac{t}{l}\right)^3 \frac{h/l+\sin\theta}{\cos^3\theta}$	
$\nu_{12}^* = -\frac{\varepsilon_2}{\varepsilon_1} = \frac{\cos^2\theta}{(h/l+\sin\theta)\sin\theta}$	$\nu_{31}^* = \nu_{32}^* = \nu_s$
$\nu_{21}^* = -\frac{\varepsilon_1}{\varepsilon_2} = \frac{(h/l+\sin\theta)\sin\theta}{\cos^2\theta}$	$\nu_{13}^* \sim \nu_{23}^* \sim 0$
$\frac{G_{12}^*}{E_s} = \left(\frac{t}{l}\right)^3 \frac{h/l+\sin\theta}{(h/l)^2(1+2h/l)\cos\theta}$	$\frac{G_{13}^*}{G_s} = \frac{G_{23}^*}{G_s} = 0.577\left(\frac{t}{l}\right)$ (regular hexagons)
Collapse stresses	
$\frac{(\sigma_{el}^*)_2}{E_s} = \frac{n^2\pi^2}{24}\frac{t^3}{lh^2}\frac{1}{\cos\theta}$	$\frac{(\sigma_{el}^*)_3}{E_s} \sim \left(\frac{t}{l}\right)^3 \frac{2}{1-\nu_s^2}\frac{l/h+2}{(h/l+\sin\theta)\cos\theta}$
$\frac{(\sigma_{pl}^*)_1}{\sigma_{ys}} = \left(\frac{t}{l}\right)^2 \frac{1}{2(h/l+\sin\theta)\sin\theta}$	$\frac{(\sigma_{pl}^*)_3}{\sigma_{ys}} = 5.6\left(\frac{t}{l}\right)^{5/3}$ plastic buckling; (regular hexagons)
$\frac{(\sigma_{pl}^*)_2}{\sigma_{ys}} = \left(\frac{t}{l}\right)^2 \frac{1}{2\cos^2\theta}$	
$\frac{(\sigma_{cr}^*)_1}{\sigma_{fs}} = \left(\frac{t}{l}\right)^2 \frac{1}{3(h/l+\sin\theta)\sin\theta}$	
$\frac{(\sigma_{cr}^*)_2}{\sigma_{fs}} = \left(\frac{t}{l}\right)^2 \frac{1}{3\cos^2\theta}$	

At a sufficiently high load, cells collapse, by elastic buckling, plastic yielding or brittle crushing, depending on the properties of the cell wall material. The elastic buckling collapse stress is related to the Euler buckling load, P_{crit}, of the vertical column BE (Fig. 3.5a) by

Table 3.2 Properties of regular hexagonal honeycombs

In-plane properties	Out-of-plane properties
$\frac{\rho^*}{\rho_s} = \frac{2}{\sqrt{3}}\frac{t}{l} = 1.15\frac{t}{l}$	
Elastic moduli	
$\frac{E_1^*}{E_s} = \frac{E_2^*}{E_s} = \frac{4}{\sqrt{3}}\left(\frac{t}{l}\right)^3 = 2.31\left(\frac{t}{l}\right)^3$	$\frac{E_3^*}{E_s} = \frac{\rho^*}{\rho_s} = 1.15\frac{t}{l}$
$\nu_{12}^* = \nu_{21}^* = 1$	$\nu_{31}^* = \nu_{32}^* = \nu_s$
$\frac{G_{12}^*}{E_s} = \frac{1}{\sqrt{3}}\left(\frac{t}{l}\right)^3 = 0.577\left(\frac{t}{l}\right)^3$	$\frac{G_{13}^*}{G_s} = \frac{G_{23}^*}{G_s} = 0.577\left(\frac{t}{l}\right)$
Collapse stresses	
$\frac{\sigma_{el}^*}{E_s} = 0.22\left(\frac{t}{l}\right)^3$	$\frac{(\sigma_{el}^*)_3}{E_s} \sim 5.2\left(\frac{t}{l}\right)^3$ for $\nu_s = 0.3$
$\frac{(\sigma_{pl}^*)_1}{\sigma_{ys}} = \frac{(\sigma_{pl}^*)_2}{\sigma_{ys}} == \frac{2}{3}\left(\frac{t}{l}\right)^2$	$\frac{(\sigma_{pl}^*)_3}{\sigma_{ys}} = 5.6\left(\frac{t}{l}\right)^{5/3}$ plastic buckling
$\frac{(\sigma_{cr}^*)_1}{\sigma_{fs}} = \frac{(\sigma_{cr}^*)_2}{\sigma_{fs}} = \frac{4}{9}\left(\frac{t}{l}\right)^2$	
Fracture toughness	
$\frac{K_{IC}^*}{\sigma_{fs}\sqrt{\pi l}} = 0.3\left(\frac{t}{l}\right)^2$	

$$\sigma_{el}^* = \frac{P_{crit}}{2lb\cos\theta} \tag{3.6}$$

where $P_{crit} = \frac{n^2\pi^2 E_s I}{h^2}$ and n is a factor describing the rotational stiffness of the node. We find that

$$\frac{(\sigma_{el}^*)_2}{E_s} = \frac{n^2\pi^2}{24}\frac{t^3}{lh^2}\frac{1}{\cos\theta} \tag{3.7}$$

For $h/l = 1$, $n = 0.686$ and for $h/l = 2$, $n = 0.806$.

The plastic collapse stress, σ_{pl}^*, can be calculated from the plastic moment to form plastic hinges in the cell walls while the brittle crushing stress, σ_{cr}^* can be calculated

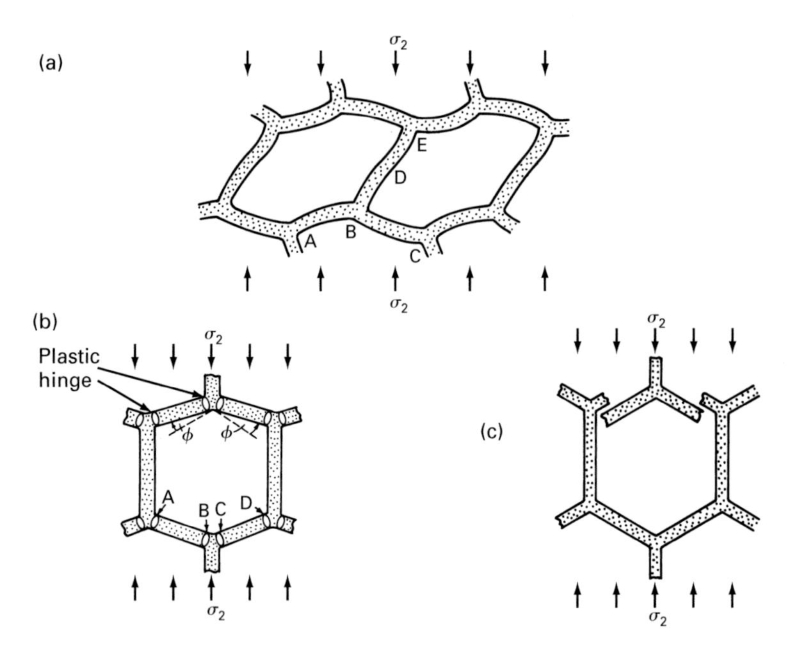

Fig. 3.5 (a) Elastic buckling deformation in an elastomeric honeycomb; (b) the formation of plastic hinges in an elastic-plastic honeycomb; (c) cell-wall fracture in an elastic-brittle honeycomb.

from the moment required to fracture the walls (Fig. 3.5b,c). The plastic collapse stress depends on the solid yield strength, σ_{ys}, while the brittle crushing strength depends on the solid modulus of rupture, σ_{fs}; both depend on $(t/l)^2$ (see Table 3.1).

The shape of the stress–strain curve in the stress plateau regime (post-collapse) either increases monotonically (Fig. 3.2a) or exhibits a slight drop just after the collapse stress and then proceeds at roughly constant stress with increasing strain (Fig. 3.2b). In the first case, cell collapse occurs uniformly throughout the specimen: increasing strain on the specimen causes further collapse in all the cells. This behavior has been observed in model silicone rubber honeycombs (Gibson *et al.*, 1982a). In the second case, cell collapse is localized: collapse initiates in a narrow band of cells which collapse nearly completely before the next, neighboring band of cells begins to collapse (Fig. 3.6). Material in the uncollapsed region remains linear elastic. With increasing strain, the band of collapsed cells propagates throughout the specimen. When all the cells have collapsed, there is a sharp increase in stress corresponding to the densification regime. Localization occurs if there are imperfections in the cellular structure or if the solid material has a non-linear softening stress–strain response; it is commonly observed in aluminum honeycombs (Gibson *et al.*, 1982a; Papka and Kyriakides, 1994; Prakash *et al.*, 1996; 1998a) and in arrays of cylindrical tubes (Poirier *et al.*, 1992; Papka and Kyriakides, 1998b). Finite element analysis of localization in aluminum honeycombs under uniaxial compression and in an array of polycarbonate cylinders in both uniaxial and biaxial compression gives a good description of the experimental results (Papka and Kyriakides, 1994, 1998a, 1998b, 1999a, 1999b) (Fig. 3.7).

Finite element analysis has also been used to study the effect of a random, Voronoi honeycomb structure (Fig. 3.8a), as well as that of defects such as missing cell walls,

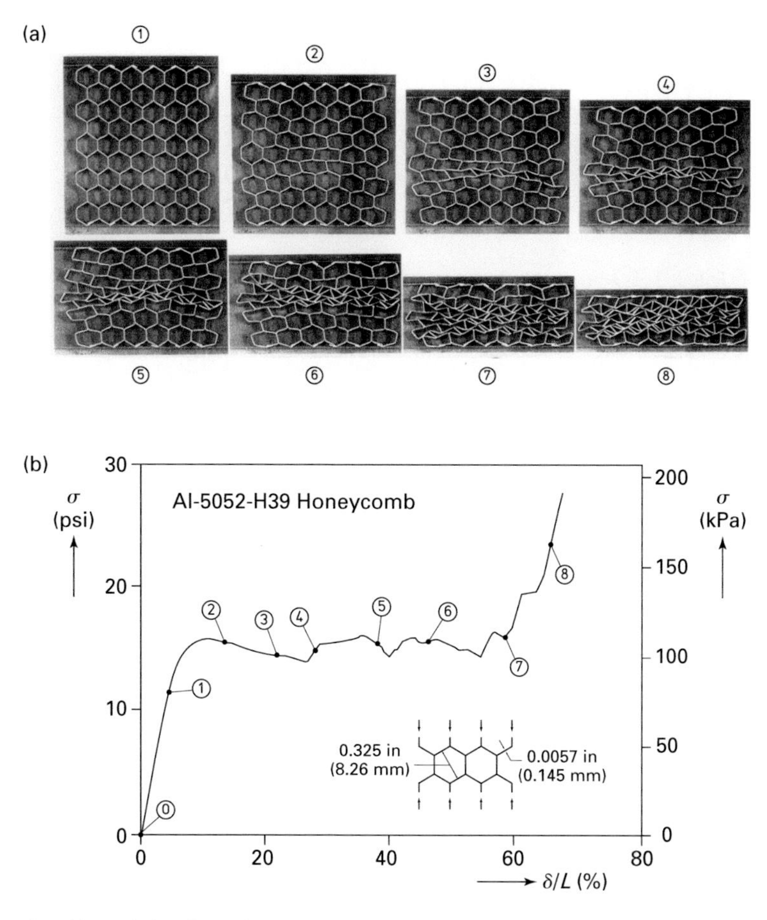

Fig. 3.6 Crushing of aluminum honeycomb: (a) sequence of deformed configurations; (b) recorded load–displacement response. (Reprinted from Papka and Kyriakides, 1994, with permission from Elsevier.)

missing clusters of cell walls and rigid inclusions, on the stiffness and compressive strength of honeycombs. The Young's modulus, shear modulus and Poisson's ratio of a random, Voronoi honeycomb are found to be nearly identical to those of a regular hexagonal honeycomb (Silva *et al.*, 1995) while the compressive strength is roughly three-quarters of that of a regular hexagonal honeycomb, most likely because there is a larger distribution of stresses within the walls of a random honeycomb, so that some are more highly stressed than those in a regular honeycomb (Silva and Gibson, 1997). Randomly removing cell walls has a dramatic effect on the Young's modulus and compressive strength: removal of 10% of the cell walls decreases both to about 40% of the initial value, for both regular hexagonal and Voronoi honeycombs (Fig. 3.8b,c) (Silva and Gibson, 1997). For comparison, reducing the relative density of an intact honeycomb by 10% by thinning the cell walls, rather than removing them, decreases both the modulus and compressive strength to about 80% of the initial values. While removing cell walls results in a reduction in the modulus and strength,

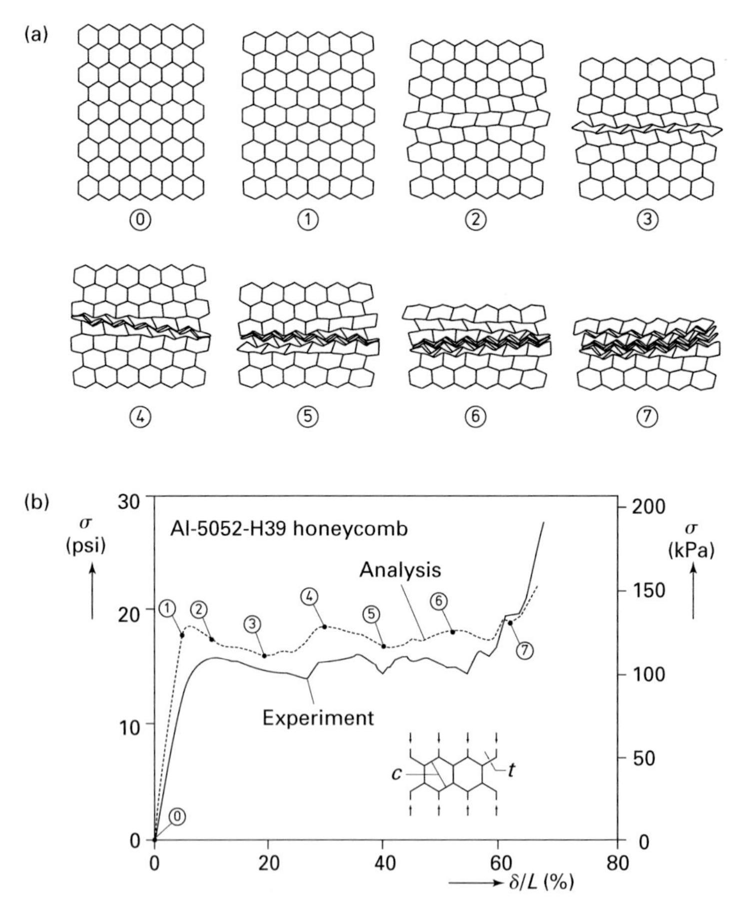

Fig. 3.7 (a) Sequence of calculated collapse configurations; (b) comparison of measured and calculated stress–strain curve for a 9 × 6 cell honeycomb specimen. (Reprinted from Papka and Kyriakides, 1994, with permission from Elsevier).

they still depend on $(\rho^*/\rho_s)^3$ and $(\rho^*/\rho_s)^2$, respectively (Guo and Gibson, 1999; Chen *et al.*, 2001). Removal of cell walls influences the localization of failure so that the initial collapse band contains the cells with missing walls (Silva and Gibson, 1997; Guo and Gibson, 1999). Low volume fractions of rigid inclusions (filling the cell voids) have little effect on the elastic moduli and uniaxial and hydrostatic compressive strengths (Chen *et al.*, 2001).

The tensile fracture strength can be calculated using the methods of fracture mechanics. If, in a brittle honeycomb loaded in tension to near its fracture stress, one cell wall fails, the stress on the neighboring walls increases and they, too, will fail: the failed cluster is like a crack. Consider a brittle honeycomb containing a crack (Fig. 3.9). When it is loaded in tension, the cell walls at first bend elastically. When the forces acting on the cell wall just ahead of the crack tip are sufficient to fracture it, the crack advances. The fracture toughness can be calculated by relating the local stress on the first wall ahead of the crack tip to the remote stress and to the bending moment on the first wall ahead of the crack tip. Fracture occurs when the bending moment

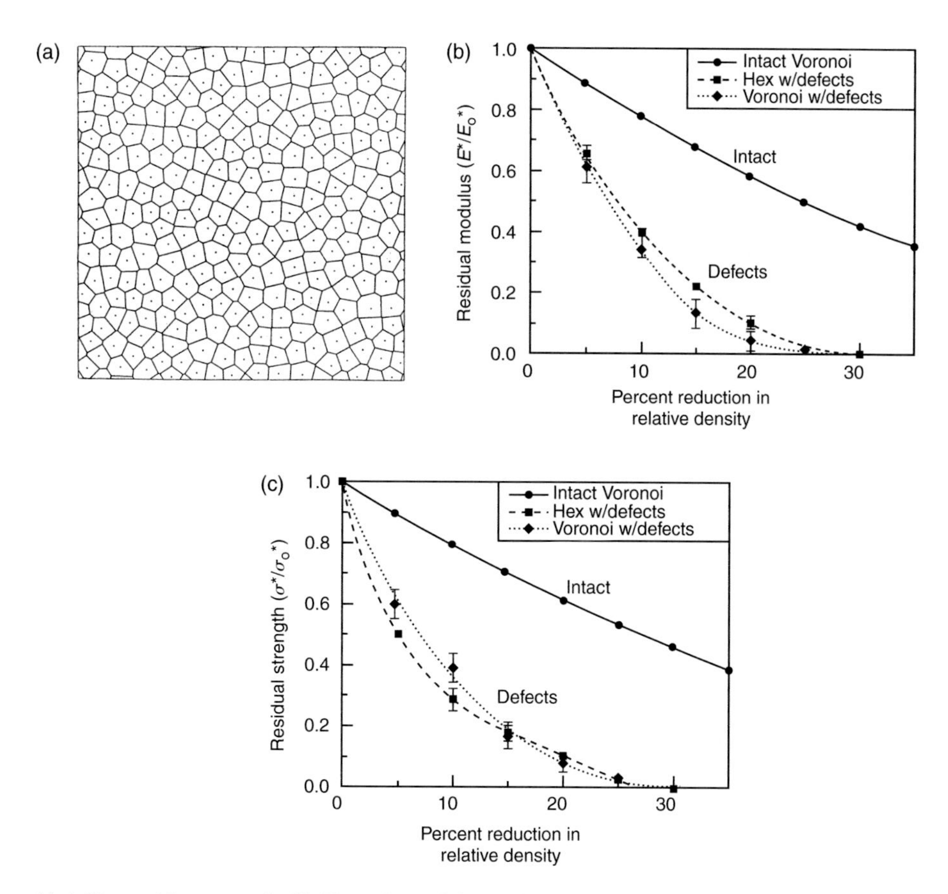

Fig. 3.8 (a) A Voronoi honeycomb, (b) Young's modulus and (c) compressive strength of honeycombs after randomly removing cell walls, normalized by the intact modulus and strength, plotted as a function of the percentage reduction in relative density. (b,c, Reprinted from Silva and Gibson (1997), with permission from Elsevier).

reaches the moment to fracture the cell wall material, which depends on the modulus of rupture of the cell wall. The fracture toughness for regular hexagonal honeycombs is given by

$$\frac{K_{IC}^*}{\sigma_{fs}\sqrt{\pi l}} = 0.3\left(\frac{t}{l}\right)^2 \tag{3.8}$$

This calculation assumes that the crack length is large relative to the cell size, that the contribution of axial forces to the internal stress in the cell wall ahead of the crack tip is negligible and that the modulus of rupture for the cell wall is constant. In addition, the relationship between the local stress on the first wall ahead of the crack tip and the bending moment cannot be calculated exactly, so that the constant of proportionality in (3.8) is approximate. These limitations are discussed in further detail in Gibson and Ashby (1997).

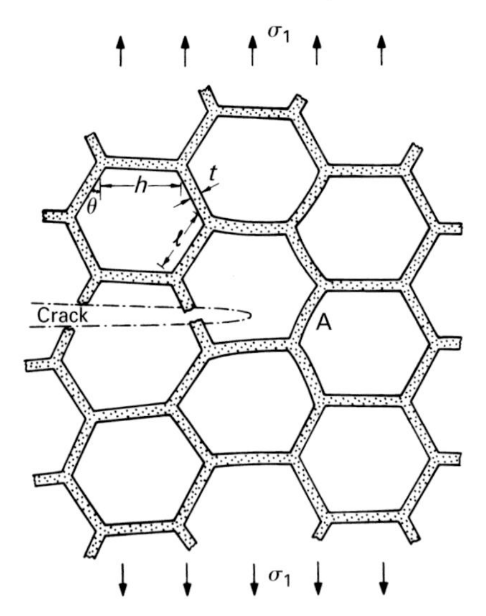

Fig. 3.9 Crack propagation leading to brittle tensile failure in a honeycomb for loading in the X_1 direction. The stress concentration at the crack tip causes the cell wall just beyond the tip to be loaded more heavily than any other.

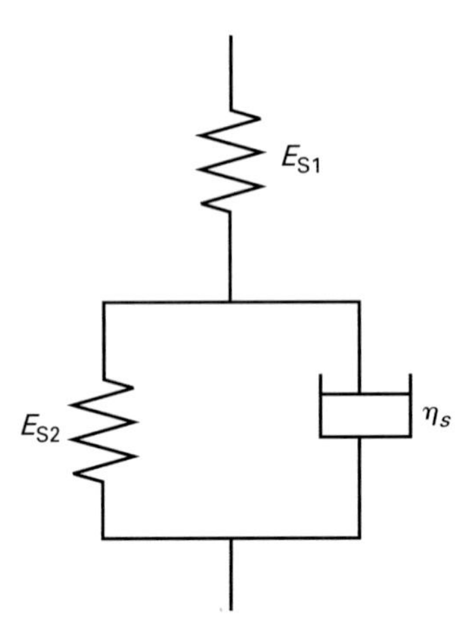

Fig. 3.10 The standard linear solid.

A honeycomb made from a linear viscoelastic material will also be linear viscoelastic. If the solid can be modeled as a standard linear solid (Fig. 3.10), with a spring of stiffness E_{s1} in series with a Voigt element (with a spring of stiffness E_{s2} in parallel with a dashpot of viscosity η_s), then the differential equation describing the viscoelastic response of a regular hexagonal honeycomb is given by

$$\dot{\varepsilon} + \frac{E_{s2}}{\eta_s}\varepsilon = \frac{\sqrt{3}l^3}{4t^3}\left\{\frac{\dot{\sigma}}{E_{s1}} + \frac{\sigma}{\eta_s}\left(\frac{E_{s1}+E_{s2}}{E_{s1}}\right)\right\} \tag{3.9}$$

The honeycomb's time-dependent response can be found by integrating this equation, subject to the appropriate boundary conditions. It is informative to examine a simple case, that of a uniaxial stress, σ. When loading is very rapid, the first term on each side of the equation dominates and the equation as a whole integrates to

$$\varepsilon = \frac{\sqrt{3}l^3}{4t^3}\frac{\sigma}{E_{s1}} \tag{3.10}$$

where E_{s1} is the unrelaxed modulus of the solid. The honeycomb is linear elastic with modulus $E^* = \sigma/\varepsilon$, which is identical to that given by (3.5) for a regular hexagon (Table 3.2). When, instead, loading is very slow, the first terms on either side of (3.9) become negligible and the honeycomb is again perfectly elastic, but with a lower modulus:

$$\varepsilon = \frac{\sqrt{3}l^3}{4t^3}\sigma\left(\frac{E_{s1}+E_{s2}}{E_{s1}E_{s2}}\right) \tag{3.11}$$

where $E_{s1}E_{s2}/(E_{s1} + E_{s2})$ is known as the relaxed modulus of the solid.

When the load is held constant, the honeycomb creeps, with time-dependent strain. Setting $\dot{\sigma} = 0$ and integrating gives

$$\varepsilon = \frac{\sqrt{3}l^3}{4t^3}\sigma\left(\frac{E_{s1}+E_{s2}}{E_{s1}E_{s2}}\right)\left[1-\exp\left(-\frac{E_{s2}}{\eta_s}t\right)\right] \tag{3.12}$$

where t is the time. The time constant for this relaxation is η_s / E_{s2} and is a property of the cell wall material only.

3.2.2 Out-of-plane loading

For loading along the prism axis, in the out-of-plane direction, the cell walls of a honeycomb initially compress axially, so that the Young's modulus simply varies with the volume fraction of solid, or with the relative density (Fig. 3.11):

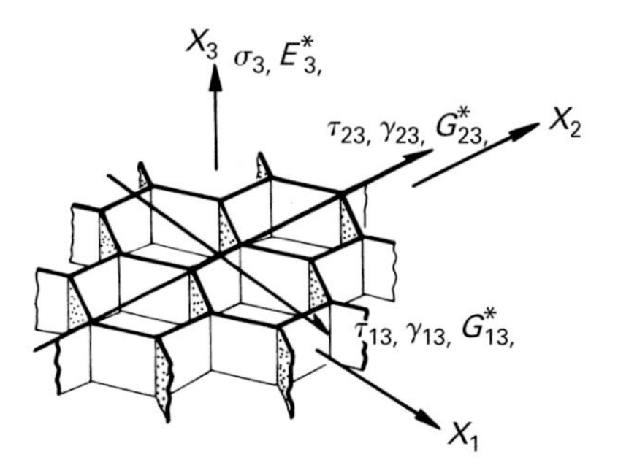

Fig. 3.11 Out-of-plane loading of a regular hexagonal honeycomb.

$$\frac{E_3^*}{E_s} = \frac{\rho^*}{\rho_s} \tag{3.13}$$

The Poisson's ratios, ν_{31}^* and ν_{32}^*, corresponding to the strain in the x_1 or x_2 direction on loading in the out-of-plane direction are simply the Poisson's ratio of the solid. The Poisson's ratios ν_{13}^* and ν_{23}^*, corresponding to the strain in the x_3 direction on loading in the x_1 or x_2 direction, are essentially zero. The out-of-plane shear moduli, G_{13}^* and G_{23}^*, involve stretching of the walls of the honeycomb and vary linearly with t/l or ρ^*/ρ_s (see Table 3.1).

If buckling is avoided, the yield strength and brittle crushing strength for loading in the x_3 direction are equal to the relative density times the yield or fracture strength of the solid, respectively. The stresses corresponding to elastic and plastic buckling are given in Table 3.1 (Gibson and Ashby, 1997).

The mechanical properties of honeycombs (apart from Poisson's ratio) depend on three parameters: the ratio of t/l, the properties of the solid cell wall materials (E_s, σ_{ys} and σ_{fs}) and the cell geometry (through h/l and θ) (Tables 3.1 and 3.2). The ratio of t/l is linearly related to the relative density, or the volume fraction of solids. The in-plane moduli and compressive strengths, dominated by bending or buckling of the honeycomb walls, vary as the relative density cubed or squared. The out-of-plane moduli and strengths, dominated by axial stretching of the walls (as long as buckling is prevented), vary linearly with the relative density. We now make use of these observations in analyzing foam behavior.

3.3 Three-dimensional cellular solids (foams)

The geometry of foams is much more complicated than that of honeycombs. The mechanisms by which foams deform and fail are similar to those in honeycombs: edge bending, buckling and yielding have been observed in polymer foams using a deformation stage in a scanning electron microscope (Gibson and Ashby, 1982b) and in aluminum foams using micro-computed tomography, for instance (Bart-Smith *et al.*, 1998; Nazarian and Muller, 2004). Here, we analyze the mechanisms of deformation and failure of foams using dimensional arguments that do not use a particular cell geometry, but instead assume that the cells in foams of different relative densities are geometrically similar. While we use a cubic cell to illustrate the argument, the same result is obtained for *any* cell geometry, so long as the mode of deformation or failure is the same.

3.3.1 Linear elasticity

In the linear elastic regime, under uniaxial stress, low-density open-cell foams deform primarily by bending of the cell edges. The Young's modulus, E^*, can be estimated as follows (Fig. 3.12). Under a transverse load, F, the bending deflection, δ, of a strut of length, l, and cross-sectional area proportional to t^2, is given by

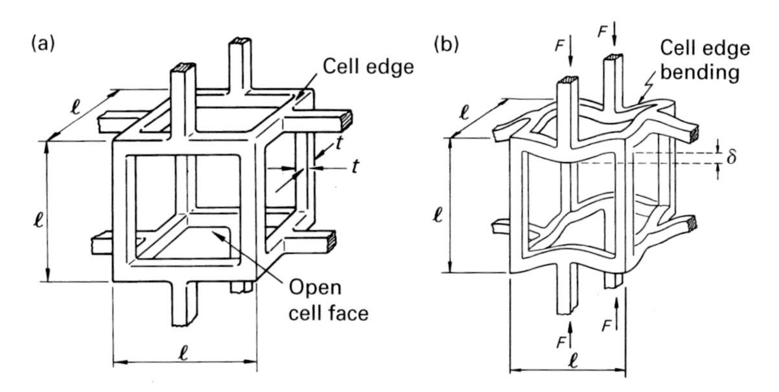

Fig. 3.12 Dimensional analysis for an open-cell foam: (a) undeformed cell; (b) linear elastic strut bending.

$$\delta \propto \frac{Fl^3}{E_s I} \tag{3.14}$$

where E_s is the Young's modulus of the solid and I is the moment of inertia (or second moment of area) and $I \propto t^4$.

The stress acting on the cell is proportional to F/l^2 and the strain is proportional to δ/l, giving

$$\frac{E^*}{E_s} \propto \left(\frac{t}{l}\right)^4 \tag{3.15}$$

The relative density of any open-cell foam, ρ^*/ρ_s, is proportional to the square of the ratio of the strut thickness to length, t/l, so that

$$\frac{E^*}{E_s} = C_1\left(\frac{\rho^*}{\rho_s}\right)^2 \tag{3.16}$$

The analysis gives the dependence of the Young's modulus on the solid modulus and the relative density, and combines all of the constants of proportionality related to the cell geometry into a single constant, C_1. By fitting eqn (3.16) to data, we find that $C_1 \sim 1$ (Gibson and Ashby, 1997). A structural analysis of a tetrakaidecaheral unit cell model with cell edges with Plateau borders finds that $C_1 = 0.98$ (Warren and Kraynik, 1997). Finite element analysis of random three-dimensional Voronoi foams with struts of circular cross-section finds that $C_1 = 0.8$ (Vajjhala *et al.*, 2000). All three approaches give similar results and all give the same dependence of modulus on the square of the relative density.

As for honeycombs, the Poisson's ratio of foams depends only on cell geometry and does not depend on relative density (assuming that bending deformations dominate). Typical values lie around 1/3, although there is a wide range in reported values.

Re-entrant foams, with negative Poisson's ratios, have been reported with values of $-0.8 < \nu < 0$ (see, for example, Lakes, 1987).

Like the Young's modulus, the shear modulus of open-cell foams is related to the bending stiffness of the cell edges, and depends on the square of the relative density. For an isotropic foam with a Poisson's ratio of 1/3, the shear modulus is

$$\frac{G^*}{E_s} = \frac{3}{8}\left(\frac{\rho^*}{\rho_s}\right)^2 \tag{3.17}$$

3.3.2 Compressive strength

As in honeycombs, the cells in foams can collapse by elastic buckling, plastic yielding or brittle crushing, depending on the nature of the cell wall material. The elastic collapse stress in an open-cell foam is proportional to the Euler buckling load divided by l^2 (Fig. 3.13a):

$$\sigma^*_{el} \propto \frac{P_{crit}}{l^2} \propto \frac{E_s t^4}{l^4} \tag{3.18}$$

or

$$\frac{\sigma^*_{el}}{E_s} = C_2\left(\frac{\rho^*}{\rho_s}\right)^2 \tag{3.19}$$

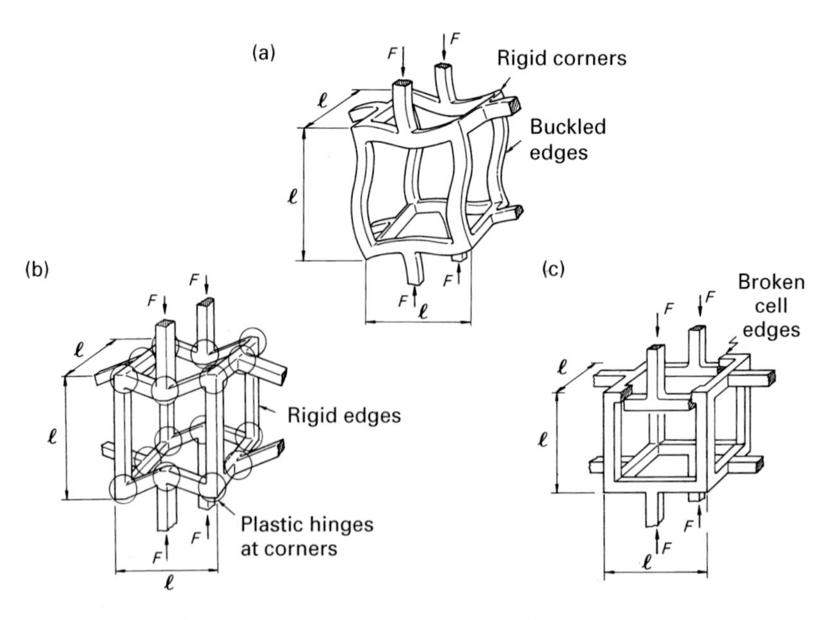

Fig. 3.13 Dimensional analysis of cell collapse in an open-cell foam by (a) elastic buckling, (b) plastic yielding and (c) brittle fracture.

Fitting (3.19) to data gives $C_2 \sim 0.05$ (Gibson and Ashby, 1997). Structural analysis of a tetrakaidecahedral unit cell gives $C_2 \sim 0.1$–0.18, depending on the cross-section of the edges, for loading in one direction; loading normal to the square faces does not produce buckling (Zhu *et al.*, 1997b).

The plastic collapse stress, σ_{pl}^*, is found by equating the applied moment, M, on a strut from a transverse force, F, to the plastic moment, M_p, required to form plastic hinges (Fig. 3.13b):

$$M \propto Fl \propto \sigma_{pl}^* l^3 \tag{3.20}$$

$$M_p \propto \sigma_{ys} t^3 \tag{3.21}$$

giving

$$\frac{\sigma_{pl}^*}{\sigma_{ys}} \propto \left(\frac{t}{l}\right)^3 = C_3 \left(\frac{\rho^*}{\rho_s}\right)^{3/2} \tag{3.22}$$

for an open-cell foam, where σ_{ys} is the yield strength of the solid cell wall material. Fitting (3.22) to data gives $C_3 \sim 0.3$ (Gibson and Ashby, 1997).

The brittle crushing strength, σ_{cr}^*, is found in a similar manner, with (Fig. 3.13c):

$$\frac{\sigma_{cr}^*}{\sigma_{fs}} = C_4 \left(\frac{\rho^*}{\rho_s}\right)^{3/2} \tag{3.23}$$

where σ_{fs} is the modulus of rupture of the solid cell wall material. Fitting (3.23) to data gives $C_4 \sim 0.2$ (Gibson and Ashby, 1997).

In closed-cell foams, stretching of the cell faces also contributes to the mechanical response. In many polymer foams, the faces are thinner than the cell edges, as polymer is drawn away from the faces and into the edges by surface tension forces during processing. The contribution of the faces to the overall mechanical response depends, then, on the fraction of the solid in the edges, ϕ, and adds a linear relative density term to the expressions already obtained for the Young's modulus (see Table 3.3) and the collapse stresses. The derivation of the properties of closed-cell foams is given in Gibson and Ashby (1997).

As with honeycombs, localization of cell collapse can occur if there are geometric imperfections or if the solid material has a non-linear, softening stress–strain curve. Strains within the locally collapsed band of cells have been measured in aluminum foams using surface strain mapping of specimens compressed to incrementally increasing global strains (Bart-Smith *et al.*, 1998) and by micro-computed tomography of specimens deformed in a stage (Nazarian and Muller, 2004). The behavior is similar to that in aluminum honeycombs: collapse initiates in a narrow band of cells and then propagates to neighboring cells while the uncollapsed region remains in the linear elastic regime (Fig. 3.14).

Table 3.3 Properties of three-dimensional cellular solids (foams)

Open-cell foams	
Young's modulus	$\frac{E^*}{E_s} = C_1\left(\frac{\rho^*}{\rho_s}\right)^2$
Shear modulus	$\frac{G^*}{E_s} = \frac{3}{8}\left(\frac{\rho^*}{\rho_s}\right)^2$
Elastic collapse strength	$\frac{\sigma^*_{el}}{E_s} = C_2\left(\frac{\rho^*}{\rho_s}\right)^2$
Plastic collapse strength	$\frac{\sigma^*_{pl}}{\sigma_{ys}} \propto \left(\frac{t}{l}\right)^3 = C_3\left(\frac{\rho^*}{\rho_s}\right)^{3/2}$
Brittle crushing strength	$\frac{\sigma^*_{cr}}{\sigma_{fs}} = C_4\left(\frac{\rho^*}{\rho_s}\right)^{3/2}$
Densification	$\varepsilon_D = 1 - 1.4\left(\frac{\rho^*}{\rho_s}\right)$
Fracture toughness	$\frac{K^*_{IC}}{\sigma_{fs}\sqrt{\pi l}} = C_5\left(\frac{\rho^*}{\rho_s}\right)^{3/2}$
Viscoelastic behavior	$\dot{\varepsilon} + \frac{E_{s2}}{\eta_s}\varepsilon = \frac{1}{C_1(\rho^*/\rho_s)^2}\left\{\frac{\dot{\sigma}}{E_{s1}} + \frac{\sigma}{\eta_s}\left(\frac{E_{s1}+E_{s2}}{E_{s1}}\right)\right\}$
Contribution of fluid flow through an open-cell foam	$\sigma^*_f = \frac{C\mu\dot{h}R^2}{hk}$
Closed-cell foams	
Modulus of closed-cell foam, accounting for edge bending, face stretching and gas pressure	$\frac{E^*}{E_s} = \phi^2\left(\frac{\rho^*}{\rho_s}\right)^2 + (1-\phi)\frac{\rho^*}{\rho_s} + \frac{p_o(1-2\nu^*)}{E_s(1-\rho^*/\rho_s)}$
Post-collapse stress for an elastomeric closed-cell foam with an enclosed gas	$\sigma^* = \sigma^*_{el} + \frac{p_o\varepsilon}{1-\varepsilon-\rho^*/\rho_s}$
Young's modulus of a liquid-filled closed-cell foam	$\frac{E^*}{E_s} = \left(\frac{\rho^*}{\rho_s}\right)\left[1+\left(\frac{\rho^*}{\rho_s}\right)\right]$

The effect of missing cell walls on the elastic moduli of foams has been modeled using finite element analysis. Three-dimensional polyhedral cells (either periodic tetrakaidecahedra or Voronoi cells) are less sensitive to defects than two-dimensional honeycombs: Fig. 3.15 shows the Young's modulus of the cellular structures with missing cell walls normalized by the intact modulus plotted as a function of the reduction in relative density from randomly removing cell walls, for both regular and random honeycombs and foams (Vajjhala *et al.*, 2000, Guo and Kim, 2002). This result is of interest in understanding resorption of trabeculae in osteoporosis (Chapter 5). Other

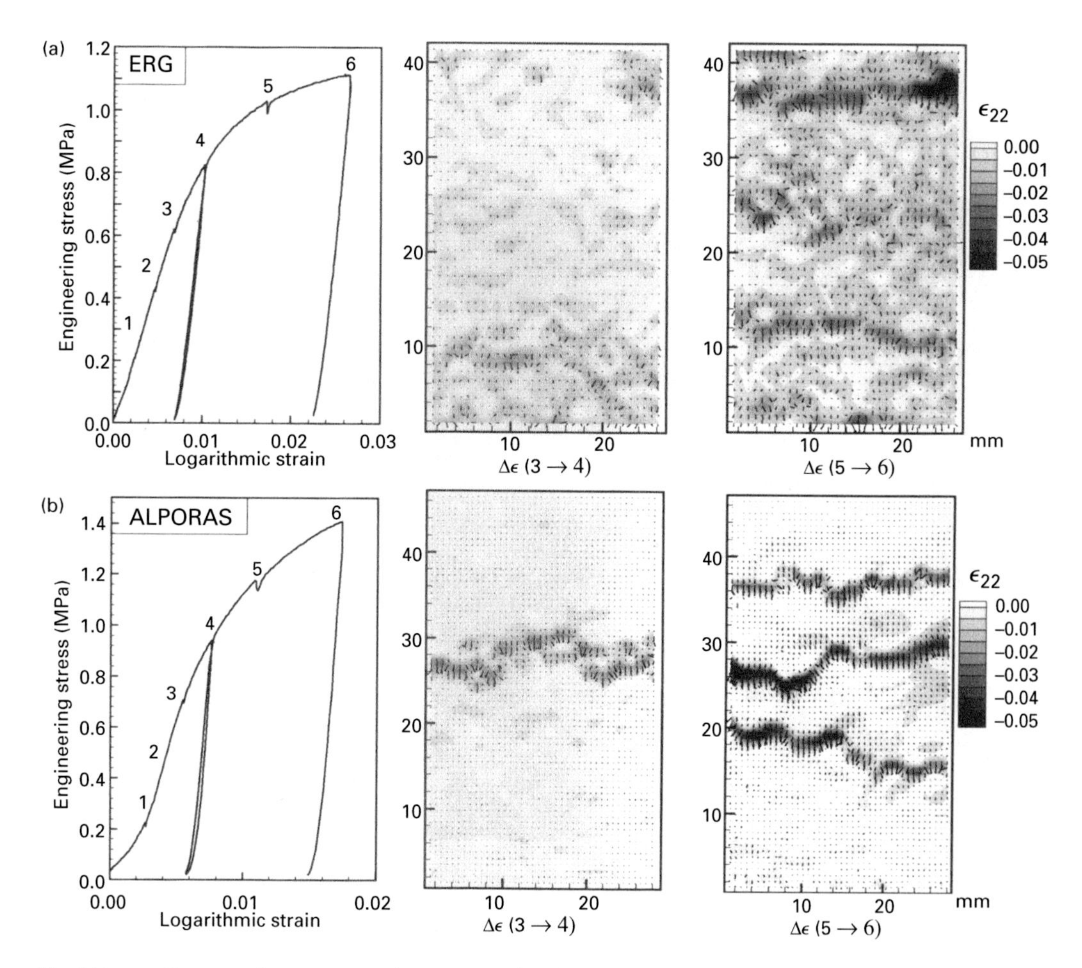

Fig. 3.14 Compressive stress–strain curves and surface strain maps for incremental strains between points 3 and 4, and 5 and 6 on the stress–strain curve. (a) Open-cell aluminum foam (ERG, Oakland, CA); (b) closed-cell aluminum foam (Alporas, Shinko Wire, Japan). (Reprinted from Bart-Smith *et al.*, 1998, with permission from Elsevier).

types of defects (curved cell walls, variations in cell wall thickness) have been analyzed by Simone and Gibson (1998a,b) and by Grenestedt and co-workers (Grenestedt, 1998; Grenestedt and Tanaka, 1999 and Grenestedt and Bassinet, 2000).

3.3.3 Densification

At large compressive strains all of the cells have collapsed and opposing cell walls touch and press against one another. When this happens, the stress–strain curve rises sharply, tending to a slope of E_s at a limiting strain of ε_D. One might expect that this limiting strain would simply be equal to the porosity $(1 - \rho^*/\rho_s)$ as this is the strain at which all the pore space has been squeezed out. In reality, the cell walls jam together at a rather smaller strain than this. Experimental data indicate that

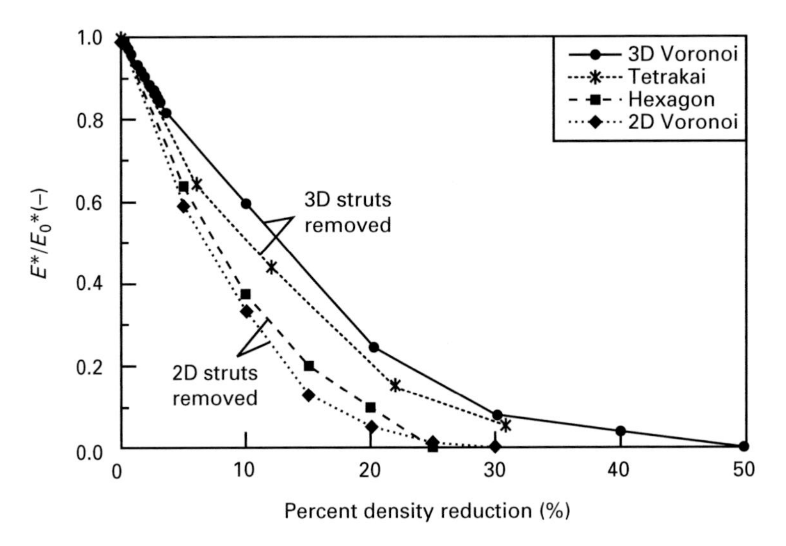

Fig. 3.15 The modulus of cellular structures with struts randomly removed normalized by that of the initial intact structure, plotted against the reduction in density from removing struts. Three-dimensional (3D) Voronoi results from Vajjhala *et al.* (2000); tetrakaidecahedron results from Guo and Kim (2002), hexagonal and two-dimensional (2D) Voronoi honeycomb results from Silva and Gibson (1997). (Reprinted from Vajjhala *et al.*, 2000, with permission of the American Society of Mechanical Engineers.)

$$\varepsilon_{\mathrm{D}} = 1 - 1.4\left(\frac{\rho^*}{\rho_{\mathrm{s}}}\right) \tag{3.24}$$

3.3.4 Fracture toughness

The fracture toughness of foams is analyzed in a manner similar to that of honeycombs. Consider a brittle foam with an edge crack (Fig. 3.16). When it is loaded in tension, the cell walls at first bend elastically. When the forces acting on the cell wall just ahead of the crack tip are sufficient to fracture it, the crack advances. The fracture toughness can be calculated by relating the local stress on the first wall ahead of the crack tip to the remote stress and to the bending moment on the first wall ahead of the crack tip. Fracture occurs when the bending moment within the cell wall reaches the moment to fracture it. The fracture toughness, K_{IC}^*, for open-cell foams is given by

$$\frac{K_{\mathrm{IC}}^*}{\sigma_{\mathrm{fs}}\sqrt{\pi l}} = C_5\left(\frac{\rho^*}{\rho_{\mathrm{s}}}\right)^{3/2} \tag{3.25}$$

where l is the cell size. Comparison of (3.25) with data suggests that $C_5 \sim 0.65$ (Gibson and Ashby, 1997). As for honeycombs, the derivation of this equation assumes that the crack length is large compared with the cell size, that the contribution of the axial

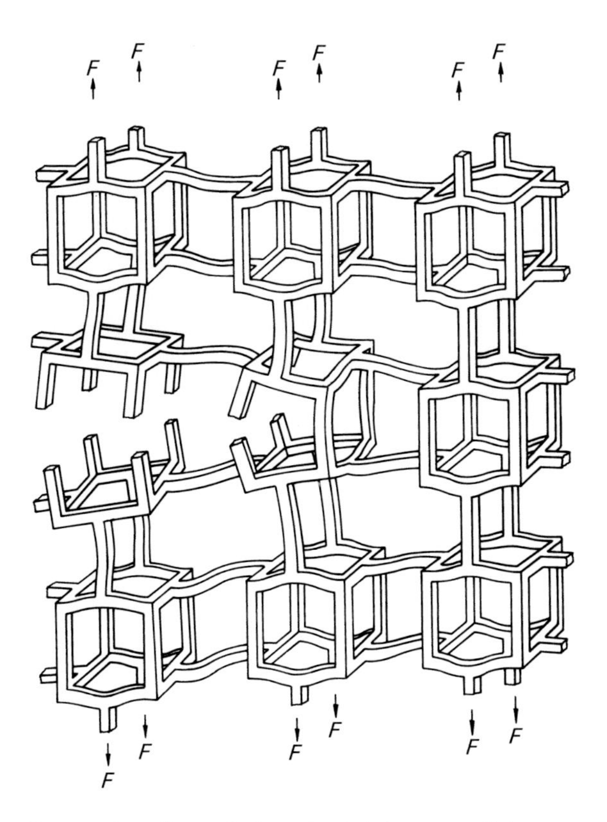

Fig. 3.16 Propagation of a crack through a brittle open-cell foam.

forces in the cell walls to the internal stress ahead of the crack tip is negligible and that the modulus of rupture of the solid, σ_{fs}, is constant.

3.3.5 Time-dependent behavior

Polymers near their glass transition temperature (as many are, at room temperature) are viscoelastic. A foam made from a linear viscoelastic polymer is itself linear viscoelastic (e.g. in a creep test, the strain at a particular time is linearly related to the applied stress). Its response to a given history of loading is found by an extension of the method developed for honeycombs in Section 3.2. The cell edges are treated as linear viscoelastic beams loaded in bending and their deflection and deflection rate are calculated, assuming that the solid behaves as a standard linear solid (Fig. 3.10). The load per beam, P, is related to the stress by $P \propto \sigma l^2$, and the strain, ε, and strain-rate, $\dot{\varepsilon}$, of the foam are related to the deflection and deflection-rate of the beam by $\varepsilon \propto \delta / l$ and $\dot{\varepsilon} \propto \dot{\delta} / l$, leading to the differential equation for linear viscoelastic deformation of an open-cell foam:

$$\dot{\varepsilon} + \frac{E_{s2}}{\eta_s}\varepsilon = \frac{1}{C_1 (\rho^* / \rho_s)^2}\left\{\frac{\dot{\sigma}}{E_{s1}} + \frac{\sigma}{\eta_s}\left(\frac{E_{s1} + E_{s2}}{E_{s1}}\right)\right\} \tag{3.26}$$

with $C_1 \sim 1$. As for honeycombs, E_{s1} is the unrelaxed modulus of the solid, $E_{s1}E_{s2}/(E_{s1}+E_{s2})$ is its relaxed modulus and η_s is its viscosity. Note that the foam properties enter only as a multiplying factor on the right-hand side of the equation; other than this, the equation is identical with that for the solid polymer itself.

The response of the foam to any given loading history is found by integrating this equation and applying the appropriate boundary conditions. But the simplicity of the equation, referred to above, means that we can use the well-known solutions for solid viscoelastic bodies (see, for example, Powell, 1983); all that is necessary is to replace the stress σ in the solution by $\sigma/C_1(\rho^*/\rho_s)^2$ (or the equivalent factor for closed-cell foams). Thus the response to a steady uniaxial stress, σ, applied for a time, t, so that $\dot{\sigma} = 0$, is

$$\varepsilon = \frac{\sigma}{C_1(\rho^*/\rho_s)^2}\left(\frac{E_{s1}+E_{s2}}{E_{s1}E_{s2}}\right)\left(1-\exp-\frac{E_{s2}}{\eta_s}t\right) \tag{3.27}$$

This has the same relaxation time as the solid, but has strains which are larger, by the factor $1/C_1(\rho^*/\rho_s)^2$, at any given time, t.

The response to a steady displacement (so that $\dot{\varepsilon} = 0$) is

$$\sigma = C_1(\rho^*/\rho_s)^2 E_{s1}\varepsilon\left\{1-\frac{E_{s1}}{E_{s1}+E_{s2}}\left(1-\exp-\frac{(E_{s1}+E_{s2})}{\eta_s}t\right)\right\} \tag{3.28}$$

which again has the same relaxation time as the solid, but has stresses which are less, by the factor $C_1(\rho^*/\rho_s)^2$, at any given time, t. Other solutions (such as that for a sinusoidally varying stress) are adapted in the same way. Note that (3.26)–(3.28) do not describe creep fracture.

Metals and ceramics creep according to a power law: the steady-state, secondary strain rate, $\dot{\varepsilon}_{ss}$, depends on the stress, σ, raised to a power (Fig. 3.17):

$$\dot{\varepsilon}_{ss} = \dot{\varepsilon}_{os}\left(\frac{\sigma}{\sigma_{os}}\right)^{n_s} \tag{3.29}$$

where $\dot{\varepsilon}_{os}$, σ_{OS} and n_s are material constants. The parameter $\dot{\varepsilon}_{os}$ includes the temperature dependence through the Arrhenius relationship, $\dot{\varepsilon}_{os} = A\exp(-Q/RT)$, where A is a constant, Q is the activation energy for the creep process, R is the ideal gas constant and T is the absolute temperature. The steady-state, secondary creep strain-rate of a foam, $\dot{\varepsilon}^*_{ss}$, can be related to that of the solid from which it is made by calculating the creep-rate of a cell wall as a deflecting beam (Gibson and Ashby, 1997):

$$\frac{\dot{\varepsilon}^*_{ss}}{\dot{\varepsilon}_{os}} = C\left(\frac{\sigma}{\sigma_{os}}\right)^{n_s}\left(\frac{\rho_s}{\rho^*}\right)^{\frac{3n_s+1}{2}} \tag{3.30}$$

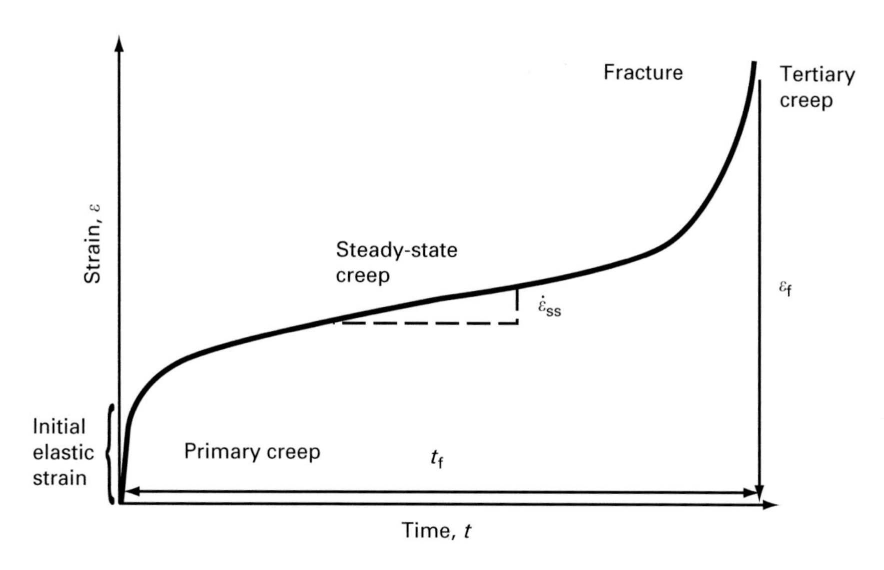

Fig. 3.17 Strain plotted against time in a creep test, showing regimes of primary, secondary and tertiary creep (Reprinted from Ashby and Jones, 1996.)

where the constant C depends on the creep exponent n_s and the cell geometry. Equation (3.30) has been confirmed in creep tests on open-cell aluminum foams (Andrews *et al.*, 1999) and in finite element simulations (Huang and Gibson, 2003). Data for open-cell aluminum foams (Andrews *et al.*, 1999) also confirms the well-known Monkman–Grant relationship that relates the time to failure, t_f, to the secondary creep strain rate, $\dot{\varepsilon}_{ss}$ (Monkman and Grant, 1956; Finnie and Heller, 1959):

$$t_f = A\dot{\varepsilon}_{ss}^{-m} \tag{3.31}$$

where A and m are constants and m is typically close to 1.

3.3.6 Fatigue

Fatigue of foams is relevant to the compressive fatigue of trabecular bone, thought to be responsible for roughly half of vertebral fractures in patients with osteoporosis. Compressive fatigue has been well documented in aluminum foams (Harte *et al.*, 1999; Sugimura *et al.*, 1999; McCullough *et al.*, 2000; Zettl *et al.*, 2000 and Zhou and Soboyejo, 2004). The main observations are as follows:

(1) Specimens progressively shorten with increasing numbers of cycles of load until a threshold is reached, at which point the compressive strain increases dramatically; compressive fatigue failure is defined as the number of cycles of loading at this threshold, N_T (Fig. 3.18a).
(2) In some closed-cell foams, the initial progressive shortening of the specimens is due to plastically buckled membranes, often in unusually large cells in the foam. The plastically buckled membranes eventually crack, leading to the formation of a local band of collapsed cells and the dramatic strain increase at N_T (Sugimura

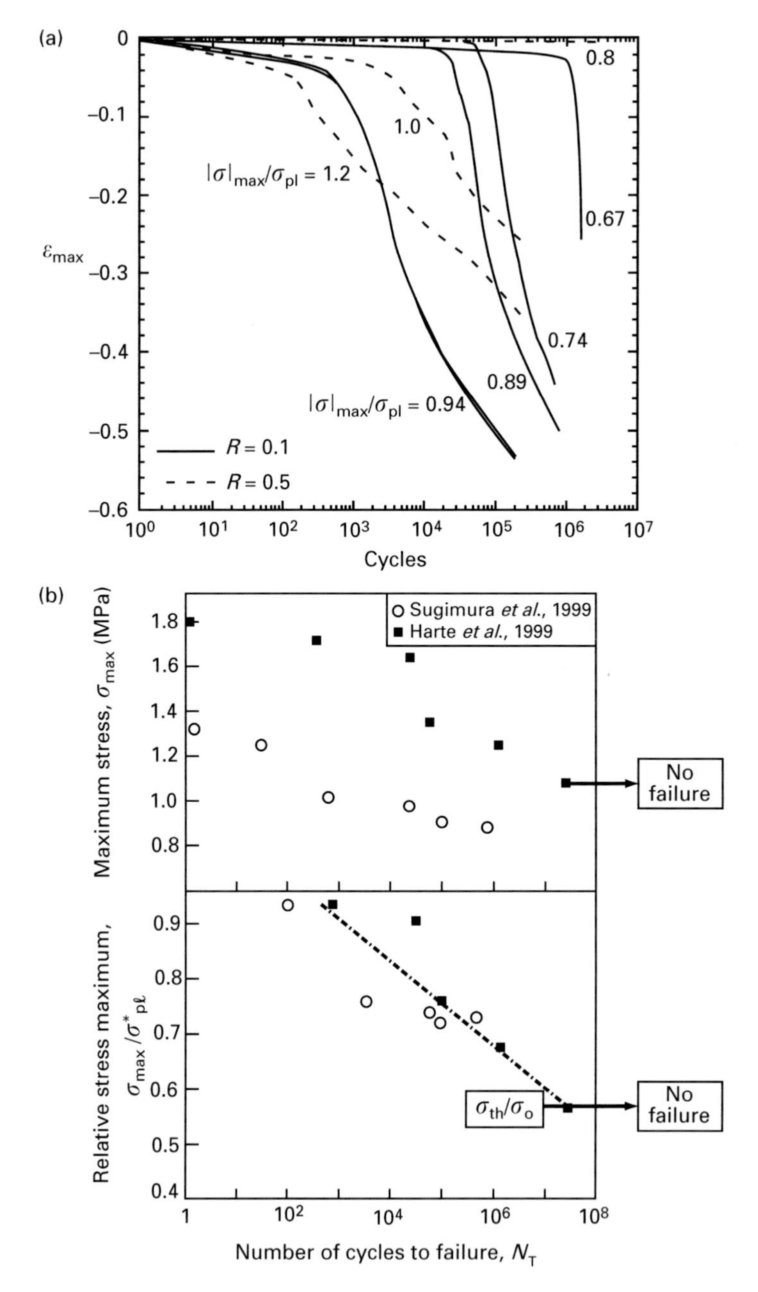

Fig. 3.18 (a) Maximum compressive strain in a fatigue cycle, ε_{max}, plotted against number of cycles of loading, showing the initial progressive shortening of the specimen and the dramatic increase in strain at the threshold, N_T. (b) S–N curves for two aluminum foams, showing that the data are described by a single line if the maximum stress in the fatigue test is normalized by the uniaxial compressive strength of the foam. (a, Reprinted from Harte *et al.*, 1999, with permission from Elsevier; b, reprinted from Sugimura *et al.*, 1999, with permission from Elsevier.)

et al., 1999). In other closed-cell foams, the initial progressive shortening of the specimens is thought to be due to material ratcheting in the cell edges and walls (McCullough *et al.*, 2000).

(3) In open-cell foams, fatigue cracks propagate across cell edges, so that at the end of the test, tetragonal joints fall out of the foam. The resulting disconnection of the foam structure gives rise to a collapse band (Harte *et al.*, 1999; Zhou and Soboyejo, 2004).

(4) In more homogeneous foams, a single band of collapsed cells forms and then progressively broadens over neighboring cells while in less homogeneous foams, several distinct collapse bands may form.

There is substantial variability in plots of cyclic stress range against number of cycles to failure (S–N curves) for the foams tested in the above studies. The variability is reduced by normalizing the data by the monotonic compressive collapse stress, σ_{pl}^* (Sugimura *et al.*, 1999) (Fig. 3.18b).

3.3.7 Fluid flow through open-cell foams

The fluid within the cells can also contribute to the mechanical response (Gibson and Ashby, 1997). In open-cell foams, there is the viscous dissipation of fluid moving through the cells. This effect depends on the fluid viscosity, μ, the strain rate, $\dot{\varepsilon}$, the strain, ε, the cell size, l, and the sample size, L. Darcy's law relates the velocity of a fluid through a porous medium to the pressure gradient, the viscosity of the fluid and the permeability, k, of the medium. Using Brace's (1977) equation relating the permeability of porous rocks to the pore size and the porosity and Gent and Rusch's (1966) equation relating the cell size of a foam at an applied compressive strain of ε to the initial size, l, Gibson and Ashby (1997) find that the contribution of the fluid to the overall stress applied to a fluid-filled foam is given by

$$\sigma_f^* = \frac{C\mu\dot{\varepsilon}}{1-\varepsilon}\left(\frac{L}{l}\right)^2 \tag{3.32}$$

A more recent analysis of the response of fluid-filled open-cell foams, that takes into account the effect on the permeability of the formation of densified bands of material within the foam after the compressive collapse strength is reached, is described by Dawson *et al.* (2008). For the case of cylindrical specimens of large radius, R, relative to their height, h, ($R/h > 4$), the analysis can be approximated to an equation of the form

$$\sigma_f^* = \frac{C\mu\dot{h}R^2}{hk} \tag{3.33}$$

For strains within the linear elastic regime, the permeability, k, is given by $k_{el} = Al_o^2(1-\varepsilon)\left(1-\frac{\rho_o^*}{\rho_s}\frac{1}{(1-\varepsilon)}\right)^3$ where A is a constant, l_o is the initial cell size and ρ_o^* is the initial foam density. At higher strains, the permeability depends on the volume fractions of foam in the linear elastic and densified regimes; the details are given by Dawson *et al.* (2007).

3.3.8 Closed-cell foams containing a gas or a liquid

In closed-cell foams containing a gas, as the volume of the cell decreases during compression, the pressure of the enclosed gas within the cells increases according to Boyle's law, contributing to the overall stress required to deform the foam. The Young's modulus for a closed-cell foam containing a gas is given by

$$\frac{E^*}{E_s} = \phi^2 \left(\frac{\rho^*}{\rho_s}\right)^2 + (1-\phi)\frac{\rho^*}{\rho_s} + \frac{p_o(1-2v^*)}{E_s(1-\rho^*/\rho_s)} \tag{3.34}$$

The first term gives the contribution of cell edge bending (where φ is the fraction of the solid contained in the cell edges), the second term gives the contribution of cell face stretching (described above) and the third term gives the contribution of the gas pressure (where p_o is the initial gas pressure in the cell and v^* is the Poisson's ratio of the foam). A gas at atmospheric pressure adds, at most, 0.1 MN/m^2 to the modulus of the foam, a contribution so small that it can be neglected for all but elastomeric foams of low relative density (less than 10%).

The effect of the gas on the rest of the stress–strain curve of a closed-cell elastomeric foam is much more pronounced. The post-collapse stress is given by:

$$\sigma = \sigma^*_{el} + \frac{p_o \varepsilon}{1-\varepsilon-\rho^*/\rho_s} \tag{3.35}$$

The post-collapse stress–strain curve rises with increasing slope. Membrane stresses, too, can contribute to the post-collapse stress–strain curve, but their contribution is typically small compared with that of the gas. (The cell walls of closed-cell foams made from materials that yield or fracture are likely to rupture before the gas pressure becomes significant, so that their post-collapse stress–strain curve is unaffected by the gas.)

A closed-cell foam filled with an incompressible liquid within the cells behaves differently. The volume of the cell is conserved by cell-wall stretching, which compensates for reductions in cell volume resulting from cell-wall bending; the stretching is of the same order as the bending deformation (Warner and Edwards, 1988; Warner *et al.*, 2000). The Young's modulus of the foam is given by the sum of the bending and stretching contributions:

$$\frac{E^*}{E_s} = \left(\frac{\rho^*}{\rho_s}\right)\left[1+\left(\frac{\rho^*}{\rho_s}\right)\right] \tag{3.36}$$

An initial pressure within the liquid pre-stresses the cell wall in tension. Additional uniaxial stress on the foam requires the same compensation of volume changes from bending deformations by cell-wall stretching as in the unpressurized case. The Young's modulus is independent of the initial cell pressure as long as the solid material itself remains linear elastic.

The cell-wall stretching affects failure, too: we expect that the compressive strength of the foam will scale with relative density in the same way as the modulus (i.e. roughly linearly). The initial pressure, p_o, of the liquid does affect the strength: it is reduced by a value equal to the initial pre-stress in the cell wall, which is roughly $p_o(\rho_s/\rho^*)$.

Table 3.3 summarizes the equations describing the behavior of foams.

3.3.9 Stretch-dominated structures

The foams analyzed so far respond to load largely by cell-edge bending. Not all cellular structures show this bending-dominated behavior. To understand this we need the *Maxwell stability criterion* (Maxwell, 1864; Deshpande *et al.*, 2001; Ashby, 2006).

The condition that a two-dimensional pin-jointed truss (meaning a structure that is hinged at its corners) made up of b struts and j frictionless joints is both statically and kinematically determinate (meaning that it is rigid and does not fold up when loaded) is

$$M = b - 2j + 3 = 0 \tag{3.37}$$

In three dimensions the equivalent equation is

$$M = b - 3j + 6 = 0 \tag{3.38}$$

If $M < 0$, the frame is a *mechanism*. It has no stiffness or strength; it collapses if loaded. If its joints are locked (Fig. 3.19a,b), preventing rotation (as they are in a cellular lattice) the bars of the frame bend when the structure is loaded, just as in Fig. 3.4(b). If, instead, $M > 0$ the frame ceases to be a mechanism (Fig. 3.19c). If it is loaded, its members carry tension or compression (even when pin-jointed), and it becomes a *stretch-dominated* structure. Locking the joints now makes little difference because slender structures are much stiffer when stretched than when bent. There is an underlying principle here: stretch-dominated structures have high structural efficiency; bending dominated structures have low.

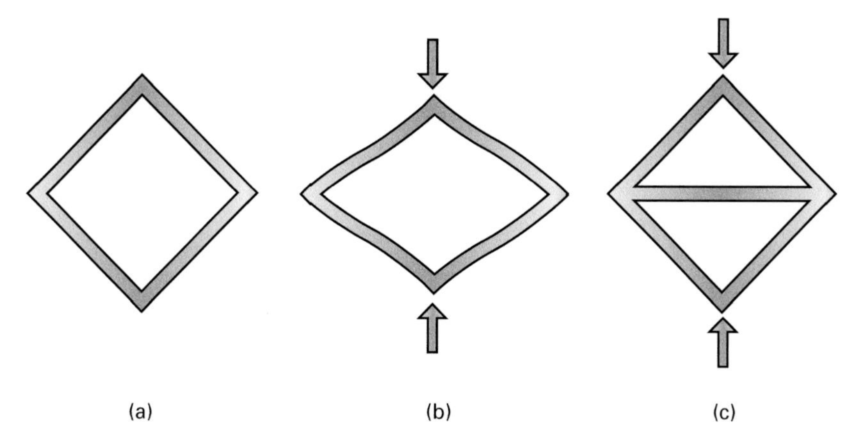

Fig. 3.19. The frame at (a) deforms by cell-edge bending when it is loaded, as in (b). The triangulated, frame at (c) is stiff when loaded because the transverse bar carries tension: its behavior is stretch-dominated.

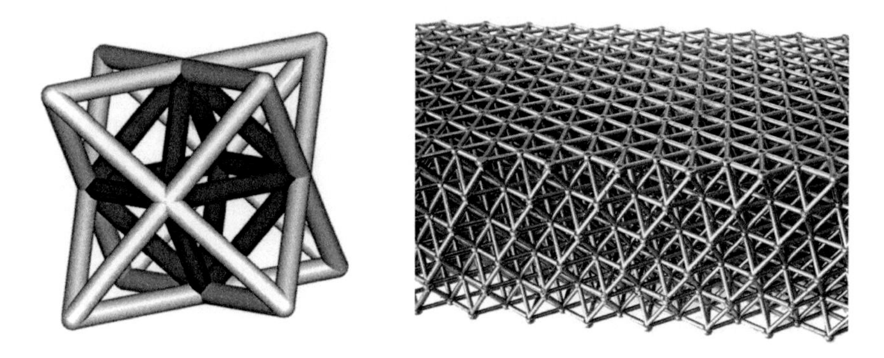

Fig. 3.20. A stretch-dominated lattice and its unit cell.

These criteria give a basis for the design of efficient micro-truss structures. For the cellular structures of Figs. 3.4 and 3.12, $M < 0$ and bending dominates. For the structure shown in Fig. 3.20, however, $M > 0$ and it behaves as an almost isotropic, stretch-dominated structure. On average one third of its bars carry tension when the structure is loaded in simple tension, regardless of the loading direction. Thus

$$\frac{E^*}{E_s} \approx \frac{1}{3}\left(\frac{\rho^*}{\rho_s}\right) \quad \text{(isotropic stretch-dominated behavior)} \quad (3.39)$$

and its collapse stress is

$$\frac{\sigma^*}{\sigma_{ys}} \approx \frac{1}{3}\left(\frac{\rho^*}{\rho_s}\right) \quad \text{(isotropic stretch-dominated behavior)} \quad (3.40)$$

This is an upper bound since it assumes that the struts yield in tension or compression when the structure is loaded. If the struts are slender, they may buckle before they yield. Then the "strength," like that of a buckling foam, (3.19), is

$$\frac{\sigma^*_{el}}{E_s} \approx 0.2\left(\frac{\rho^*}{\rho_s}\right)^2 \quad (3.41)$$

These results are summarized in Fig. 3.21, in which the modulus E^* is plotted against the density ρ^*. Stretch-dominated, prismatic microstructures have moduli that scale as ρ^*/ρ_s (slope 1); bending-dominated, foam-like microstructures have moduli that scale as $(\rho^*/\rho_s)^2$ (slope 2). Given that the density can by varied through a wide range, this allows great scope for material design. Note how the use of microscopic shape has expanded the occupied area of $E - \rho$ space.

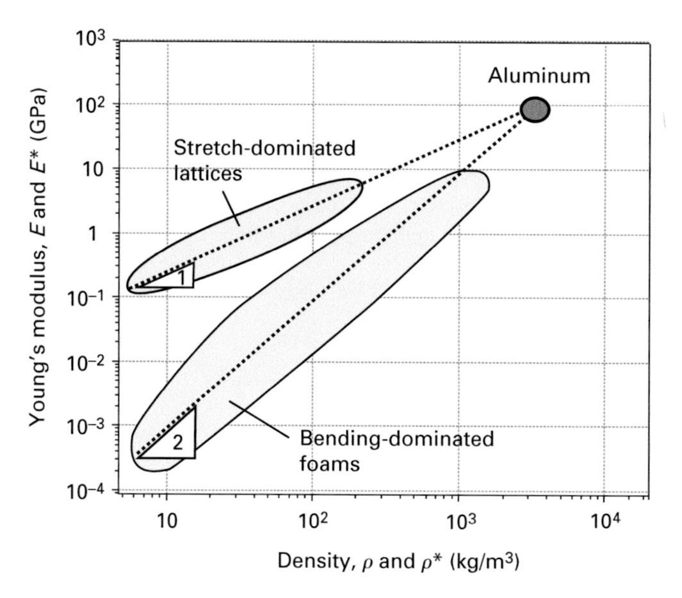

Fig. 3.21 The contrast between bend- and stretch-dominated behavior, shown here for aluminum foams and lattices. Real aluminum foams and lattices have properties that occupy the shaded regions.

3.4 The mechanics of structural members with cellular materials

3.4.1 Sandwich structures

Consider a sandwich beam under a load P in three-point bending (Fig. 3.22). The span is ℓ, the width, b, the core thickness, c, and the face thickness, t. The central deflection is δ. The density and Young's modulus of the faces are ρ_f and E_f, respectively. Those of the core are ρ_c^* and E_c^*; the shear modulus of the core is G_c^*. For a foam core, the moduli are related to the core density by (3.16) and (3.17). The normal stresses in the face and core are σ_f and σ_c, respectively, while the shear stresses in the face and core are τ_f and τ_c, respectively.

The faces of the sandwich beam largely carry the normal stresses while the core carries nearly all the shear stresses (Fig. 3.23) (Allen, 1969). The deflection of the beam is the sum of its bending and shear deflections, calculated from the equivalent flexural rigidity, $(EI)_{eq}$, and the equivalent shear rigidity, $(AG)_{eq}$, respectively. The equivalent flexural rigidity is found from the parallel axis theorem; for a rectangular cross-section:

$$(EI)_{eq} = \frac{E_f b t^3}{6} + \frac{E_c^* b c^3}{12} + \frac{E_f b t d^2}{2} \tag{3.42}$$

where $d = c + t$. The first and second terms describe the bending stiffnesses of the face and core about their own centroids: together they give the stiffness of the beam if the faces were not bonded to the core. In optimal sandwich design, both are small compared to the third term, which describes the bending stiffness of the faces about the

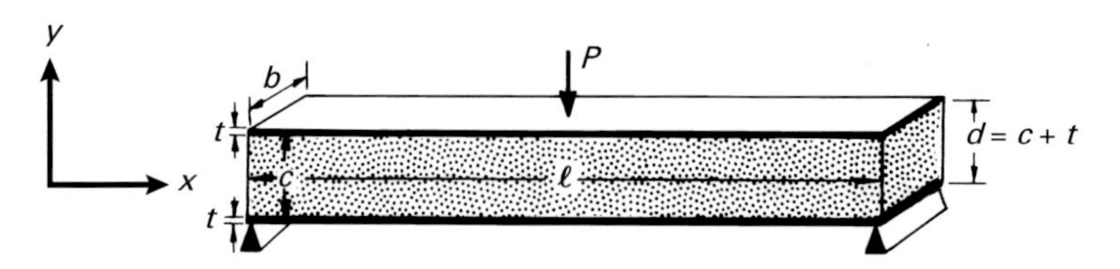

Fig. 3.22 Schematic of a sandwich beam loaded in three-point bending.

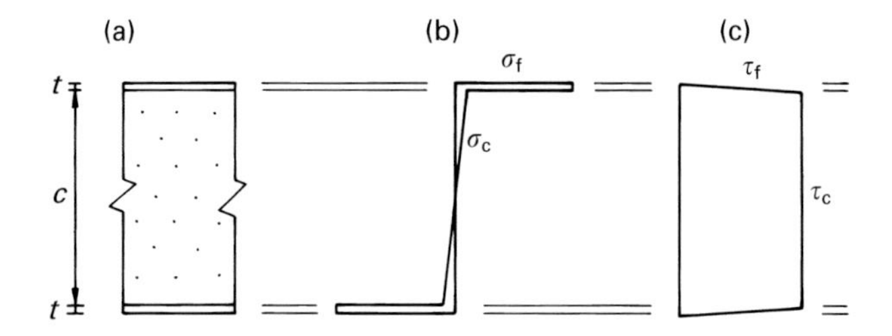

Fig. 3.23 Approximate stress distributions in a rectangular sandwich beam: (a) beam cross-section; (b) normal stress distribution; (c) shear distribution.

centroid of the beam itself. Typically, the faces of the core are thin compared with the core, so that $d \sim c$. Then, to a good approximation (assuming that $E_f \gg E_c^*$),

$$(EI)_{eq} = \frac{E_f btc^2}{2} \tag{3.43}$$

The equivalent shear rigidity is

$$(AG)_{eq} = \frac{bd^2 G_c^*}{c} \sim bcG_c^* \tag{3.44}$$

The deflection δ is the sum of the bending and shear components:

$$\delta = \delta_b + \delta_s = \frac{P\ell^3}{B_1 (EI)_{eq}} + \frac{P\ell}{B_2 (AG)_{eq}} \tag{3.45}$$

The constants B_1 and B_2 relate to the geometry of the loading configuration. Values for several common loading configurations are given in Table 3.4; for three-point bending with a central concentrated load, $B_1 = 48$ and $B_2 = 4$.

Sandwich beams can fail by a variety of mechanisms (Fig. 3.24). The stress in the tensile face can reach the strength of the face material (e.g. by yielding). The stress in the compressive face can reach the local buckling or wrinkling stress. The core can fail in shear. And the face can delaminate from the core. Here we consider the first three modes (Allen, 1969), assuming that the bond between the faces and the core is sufficiently strong that delamination is avoided.

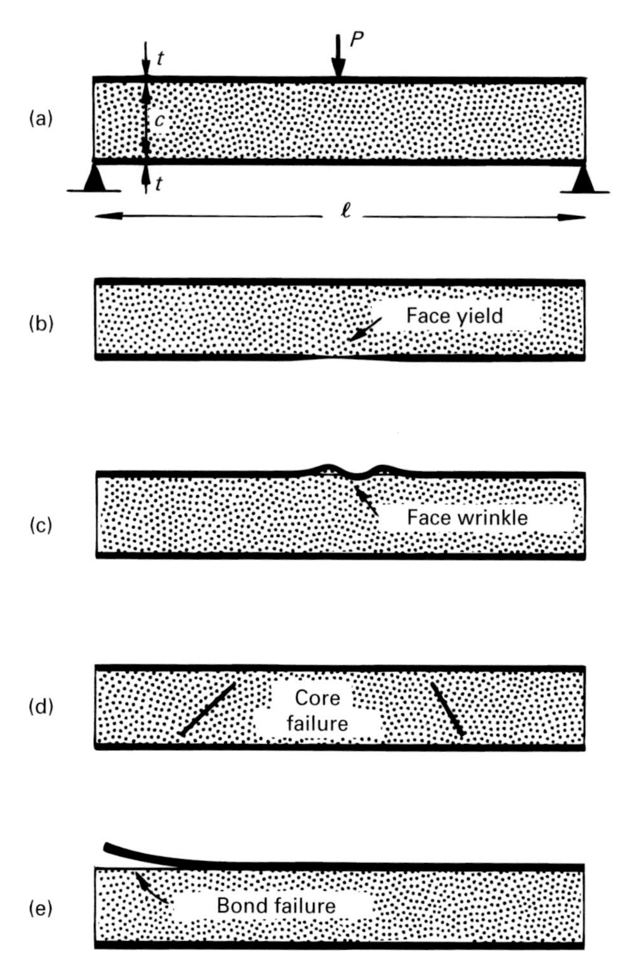

Fig. 3.24 Failure modes for sandwich panels: (a) the loading geometry; (b) face yielding; (c) face wrinkling; (d) core failure; (e) debonding.

The normal stress in the face, at a cross-section with a bending moment, M, is given by

$$\sigma_f = \frac{MyE_f}{(EI)_{eq}} = \frac{M}{btc} \tag{3.46}$$

where y is the vertical distance from the neutral axis in the center of the beam. The stress is maximum when $y = (c/2) + t \sim c/2$, for thin faces relative to the core, and when $M = M_{max}$. The maximum moment can be related to the applied load, P, and the span of the beam, l, by $M_{max} = Pl/B_3$. Values of B_3 for common loading configurations are given in Table 3.4. Face tensile failure occurs when this stress reaches the yield strength of the face material, σ_{yf}. Face yielding then occurs when

Table 3.4 Constants for bending and failure of beams

Mode of loading (all beams of length l)	B_1	B_2	B_3	B_4
	$\delta_b = \frac{Pl^3}{B_1 (EI)_{eq}}$	$\delta_s = \frac{Pl}{B_2 (AG)_{eq}}$	$M = \frac{Pl}{B_3}$	$Q = \frac{P}{B_4}$
Cantilever, end load, P	3	1	1	1
Cantilever, uniformly distributed load, $^{\dagger}q = P/l$	8	2	2	1
Three-point bend, central load, P	48	4	4	2
Three-point bend, uniformly distributed load, $q = P/l$	$\frac{384}{5}$	8	8	2
Ends built in, central load, P	192	4	8	2
Ends built in, uniformly distributed load, $q = P/l$	384	8	12	2

† q is a distributed load per unit length.

$$\frac{P\ell}{B_3 btc} = \sigma_{yf} \tag{3.47}$$

Face wrinkling occurs when the normal stress in the compressive face reaches the local instability stress. Allen (1969) gives the result; wrinkling occurs when the compressive stress in the face is

$$\sigma_f = 0.57 E_f^{1/3} E_c^{*\,2/3} \tag{3.48}$$

for Poisson's ratio of the core of 0.33. For a foam core, the Young's modulus is related to the relative density by (3.16), giving:

$$\frac{P\ell}{B_3 btc} = 0.57 E_f^{1/3} E_s^{2/3} \left(\rho_c^* / \rho_s\right)^{4/3} \tag{3.49}$$

Finally, core failure occurs when the shear stress in the core equals the shear strength of the core:

$$\frac{P}{B_4 bc} = C_{11} \left(\frac{\rho_c^*}{\rho_s}\right)^{3/2} \sigma_{ys} \tag{3.50}$$

where C_{11} is a constant of proportionality for the foam shear strength (assuming failure by plastic yielding) and B_4 is a constant relating the maximum shear force in the beam to the applied load Q; values are given in Table 3.4 for common loading geometries.

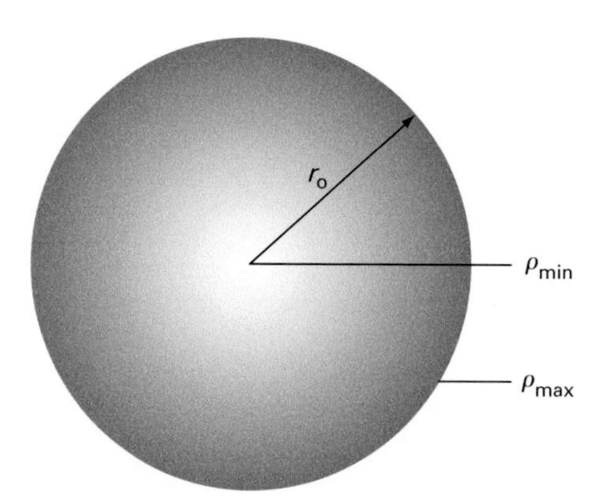

Fig. 3.25 A circular cross-section with a radial density gradient.

3.4.2 Circular sections with a radial density gradient

Consider a member with a circular cross-section of radius, r_o, made of a cellular material with a radial variation in density (Fig. 3.25). The minimum density at the center of the cross-section is ρ_{min}, the maximum density at the periphery is ρ_{max}, and the density at a radius, r, is ρ. The radial variation in density can be represented by

$$\frac{\rho - \rho_{min}}{\rho_{max} - \rho_{min}} = \left(\frac{r}{r_o}\right)^n \tag{3.51}$$

(In Chapter 6, we will see that this is a good approximation for palm and bamboo stems.) We define R as the ratio of the minimum to the maximum density:

$$R = \frac{\rho_{min}}{\rho_{max}} \tag{3.52}$$

so that

$$\frac{\rho(r)}{\rho_{max}} = (1-R)\left(\frac{r}{r_o}\right)^n + R \tag{3.53}$$

The equivalent density of a uniform cross-section, $\bar{\rho}$, is

$$\frac{\bar{\rho}}{\rho_{max}} = \frac{1}{\pi r_o^2}\int_0^{r_o} \frac{\rho(r)}{\rho_{max}} 2\pi r dr = \frac{2+nR}{2+n} \tag{3.54}$$

The Young's modulus of the cellular material is

$$E = C\left(\frac{\rho}{\rho_{max}}\right)^m \tag{3.55}$$

The flexural rigidity for an equivalent section of uniform density, $\bar{\rho}$, is:

$$(EI)_{\text{uniform}} = C\left(\frac{2+nR}{2+n}\right)^{\text{m}} \frac{\pi r_o^4}{4} \tag{3.56}$$

The flexural rigidity for the section with the radial density gradient is

$$(EI)_{\text{rad grad}} = \int_0^{r_o} C\left[(1-R)\left(\frac{r}{r_o}\right)^{\text{n}} + R\right]^{\text{m}} \pi r^3 dr \tag{3.57}$$

3.4.3 Cylindrical shells filled with uniform compliant cores

We first review standard results for a hollow cylindrical shell (without a core (Fig. 3.26a)). In uniaxial compression, a hollow, thin-walled cylinder of radius, a, and wall thickness, t, made from an isotropic material with Young's modulus, E, and Poisson's ratio, ν, buckles axisymmetrically at a critical stress of (Timoshenko and Gere, 1961)

$$\sigma_0 = \frac{Et}{a\sqrt{3(1-\nu^2)}} \tag{3.58}$$

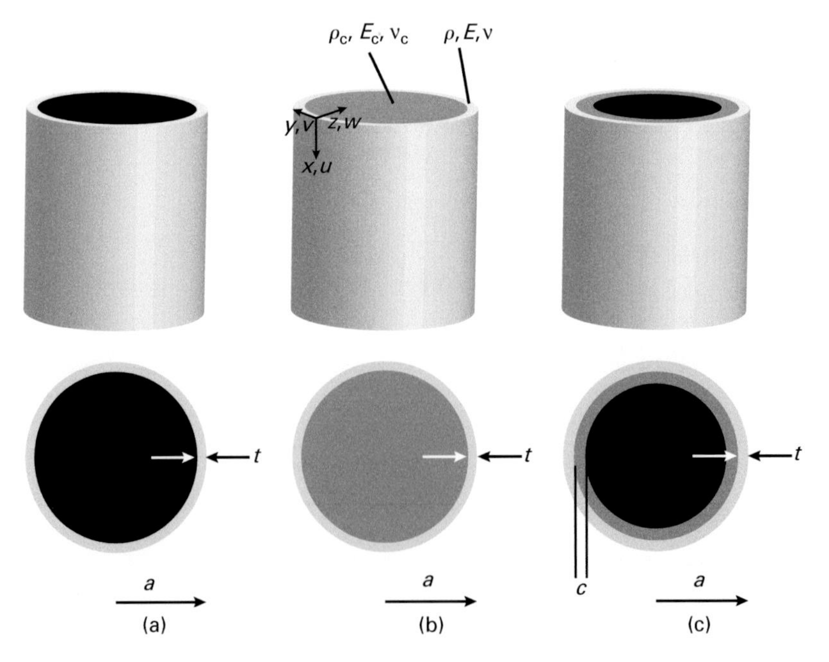

Fig. 3.26 (a) A hollow cylinder of radius, a, and wall thickness, t. (b) A cylinder filled with a compliant core. (c) A cylinder with a compliant core of thickness, c, with a central borehole.

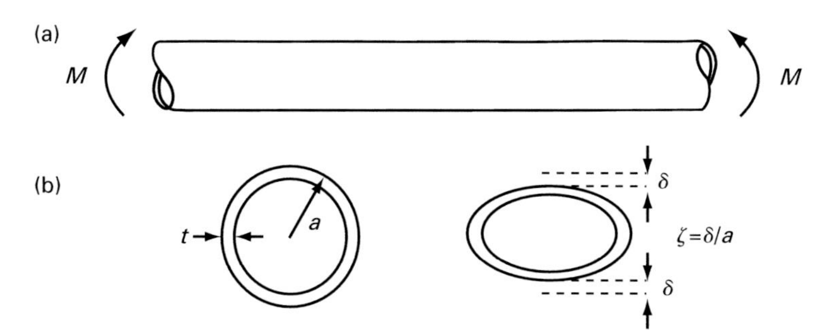

Fig. 3.27 (a) A hollow cylindrical shell in pure bending. (b) Ovalization of the circular cross-section. The degree of ovalization is $\zeta = \delta/a$. (Reprinted from Karam and Gibson, 1995a, with permission of Elsevier.)

or, for $\nu = 1/3$, $\sigma_o \approx 0.6(Et/a)$.

In pure bending of an infinitely long hollow cylindrical shell, the circular cross-section ovalizes, reducing its moment of inertia. The ovalization is characterized by $\zeta = \delta/a$ (Fig. 3.27). As the curvature increases, the ovalization increases, so that the bending moment reaches a theoretical maximum at (Brazier, 1927):

$$M_{\text{Brazier}} = \frac{2\sqrt{2}\pi Eat^2}{9\sqrt{1-\nu^2}} \approx 1.06Eat^2 \tag{3.59}$$

Local buckling of a cylindrical shell in bending, corresponding to a bifurcation point, occurs when the maximum compressive stress in the shell equals the stress to cause axisymmetric buckling under uniaxial compression (3.58); taking account of ovalization, Calladine (1983) found that:

$$M_{\text{lb}} = \frac{0.939Eat^2}{\sqrt{1-\nu^2}} \approx 1.0Eat^2 \tag{3.60}$$

The moment required for local buckling is always lower than the Brazier moment.

Cylindrical shells fully filled with a compliant core

The stress at which a cylindrical shell filled with a compliant core buckles under uniaxial compressive loading can be found by considering the core to act as an elastic foundation supporting the outer shell (Fig. 3.26b). The shell has a radius, a, and a thickness, t. The Young's modulus, Poisson's ratio and density of the outer shell are E, ν and ρ, while those of the core are E_c, ν_c and ρ_c. The critical stress for axisymmetric elastic buckling under uniaxial compression is calculated by modifying Timoshenko and Gere's (1961) results for the symmetric deformation and axisymmetric buckling of a hollow cylindrical shell to account for the compliant core as a two-dimensional elastic foundation stabilizing a longitudinal strip of the shell.

The uniaxial buckling stress is (Karam and Gibson, 1995a):

$$\sigma_{cr} = \frac{Et}{a} f_1 \tag{3.61}$$

$$\text{where } f_1 = \left[\frac{1}{12(1-\nu^2)} \frac{(a/t)}{(\lambda_{cr}/t)^2} + \frac{(\lambda_{cr}/t)^2}{(a/t)} + \frac{2}{(3-\nu_c)(1+\nu_c)} \frac{E_c}{E} \left(\frac{\lambda_{cr}}{t}\right)\left(\frac{a}{t}\right) \right], \tag{3.61a}$$

and λ_{cr} is the half buckled wavelength divided by π. The solution for λ_{cr} is given by Karam and Gibson (1995a); it is plotted against a/t in Fig. 3.28, for various values of E_c/E. Two limits are of particular interest. If the core is sufficiently stiff, then the solution for λ_{cr} corresponds to wrinkling of a flat sheet on an elastic foundation (Allen, 1969):

$$\frac{\lambda_{cr}}{t} = \left[\frac{(3-\nu_c)(1+\nu_c)}{12(1-\nu^2)} \right]^{1/3} \left[\frac{E}{E_c} \right]^{1/3} \approx 0.69 \left(\frac{E}{E_c} \right)^{1/3} \tag{3.62a}$$

independently of a/t (the horizontal region of the dashed lines for various E_c/E in Fig. 3.28). In the second limit, if the core is too compliant, it fails to act as an elastic foundation and the outer shell behaves like a hollow cylindrical shell. In this case,

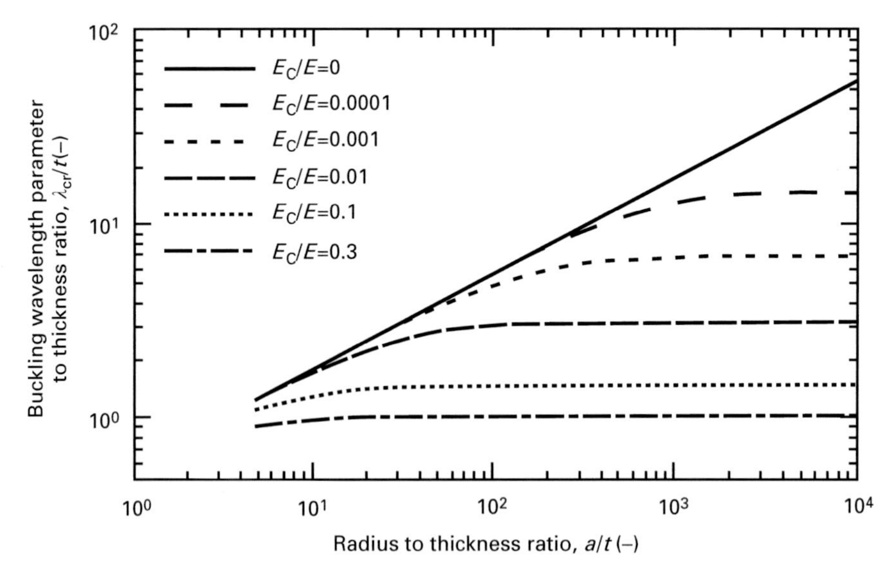

Fig. 3.28 Buckling wavelength parameter, λ_{cr}, normalized by the shell thickness, t, plotted against the shell radius, a, normalized by t. The solid line represents the case of the hollow shell (3.62b), while the horizontal dashed lines to the right represent (3.62a). (Reprinted from Karam and Gibson, 1995a, with permission of Elsevier.)

$$\frac{\lambda_{cr}}{t} = \frac{1}{\left[12\left(1-\nu^2\right)\right]^{1/4}}\left(\frac{a}{t}\right)^{1/2} \approx 0.55\left(\frac{a}{t}\right)^{1/2} \tag{3.62b}$$

independently of the shell and core moduli (solid line in Fig. 3.28). There is a gradual transition between the behavior of the hollow shell and that of the shell with a core that acts as an elastic foundation. As we shall see in Chapter 6, some animal quills and plant stems have cores that act as an elastic foundation, so that (3.62a) is applicable, while others do not.

The performance of a thin walled cylindrical shell with a compliant elastic core can be evaluated by comparing its buckling resistance with that of an equivalent hollow shell of equal diameter and mass. The thickness of the equivalent hollow shell, t_{eq}, found by equating the masses, is

$$t_{eq} = t + \left(\frac{\rho_c}{\rho}\right)\left(\frac{a}{2}\right) \tag{3.63}$$

The critical uniaxial compressive buckling load for the shell filled with the compliant core, P_{cr}, divided by that of the hollow shell, P_{eq}, is

$$\frac{P_{cr}}{P_{eq}} = \sqrt{3\left(1-\nu^2\right)}\frac{\left(1+\frac{a}{2t}\frac{E_c}{E}\right)f_1}{\left(1+\frac{a}{2t}\frac{\rho_c}{\rho}\right)^2} \tag{3.64}$$

In bending, local buckling occurs when the maximum normal stress on the outer skin of the cylinder equals the uniaxial buckling stress, σ_{cr} (3.61). The local buckling moment is given by (Karam and Gibson, 1995a; Cheng, 1996)

$$M_{lb} = \frac{\pi E a t^2\sqrt{\zeta_{lb}}}{\sqrt{1-\nu^2}}\left(1-\frac{3}{2}\zeta_{lb}\right)\left(1+\frac{\alpha a}{4t}\right)^{1/2}\left(1+\frac{2\alpha\beta a^3\left(1-\nu^2\right)}{3t^3}\right)^{1/2} \tag{3.65}$$

where $\alpha = \frac{E_c}{E}$, $\beta = \frac{3-5\nu_c}{\left(1+\nu_c\right)\left(1-2\nu_c\right)} = 3$ (for $\nu = 1/3$), and the ovalization at the point of local buckling, ζ_{lb}, is found from the solution to

$$\frac{\zeta_{lb}^{1/2}\left(1-\zeta_{lb}\right)}{\sqrt{1-\nu^2}\left(1-3\zeta_{lb}\right)} = f_1\frac{\left(1+\frac{\alpha a}{4t}\right)^{1/2}}{\left[1+\frac{2\alpha\beta a^3\left(1-\nu^2\right)}{3t^3}\right]^{1/2}} \tag{3.66}$$

where f_1 is given by (3.61). We note that the first bracketed factor in (3.65) differs slightly from the result in Karam and Gibson (1995a) due to an error in their expression for the strain energy (Cheng, 1996). As for the hollow cylinder, the local buckling moment for the cylinder with the compliant core is always lower than the Brazier moment (Karam and Gibson, 1995a).

The ratio of the local buckling moment of a cylindrical shell filled with a compliant core to that of an equivalent hollow cylindrical shell of the same mass and diameter is

$$\frac{M_{\mathrm{lb}}}{M_{\mathrm{eq}}}=\frac{\pi\sqrt{\zeta_{\mathrm{lb}}}\left(1-\frac{3}{2}\zeta_{\mathrm{lb}}\right)\left(1+\frac{\alpha a}{4t}\right)^{1/2}\left(1+\frac{2\alpha\beta a^3\left(1-\nu^2\right)}{3t^3}\right)^{1/2}}{0.939\left[1+\left(\frac{a}{2t}\right)\left(\frac{\rho_{\mathrm{c}}}{\rho}\right)\right]^2} \tag{3.67}$$

Cylindrical shells partially filled with a compliant core

The buckling analyses described above treat the core as an elastic foundation resisting buckling of the shell. The stresses in the core are maximum at the interface between the shell and the core and decay radially into the core. The decay can be approximated by that for a buckled flat strip on an elastic half-space (Allen, 1969). It is straightforward to show that the stresses decay to about 5% of the maximum value at a radial depth of 1.6 half buckling wavelengths or $5\lambda_{\mathrm{cr}}/t$ (Karam and Gibson, 1995a). Core material interior to this does not resist any load and can be removed without reducing the uniaxial compressive buckling stress (3.61) or the local buckling moment in bending (3.65) (Fig. 3.26c).

The uniaxial compressive stress at which axisymmetric buckling occurs is unchanged by the presence of a central borehole (3.61). The buckling load is altered slightly by the smaller area of the core. The buckling load of a cylindrical shell with a compliant core and a central borehole is

$$P_{\mathrm{cr}}=2\pi a t\sigma_{\mathrm{cr}}\left[1+\frac{c}{t}\frac{E_{\mathrm{c}}}{E}\left(1-\frac{c/t}{2a/t}\right)\right] \tag{3.68}$$

where t is again the thickness of the outer shell and c is the thickness of the compliant core (Fig. 3.26c). The equivalent thickness of a hollow shell of the same mass and diameter as the partially filled cylindrical shell is:

$$t_{\mathrm{eq}}=t\left[1+\frac{c}{t}\frac{\rho_{\mathrm{c}}}{\rho}\left(1-\frac{c/t}{2a/t}\right)\right] \tag{3.69}$$

The ratio of the axial compressive buckling load of the partially filled shell to that of the hollow shell is (assuming $\nu=\nu_{\mathrm{c}}=1/3$)

$$\frac{P_{\mathrm{cr}}}{P_{\mathrm{eq}}}=\frac{\sqrt{3\left(1-\nu^2\right)}\left[1+\frac{c}{t}\frac{E_c}{E}\left(1-\frac{c/t}{2a/t}\right)\right]f_1}{\left[1+\frac{c}{t}\frac{\rho_{\mathrm{c}}}{\rho}\left(1-\frac{c/t}{2a/t}\right)\right]^2} \tag{3.70}$$

The local buckling moment in bending is the same as for the completely filled core (3.65), with the following modifications to account for the reduced thickness of the core in calculating the moment of inertia and area of the section (Cheng, 1996): $1+\frac{\alpha a}{4t}$ is modified to $1+\frac{\alpha a}{4t}\left[1-\left(1-\frac{c/t}{a/t}\right)^4\right]$ and $1+\frac{2\alpha\beta a^3(1-\nu^2)}{3t^3}$ is modified to $1+\frac{2\alpha\beta a^3(1-\nu^2)}{3t^3}\left[1-\left(1-\frac{c/t}{a/t}\right)^2\right]$.

Defining $F' = \left[1-\left(1-\frac{c/t}{a/t}\right)^4\right]$ and $G'\left[1-\left(1-\frac{c/t}{a/t}\right)^2\right]$, the local buckling moment is then (Cheng, 1996):

$$M_{\text{lb}} = \frac{\pi E a t^2 \sqrt{\zeta_{\text{lb}}}}{\sqrt{1-\nu^2}}\left(1-\frac{3}{2}\zeta_{\text{lb}}\right)\left(1+\frac{\alpha a}{4t}F'\right)^{1/2}\left(1+\frac{2\alpha\beta a^3(1-\nu^2)}{3t^3}G'\right)^{1/2} \tag{3.71}$$

The ovalization at the point of local buckling, ζ_{lb}, is then, from (3.66) modified by F' and G',

$$\frac{\zeta_{\text{lb}}^{1/2}(1-\zeta_{\text{lb}})}{\sqrt{1-\nu^2}\,(1-3\zeta_{\text{lb}})} = f_1 \frac{\left(1+\frac{\alpha a}{4t}F'\right)^{1/2}}{\left[1+\frac{2\alpha\beta a^3(1-\nu^2)}{3t^3}G'\right]^{1/2}} \tag{3.72}$$

The ratio of the local buckling moment of the partially filled shell to that of the equivalent hollow shell is

$$\frac{M_{\text{lb}}}{M_{\text{eq}}} = \frac{\pi\sqrt{\zeta_{\text{lb}}}\left(1-\frac{3}{2}\zeta_{\text{lb}}\right)\left(1+\frac{\alpha a}{4t}F'\right)^{1/2}\left(1+\frac{2\alpha\beta a^3(1-\nu^2)}{3t^3}G'\right)^{1/2}}{0.939\left[1+\frac{c}{t}\frac{\rho_c}{\rho}\left(1-\frac{c/t}{2a/t}\right)\right]^2} \tag{3.73}$$

The results are summarized in Table 3.5. Data for buckling of model rubber cylinders with and without compliant cores in uniaxial compression and bending are described by Karam and Gibson (1995b). The above equations give a good description of the experimental results.

3.5 Material design in engineering and nature

There are established strategies for selecting materials for engineering design, guiding the search for the best match between the attribute-profiles of the materials and those required by the design. Materials are chosen to perform one or more *functions* – carrying

Table 3.5 Summary of buckling equations for a cylindrical shell with a core that acts as an elastic foundation $\nu = \nu_c = 1/3$

Fully filled shell	Partially filled shell
Uniaxial compression	
$\sigma_{cr} = \frac{Et}{a} f_1 \quad \frac{\lambda_{cr}}{t} = 0.69\left(\frac{E}{E_c}\right)^{1/3}$	$P_{cr} = 2\pi a t \sigma_{cr}\left[1+\frac{c}{t}\frac{E_c}{E}\left(1-\frac{c/t}{2a/t}\right)\right]$
$\frac{P_{cr}}{P_{eq}} = 1.63\frac{\left(1+\frac{a}{2t}\frac{E_c}{E}\right)f_1}{\left(1+\frac{a}{2t}\frac{\rho_c}{\rho}\right)^2}$	$\frac{P_{cr}}{P_{eq}} = \frac{1.63\left[1+\frac{c}{t}\frac{E_c}{E}\left(1-\frac{c/t}{2a/t}\right)\right]f_1}{\left[1+\frac{c}{t}\frac{\rho_c}{\rho}\left(1-\frac{c/t}{2a/t}\right)\right]^2}$
Local buckling in bending	
$M_{lb} = 3.33Eat^2\sqrt{\zeta_{lb}}\left(1-\frac{3}{2}\zeta_{lb}\right) \times\left(1+\frac{a}{4t}\frac{E_c}{E}\right)^{1/2}\left(1+1.77\left(\frac{a}{t}\right)^3\left(\frac{E_c}{E}\right)\right)^{1/2}$	$M_{lb} = 3.33Eat^2\sqrt{\zeta_{lb}}\left(1-\frac{3}{2}\zeta_{lb}\right)\left(1+\frac{a}{4t}\frac{E_c}{E}F'\right)^{1} \times\left(1+1.77\left(\frac{a}{t}\right)^3\left(\frac{E_c}{E}\right)G'\right)^{1/2}$
$\frac{M_{lb}}{M_{eq}} = \frac{3.35\sqrt{\zeta_{lb}}\left(1-\frac{3}{2}\zeta_{lb}\right)\left(1+\frac{a}{4t}\frac{E_c}{E}\right)^{1/2}\left(1+1.77\left(\frac{a}{t}\right)^3\left(\frac{E_c}{E}\right)\right)^{1/2}}{\left[1+\left(\frac{a}{2t}\right)\left(\frac{\rho_c}{\rho}\right)\right]^2}$	$\frac{M_{lb}}{M_{eq}} = \frac{3.35\sqrt{\zeta_{lb}}\left(1-\frac{3}{2}\zeta_{lb}\right)\left(1+\frac{a}{4t}\frac{E_c}{E}F'\right)^{1/2}\left(1+1.77\left(\frac{a}{t}\right)^3\left(\frac{E_c}{E}\right)G'\right)^{1/2}}{\left[1+\frac{c}{t}\frac{\rho_c}{\rho}\left(1-\frac{c/t}{2a/t}\right)\right]^2}$
$f_1 = \left[\frac{0.20(a/t)}{(E/E_c)^{2/3}} + \frac{0.48(E/E_c)^{2/3}}{(a/t)} + 0.39\left(\frac{E_c}{E}\right)^{2/3}\left(\frac{a}{t}\right)\right]$	

a given load, for example. They must do so while meeting a number of *constraints*: that they carry the load without failing, that they tolerate the environment in which they must operate, and so on. Among those that meet the constraints the best choices are those that minimize one or more *objectives*, which, in engineering design, are frequently those of minimizing cost, or mass, or volume, or carbon footprint. In making the choice the designer has, at his or her disposal, a number of *free variables*, parameters that can be varied to best meet the objectives. Function, constraints, objectives and free variables (Table 3.6) define the selection problem. The choice is made by *screening out* all materials that fail to meet the constraints and then *ranking* the survivors by their ability to minimize the objectives. The most highly ranked candidate is the best choice. There are well-developed methods for dealing with multiple constraints and the trade-offs that become necessary if there is more than one objective.

It is important to be clear about the distinction between constraints and objectives. A constraint is an essential condition that must be met, usually expressed as an upper or lower limit on a material property. An objective is a quantity for which an extreme value (a maximum or minimum) is sought. Table 3.7 lists common engineering constraints and objectives.

Table 3.6 Function, constraints, objectives and free variables

Function	What does the component do?
Constraints	What non-negotiable conditions must be met?
Objective	What is to be maximized or minimized?
Free variables	What parameters of the problem is the designer free to change?

Table 3.7 Some constraints and objectives in engineering

Common constraints	Common objectives	Free variables
Meet a target value of:	*Minimize*:	*Design of*:
Stiffness	Cost	Certain dimensions
Strength	Mass	Shape and configuration at
Toughness	Volume	macro and mm scale
Thermal conductivity	Thermal losses	
Service temperature	Electrical losses	
Must be:	Resource depletion	
Electrically conducting	Energy consumption	
Optically transparent	Carbon emissions	
Corrosion resistant	Waste	

Using constraints and objectives in engineering

Constraints are gates: meet the constraint and you pass through the gate, fail to meet it and you are shut out. Screening does just that: it eliminates candidates that cannot do the job at all because one or more of their attributes lie outside the limits set by the constraints.

To rank the materials that survive the screening step we need criteria of excellence – what we have called objectives. The middle column of Table 3.7 lists common engineering objectives: they are measures of performance. Performance is sometimes limited by a single property, sometimes by a combination of them. Thus the best materials to minimize thermal losses (an objective) are the ones with the smallest values of the thermal conductivity, λ, provided, of course, that they also meet all other constraints imposed by the design. Here the objective is met by minimizing a single property. Often, though, it is not one, but a group of properties that are relevant. Thus the best materials for a light stiff tie-rod are those with the smallest value of the group, ρ/E, where ρ is the density and E is the Young's modulus. Those for a strong beam of lowest mass are those with the lowest value of $\rho/\sigma_y^{2/3}$ where σ_y is its yield strength. The property or property-group that maximizes performance for a given design is called its *material index*.

Table 3.8 lists indices for stiffness and strength-limited design for three generic components (a tie, a beam and a panel, for each of four objectives), along with indices for elastic energy storage, elastic hinges that allow large, recoverable deformations without breaking, impact energy absorption and resistance to crack growth. Selecting materials with the objective of minimizing volume uses as little material as possible,

Table 3.8 Indices for stiffness- and strength-limited design

Constraint	Configuration	Minimum volume – *minimize*:	Minimum mass – *minimize*:	Minimum embodied energy – *minimize*:	Minimum material cost – *minimize*:
Stiffness	Tie	$1/E$	ρ/E	$H_m\rho/E$	$C_m\rho/E$
	Beam	$1/E^{1/2}$	$\rho/E^{1/2}$	$H_m\rho/E^{1/2}$	$C_m\rho/E^{1/2}$
	Panel	$1/E^{1/3}$	$\rho/E^{1/3}$	$H_m\rho/E^{1/3}$	$C_m\rho/E^{1/3}$
Strength	Tie	$1/\sigma_y$	ρ/σ_y	$H_m\rho/\sigma_y$	$C_m\rho/\sigma_y$
	Beam	$1/\sigma_y^{2/3}$	$\rho/\sigma_y^{2/3}$	$H_m\rho/\sigma_y^{2/3}$	$C_m\rho/\sigma_y^{2/3}$
	Panel	$1/\sigma_y^{1/2}$	$\rho/\sigma_y^{1/2}$	$H_m\rho/\sigma_y^{1/2}$	$C_m\rho/\sigma_y^{1/2}$
Elastic energy storage		E/σ_y^2	$\rho E/\sigma_y^2$	$H_m\rho E/\sigma_y^2$	$C_m\rho E/\sigma_y^2$
Bend without breaking		E/σ_y	$\rho E/\sigma_y$	$H_m\rho E/\sigma_y$	$C_m\rho E/\sigma_y$
Absorb impact energy		$1/J_c$	ρ/J_c	$H_m\rho/J_c$	$C_m\rho/J_c$
Resist crack growth		$1/K_{IC}$	ρ/K_{IC}	$H_m\rho/K_{IC}$	$C_m\rho/K_{IC}$

ρ = density; E = Young's modulus; σ_y = yield strength; J_c = toughness; K_{IC} = fracture toughness; H_m = embodied energy per unit mass; C_m = cost per unit mass
For materials that do not yield, a failure strength can be substituted for the yield strength.

Table 3.9 Indices for thermal design

Objective	Minimum steady-state heat loss – *minimize*:	Minimum thermal inertia – *minimize*:	Minimum heat loss in a thermal cycle – *minimize*:
	λ	$C_p\rho$	$(\lambda C_p\rho)^{1/2}$

conserving resources. Selection with the objective of minimizing mass is central to the eco-design of transport systems (or indeed of anything that moves) because fuel consumption for transport scales with weight. Selection with the objective of minimizing embodied energy is important when large quantities of material are used, as they are in construction of buildings, bridges, roads and other infrastructure. The fourth column, selection with the objective of minimizing cost, is always with us. Table 3.9 lists indices for thermal design. The first is a single property, the thermal conductivity λ; materials with the lowest values of λ minimize heat loss at steady state, that is, when the temperature gradient is constant. The other two guide material choice when the temperature fluctuates.

There are many such indices, each associated with maximizing some aspect of performance. They provide criteria of excellence that allow ranking of materials by their ability to perform well in the given application. Their derivation is described more fully in the appendix to this chapter.

3.6 Summary

The mechanics of honeycombs and foams is now a mature subject. The stiffness, strength, toughness, visco-elasticity, creep and permeability to fluids are all well understood. Robust models, reviewed in the first part of this chapter, describe the experimental data, given in Gibson and Ashby (1997), well. The key results for honeycombs are summarized in Tables 3.1 and 3.2, those for foams in Table 3.3. We draw on these later in the book to interpret the mechanical response of natural materials.

Cellular solids in nature often appear as part of a structure: the core of a sandwich-like configuration (leaf, for example) or as the internal filling of tubular structures (stem, for instance). The mechanics of these, too, is well understood. The second part of the chapter reviews models for the behavior of sandwich structures, cylindrical structures with a radial density gradient and filled cylindrical tubes.

Established methods exist for selecting materials to meet the constraints of engineering design while minimizing the mass or volume of material that is used. The chapter ends by summarizing how these work, and the material indices that allow optimized choice.

Appendix: Deriving material indices

This appendix describes how material indices are derived. You can find out more about them and their use in Ashby (2005) and Ashby *et al.* (2007).

(a) Minimizing mass: a light, stiff tie-rod

A material is sought for a cylindrical tie-rod that must be as light as possible (Fig. 3.29a). Its length L_o is specified and it must carry a tensile force F without extending elastically by more than δ. Its stiffness must be at least $S^* = F/\delta$. We are free to choose the cross-sectional area, A, and, of course, the material. The design requirements, translated, are listed in Table 3.10.

We first seek an equation that describes the quantity to be minimized, here the mass m of the tie. This equation, the *objective function*, is

$$m = A L_o \rho \tag{3.74}$$

where ρ is the density of the material from which it is made. We can reduce the mass by reducing the cross-section, but there is a constraint: the section-area A must be sufficient to provide a stiffness of S^*, which, for a tie, is

Table 3.10 Design requirements for the light stiff tie

Function	Tie rod
Constraints	Stiffness S^* specified (*a functional constraint*)
	Length L_0 specified (*a geometric constraint*)
Objective	Minimize mass
Free variables	Choice of material
	Choice of cross-sectional area, A

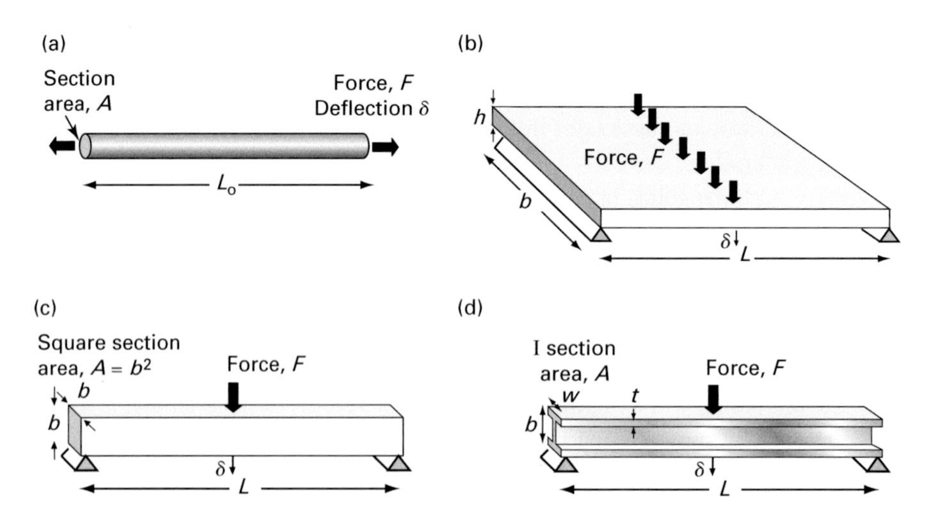

Fig. 3.29 Generic components: (a) a tie, a tensile component; (b) a panel, loaded in bending; (c,d) beams, loaded in bending.

$$S^* = \frac{AE}{L_o} \geq S^* \tag{3.75}$$

where E is the Young's modulus. If the material has a low modulus a large area A is needed to give the necessary stiffness; if E is high, a smaller A is needed. But which gives the lower mass? To find out, we must eliminate the free variable A between these two equations, giving

$$m = S^* L_o^2 \left(\frac{\rho}{E}\right) \tag{3.76}$$

Both S^* and L_o are specified. The lightest tie that will provide a stiffness S^* is that made of the material with the smallest value of the index

$$M_{t_1} = \frac{\rho}{E} \tag{3.77}$$

Table 3.11 Design requirements for the light stiff panel

Function	Panel
Constraints	Stiffness S^* specified (*a functional constraint*)
	Length L and width b specified (*a geometric constraint*)
Objective	Minimize mass
Free variables	Choice of material
	Choice of panel thickness, h

provided that they also meet all other constraints of the design. If the constraint is not stiffness but strength the index becomes

$$M_{t_2} = \frac{\rho}{\sigma_y} \tag{3.78}$$

where σ_y is the yield strength. That means that the best choice of material for the lightest tie that can support a load F without yielding is that with the smallest value of this index.

The mode of loading that most commonly dominates in both engineering and in nature is not tension, but bending – think of floor joists of buildings, of wing spars of aircraft, the branches of trees, the loading in arms and backbone. The index for bending differs from that for tension, and this (significantly) changes the optimal choice of material. We start by modeling a panel, specifying stiffness and seeking to minimize the mass of material needed to make it.

(b) Minimizing mass: a light, stiff panel

A panel is a flat slab, like a table top. Its length L and width b are specified but its thickness h is free. It is loaded in bending by a central load F (Fig. 3.29b). The stiffness constraint requires that it must not deflect more than δ. The objective is to achieve this with minimum mass, m. Table 3.11 summarizes the design requirements.

The objective function for the mass of the panel is the same as that for the tie: $m = AL\rho = bhL\rho$. Its bending stiffness S must be at least S^*:

$$S = \frac{C_1 EI}{L^3} \geq S^* \tag{3.79}$$

Here C_1 is a constant that depends only on the distribution of the loads and I is the second moment of area, which, for a rectangular section, is

$$I = \frac{bh^3}{12} \tag{3.80}$$

We can reduce the mass by reducing h, but only so far that the stiffness constraint is still met. Using the last two equations to eliminate h in the objective function gives

$$m = \left(\frac{12\,S^*}{C_1\,b}\right)^{1/3}\left(bL^2\right)\left(\frac{\rho}{E^{1/3}}\right) \tag{3.81}$$

The quantities S^*, L, b and C_1 are all specified; the only freedom of choice left is that of the material. The best materials for a light, stiff panel are those with the smallest values of

$$M_{P_1} = \frac{\rho}{E^{1/3}} \tag{3.82}$$

Repeating the calculation with a constraint of strength rather than stiffness leads to the index

$$M_{P_2} = \frac{\rho}{\sigma_y^{1/2}} \tag{3.83}$$

These don't look much different from the previous indices, ρ/E and ρ/σ_y, but they are; they lead to different choices of material, as we shall see in a moment. For now, note the procedure. The in-plane dimensions of the panel were specified but we were free to vary the thickness h. The objective is to minimize its mass, m. Use the stiffness constraint to eliminate the free variable, here h. Then read off the combination of material properties that appears in the objective function – the equation for the mass. It sounds easy, and it is – so long as you are clear from the start what the constraints are, what you are trying to maximize or minimize, and which parameters are specified and which are free.

Now for another bending problem, in which the freedom to choose shape is rather greater than for the panel.

(c) Minimizing mass: a light, stiff beam

Beams come in many shapes: solid rectangles, cylindrical tubes, I-beams and more. Some of these have too many free geometric variables to apply the method above directly. However, if we constrain the shape to be *self-similar* (such that all dimensions change in proportion as we vary the overall size), the problem becomes tractable again. We therefore consider beams in two stages: first, to identify the optimum materials for a light, stiff beam of a prescribed simple shape (a square section); then, second, we explore how much lighter it could be made, for the same stiffness, by using a more efficient shape.

Consider a beam of square section $A = b \times b$ that may vary in size but the square shape is retained. It is loaded in bending over a span of fixed length L with a central load F (Fig. 3.29c). The stiffness constraint is again that it must not deflect more than δ

Table 3.12 Design requirements for the light stiff beam

Function	Beam
Constraints	Stiffness S^* specified (*a functional constraint*) Length L (*a geometric constraint*) Section shape square
Objective	Minimize mass
Free variables	Choice of material Area of cross-section, A

under the load F, with the objective that the beam should again be as light as possible. Table 3.12 summarizes the design requirements.

Proceeding as before, the objective function for the mass is: $m = AL\rho = b^2 L\rho$. The bending stiffness S of the beam must be at least S^*:

$$S = \frac{C_1 EI}{L^3} \geq S^* \tag{3.84}$$

where C_1 is a constant – we don't need its value. The second moment of area, I, for a square section beam is

$$I = \frac{b^4}{12} = \frac{A^2}{12} \tag{3.85}$$

For a given length L, the stiffness S^* is achieved by adjusting the size of the square section. Now eliminating b (or A) in the objective function for the mass gives

$$m = \left(\frac{12 S^* L^3}{C_1}\right)^{1/2} (L) \left(\frac{\rho}{E^{1/2}}\right) \tag{3.86}$$

The quantities S^*, L and C_1 are all specified or constant – the best materials for a light, stiff beam are those with the smallest values of the index M_b where

$$M_{b_1} = \frac{\rho}{E^{1/2}} \tag{3.87a}$$

Repeating the calculation with a constraint of strength rather than stiffness leads to the index

$$M_{b_2} = \frac{\rho}{\sigma_y^{2/3}} \tag{3.87b}$$

This analysis was for a square beam, but the result in fact holds for any shape, so long as the shape is held constant. This is a consequence of (3.84) – for a given shape, the

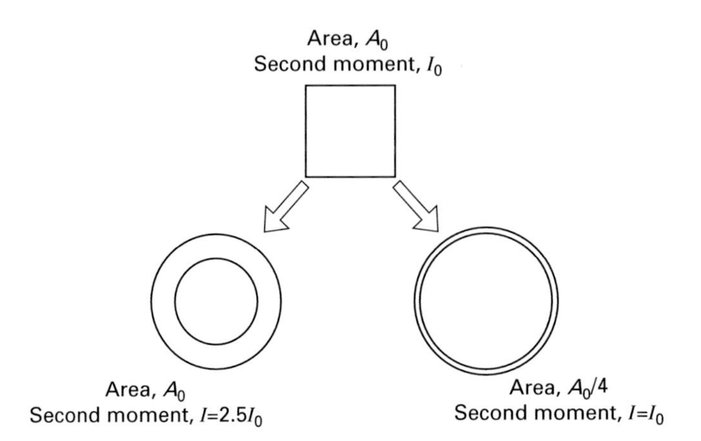

Fig. 3.30 The effect of section shape on bending stiffness, EI: a square-section beam, compared, left, with a tube of the same area (but 2.5 times stiffer) and, right, a tube with the same stiffness (but four times lighter).

second moment of area I can always be expressed as a constant times A^2, so changing the shape just changes the constant C_1 in (3.86), not the resulting index.

As noted above, real beams have section shapes that improve their efficiency in bending, requiring less material to get the same stiffness. By shaping the cross-section it is possible to increase I without changing A. This is achieved by locating the material of the beam as far from the neutral axis as possible, as in thin-walled tubes or I-beams (Fig. 3.29d). Some materials are more amenable than others to being made into efficient shapes. Comparing materials on the basis of the index in M_b therefore requires some caution – materials with lower values of the index may "catch up" by being made into more efficient shapes. So we need to get an idea of the effect of shape on bending performance.

Figure 3.30 shows a solid square beam, of cross-sectional area A. If we turned the same area into a tube, as shown in the right of the figure, the mass of the beam is unchanged. The second moment of area, I, however, is now much greater – and so is the stiffness (3.84). We define the ratio of I for the shaped section to that for a solid square section with the same area (and thus mass) as the *shape factor* Φ. The more slender the shape the larger is Φ, but there is a limit – make it too thin and the flanges will buckle – so there is a maximum shape factor for each material that depends on its properties. Table 3.13 lists some typical values.

Shaping is used to make structures lighter: it is a way to get the same stiffness with less material. The mass ratio is given by the reciprocal of the square root of the maximum shape factor, $\Phi^{-1/2}$ (because C_1, which proportional to the shape factor, appears as $(C_1)^{-1/2}$ in equation (3.86)). Table 3.13 lists the factor by which a beam can be made lighter, for the same stiffness, by shaping. Metals and composites can all be improved significantly (though the metals do a little better), but wood has more limited potential because it is more difficult to shape it into efficient, thin-walled shapes. So, when comparing materials for light, stiff beams using the index in (3.87a) and (3.87b), the performance of wood is not as good as it looks because other materials can be made

Table 3.13 The effect of shaping on stiffness and mass of beams in different structural materials

Material	Typical maximum shape factor (stiffness relative to that of a solid square beam)	Typical mass ratio by shaping (relative to that of a solid square beam)
Steels	64	1/8
Al alloys	49	1/7
Composites (GFRP, CFRP)	36	1/6
Wood	9	1/3

into more efficient shapes. Composites (particularly CFRP) have attractive (i.e. low) values of all the indices M_t, M_p and M_b, but this advantage relative to metals is reduced a little by the effect of shape.

References

Allen HG (1969) *Analysis and Design of Structural Sandwich Panels*. Oxford: Pergamon Press.

Andrews EW, Gibson LJ and Ashby MF (1999) The creep of cellular solids. *Acta Mater* **47**, 2853–63.

Ashby MF and Jones DRH (1996) *Engineering Materials: An Introduction to their Properties and Applications*. Oxford: Pergamon Press.

Ashby MF (2005) *Materials Selection in Mechanical Design*, 3rd edn. Oxford: Butterworth Heinemann.

Ashby MF (2006) The properties of foams and lattices. *Phil. Trans. Royal Soc.* **A364**, 15–30.

Ashby MF, Shercliff HR and Cebon D (2007) *Materials: Engineering, Science, Processing and Design*. Oxford: Butterworth Heinemann.

Bart-Smith H, Bastawros A-F, Mumm DR, Evans AG, Sypeck DJ and Wadley HNG (1998) Compressive deformation and yielding mechanisms in cellular Al alloys determined using X-ray tomography and surface strain mapping. *Acta Mat.* **46**, 3583–92.

Brace WF (1977) Permeability from resistivity and pore shape. *J. Geophysical Research* **82**, 3343–9.

Brazier LG (1927) On the flexure of thin cylindrical shells and other thin sections. *Proc. R. Soc. Lond.* **A116**, 104–14.

Calladine CR (1983) *Theory of Shell Structures*. Cambridge: Cambridge University Press.

Chen C, Lu TJ and Fleck NA (2001) Effect of inclusions and holes on the stiffness and strength of honeycombs. *Int. J. Mech. Sci.* **43**, 487–504.

Cheng P (1996) Weight optimization of cylindrical shells with cellular cores, SM thesis. Department of Civil and Environmental Engineering, MIT.

Dawson MA, Germaine JT and Gibson LJ (2007) Permeability of open-cell foams under compressive strain. *Int. J. Solids Struct.* **44**, 5133–45.

Dawson MA, McKinley GH and Gibson LJ (2008) The dynamic compressive response of open-cell foam impregnated with a Newtonian fluid. *J. Appl. Mech.* **75**, 041015.

Deshpande VS, Ashby MF and Fleck NA (2001) Foam topology bending versus stretching dominated architectures. *Acta Mater.* **49**, 1035–40.

Finnie I and Heller WR (1959) *Creep of Engineering Materials*. New York: McGraw Hill.

Gent A and Rusch K (1966) Permeability of open-cell foamed materials. *J. Cellular Plastics*, **2**, 46–51.

Gibson LJ, Ashby MF, Schajer GS and Robertson CL (1982a) The mechanics of two-dimensional cellular materials. *Proc. Roy. Soc. Lond.* **A382**, 25–42.

Gibson LJ and Ashby MF (1982b) The mechanics of three-dimensional cellular materials. *Proc. Roy. Soc. Lond.* **A382**, 43–59.

Gibson LJ and Ashby MF (1997) *Cellular Solids: Structure and Properties*, 2nd edn. Cambridge: Cambridge University Press.

Grenestedt JL (1998) Influence of wavy imperfections in cell walls on elastic stiffness of cellular solids. *J. Mech. Phys. Solids* **46**, 29–50.

Grenestedt JL and Tanaka K (1999) Influence of cell shape variations on elastic stiffness of closed cell cellular solids. *Scripta Materialia* **40**, 71–7.

Grenestedt JL and Bassinet F (2000) Influence of cell wall thickness variations on elastic stiffness of closed cell cellular solids. *Int. J. Mech. Sci.* **42**, 1327–38.

Guo XE and Gibson LJ (1999) Behaviour of intact and damaged honeycombs: a finite element study. *Int. J. Mech. Sci.* **41**, 85–105.

Guo XE and Kim (2002) Mechanical consequence of trabecular bone loss and its treatment: a three-dimensional model simulation. *Bone* **30**, 404–11.

Harte A-M, Fleck NA and Ashby MF (1999) Fatigue failure of an open cell and a closed cell aluminum alloy foam. *Acta Mat.* **47**, 2511–24.

Huang J-S and Gibson LJ (2003) Creep of open-cell Voronoi foams. *Mat. Sci. Eng.* **A339**, 220–6.

Karam GN and Gibson LJ (1995a) Elastic buckling of cylindrical shells with elastic cores I: Analysis Int. *J. Solids Struct.* **32**, 1259–83.

Karam GN and Gibson LJ (1995b) Elastic buckling of cylindrical shells with elastic cores II: Experiments. *Int. J. Solids Struct.* **32**, 1285–306.

Ko WL (1965) Deformations of foamed elastomers. *J. Cellular Plastics* **1**, 45.

Lakes R (1987) Foam structures with a negative Poisson's ratio. *Science* **235**, 1038–40.

Maxwell JC (1864) On the calculation of the equilibrium and stiffness of frames. *Philosophical Magazine* **27**, 294–9.

McCullough KYG, Fleck NA and Ashby MF (2000) The stress-life fatigue behaviour of aluminum alloy foams. *Fatigue Fract. Eng. Mat. Struct.* **23**, 199–208.

Monkman FC and Grant NJ (1956) *Proc. ASTM STP*, **56**, 593.

Nazarian A and Muller R (2004) Time-lapsed microstructural imaging of bone failure. *J. Biomech.* **37**, 55–65.

Papka SD and Kyriakides S (1994) In-plane compressive response and crushing of honeycomb. *J. Mech. Phys. Solids* **42**, 1499–532.

Papka SD and Kyriakides S (1998a) Experiments and full-scale numerical simulations of in-plane crushing of a honeycomb. *Acta Mat.* **46**, 2765–76.

Papka SD and Kyriakides S (1998b) In-plane crushing of a polycarbonate honeycomb. *Int. J. Solids Struct.* **35**, 239–67.

Papka SD and Kyriakides S (1999a) Biaxial crushing of honeycombs – part I: experiments. *Int. J. Solids Struct.* **36**, 4367–96.

Papka SD and Kyriakides S (1999b) Biaxial crushing of honeycombs – Part II: analysis. *Int. J. Solids Struct.* **36**, 4397–423.

Patel MR and Finnie I (1970) Structural features and mechanical properties of rigid cellular plastics. *J. Mat.* **5**, 909.

Poirier C, Ammi M, Bideau D and Troadec JP (1992) Experimental study of the geometrical effects in the localization of deformation. *Phys. Rev. Letters* **68**, 216–19.

Powell PC (1983). *Engineering with Polymers*. London: Chapman and Hall.

Prakash O, Bichebois P, Brechet Y, Louchet F and Embury JD (1996) A note on the deformation behaviour of two-dimensional model cellular structures. *Phil. Mag.* **A73**, 739–51.

Silva MJ, Hayes WC and Gibson LJ (1995) The effects of non-periodic microstructure on the elastic properties of two-dimensional cellular solids. *Int J. Mech. Sci.* **37**, 1161–77.

Silva MJ and Gibson LJ (1997) The effects of non-periodic microstructure and defects on the compressive strength of two-dimensional cellular solids. *Int J. Mech. Sci.* **39**, 549–63.

Simone AE and Gibson LJ (1998a) Effects of solid distribution on the stiffness and strength of metallic foams. *Acta Materialia* **46**, 2139–50.

Simone AE and Gibson LJ (1998b) Effects of cell face curvature and corrugations on the stiffness and strength of metallic foams. *Acta Materialia* **46**, 3929–35.

Sugimura Y, Rabiei A, Evans AG, Harte AM and Fleck NA (1999) Compression fatigue of a cellular Al alloy. *Mat. Sci. Eng.* **A269**, 38–48.

Timoshenko SP and Gere JM (1961) *Theory of Elastic Stability*, 2nd edn. New York: McGraw-Hill.

Ulrich D, Van Rietbergen B, Laib A, Ruegsegger P (1999) The ability of three dimensional structural indices to reflect mechanical aspects of trabecular bone. *Bone* **25**, 55–60.

Vajjhala S, Kraynik AM, Gibson LJ (2000) A cellular solid model for modulus reduction due to resorption of trabeculae in bone. *J. Biomech. Eng.* **122**, 511–15.

Van der Burg MWD, Shulmeister V, van der Geissen E, Marissen R (1997) On the linear elastic properties of regular and random open-cell foam models. *J. Cellular Plastics* **33**, 31–54.

Van Rietbergen B, Weinans H, Huiskes R and Odgaard A (1995) A new method to determine trabecular bone elastic properties and loading using micromechanical finite element models. *J. Biomech.* **28**, 69–81.

Warner M and Edwards SF (1988) A scaling approach to elasticity and flow in solid foams. *Europhys. Lett.* **5**, 623–8.

Warner M, Thiel BL and Donald AM (2000) The elasticity and failure of fluid-filled cellular solids: Theory and experiment. *Proc. Nat. Acad. Sci.* **97**, 1370–5.

Warren WE and Kraynik AM (1997) Linear elastic behavior of a low-density Kelvin foam with open cells. *J. Appl. Mech.* **64**, 787–94.

Zettl B, Mayer H, Stanzl-Tschegg SE and Degischer HP (2000) Fatigue properties of aluminum foams at high numbers of cycles. *Mat. Sci. Eng.* **A292**, 1–7.

Zhou J and Soboyejo WO (2004) Compression-compression fatigue of open cell aluminum foams: macro/micro-mechanisms and the effects of heat treatment. *Mat. Sci. Eng.* **A369**, 23–35.

Zhu HX, Knott JF and Mills NJ (1997a) Analysis of the elastic properties of open-cell foams with tetrakaidecahedral cells. *J. Mech. Phys. Solids* **45**, 319–43.

Zhu HX, Mills NJ and Knott JF (1997b) Analysis of the high strain compression of open-cell foams. *J. Mech. Phys. Solids* **45**, 1875–904.

Part II

Cellular materials in nature

4 Honeycomb-like materials in nature

4.1 Introduction

The queen of cellular structures, of course, is the honeycomb of the bee (Fig. 4.1). Each layer has near-perfect hexagonal cells, which, when filled, are sealed off and a new layer, offset from the first, is started. There are two possible explanations for the reason that honeycomb is composed of hexagons, rather than any other shape. One is that the hexagon tiles the plane with minimum surface area. Thus a hexagonal structure uses the least material to create a lattice of cells within a given volume. Another, given by D'Arcy Thompson (1961), is that the shape simply results from the process of individual bees putting cells together: somewhat analogous to the boundary shapes created in a mass of soap bubbles. In support of this he notes that queen cells, which are constructed singly, are irregular and lumpy with no apparent attempt at efficiency.

Wood and cork both have cellular structures similar to that of a honeycomb. Although wood has several types of cells of somewhat different geometries, and the cellular structure of cork is interrupted by small air channels, called lenticels, the overall mechanical behavior of both wood and cork can be understood by applying the models for honeycomb materials, described in Chapter 3.

4.2 Wood

4.2.1 Introduction

Wood is one of the oldest and most widely used structural materials available – the word "material" derives from the Latin *materies*, *materia*: the trunk of a tree. Today, the world production of wood is roughly 1500 million metric tons/year, similar to that of steel (United Nations Food and Agricultural Organization, 2005; International Iron and Steel Institute, 2008), and, if measured by volume, that of wood exceeds that of steel by a factor of ten. Roughly 20% of the total production is used structurally: for beams, joists, supports and panels that bear load. Then the properties that interest the designer are the moduli, the crushing strength and the toughness. These properties vary enormously from one wood to another: oak is more than ten times stiffer, stronger and tougher than balsa. And wood can be very anisotropic, too: some species are more than 50 times stiffer when loaded along the

Fig. 4.1 A bee's honeycomb, showing the cell size in relation to the size of the bee. (Image courtesy of Matthew Rader of Wills Point TX.)

grain than across it. In this section we review the structure and mechanical properties of wood and show how its stiffness, strength and toughness can be explained by extensions of the models for honeycombs developed in Chapter 3. We show that the properties depend primarily, like those of all cellular solids, on the properties of the cell wall, on the relative density and on the shape of the cells. Other factors (e.g. temperature and moisture content) play a secondary, though still important, role.

4.2.2 Structure of wood

If a sample of wood is cut at a sufficient distance from the center of the tree that the curvature of the growth rings can be neglected, its properties are orthotropic. It has three orthogonal planes of symmetry: radial, tangential and axial (Fig. 4.2). The stiffness and strength are greatest in the axial direction, that is, parallel to the trunk of the tree; in the radial and tangential directions, they are less by a factor of ½ to 1/20, depending on the species.

These differences all relate to the structure of wood. At one scale (that of millimeters), wood is a cellular solid: cell walls, often with the shape of hexagonal prisms, enclose pore space. Three features characterize the microstructure (Dinwoodie, 1981; Bodig and Jayne, 1982; Kettunen, 2006):

(1) the highly elongated cells that make up the bulk of the wood, called *tracheids* in softwoods and *fibers* in hardwoods;
(2) the *rays*, made up of radial arrays of smaller, more rectangular, parenchyma cells;
(3) the *sap channels*, which are enlarged cells with thin walls and large pore spaces which conduct fluids up the tree.

There are, of course, structural differences between softwoods and hardwoods. The rays in softwoods are narrow and extend only a few cells in the axial direction, while

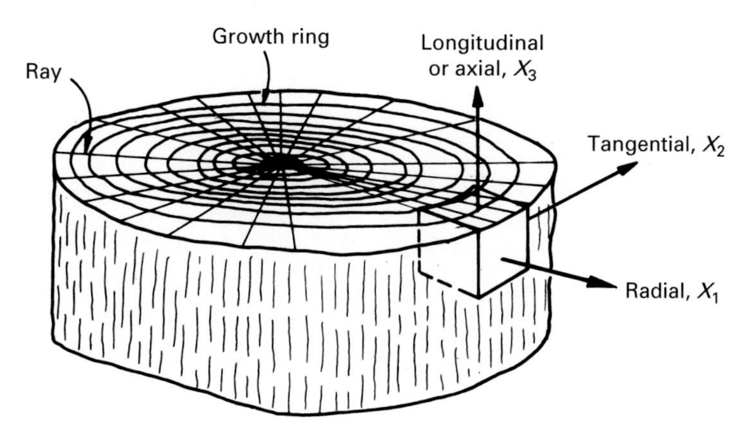

Fig. 4.2 A section through the trunk of a tree, showing the axial, radial and tangential directions. (Reprinted from Easterling *et al.*, 1982.)

those in hardwoods are wider and extend hundreds of cells in the axial direction. In softwoods the sap channels make up less than 3% of the wood volume while in hardwoods they can account for as much as 55%. The growth rings in softwoods are made up of alternating circumferential bands of thick- and thin-walled tracheids, while those in ring-porous hardwoods are characterized by bands of large- and small-diameter sap channels. Diffuse-porous hardwoods have a uniform distribution of sap channels of the same size; they do not exhibit any characteristic growth rings on a microscopic scale. All of these features can be seen in Fig. 4.3, which shows the structures of a representative softwood (cedar) and a hardwood (oak). The volume fractions and dimensions of the cells in wood are listed in Table 4.1.

At a finer scale (that of microns) wood is a fiber-reinforced-composite. The cell walls are made up of fibers of largely crystalline cellulose embedded in a matrix of amorphous hemicellulose and lignin (Fig. 2.7). The arrangement of the cellulosic fibers in the wall is complicated but important because it accounts for part of the great anisotropy of wood – the difference in properties along and across the grain. It is helpful (though a simplification) to think of the cell walls as helically wound, like the shaft of a racquet, with the fiber direction nearer the cell axis than across it. This gives the cell wall a modulus and a strength which are large parallel to the axis of the cell, and smaller, by a factor of about 3, across it. But this accounts for only a part of the large anisotropy of wood. The rest is related to the cell shape: elongated cells are stiffer and stronger when loaded along the long axis of the cell than when loaded across it. Details are given below.

Although woods differ enormously in their density and mechanical properties, the properties of the cell wall are, as a rough approximation, the same for all woods. Woods as different as balsa and beech have cell walls with a density near 1500 kg/m^3 and moduli and strengths which have values close to those listed in Table 4.2. The relative density, ρ^*/ρ_s, the density of the wood divided by that of the cell-wall material, can be as low as 0.05 for balsa and as high as 0.80 for lignum vitae.

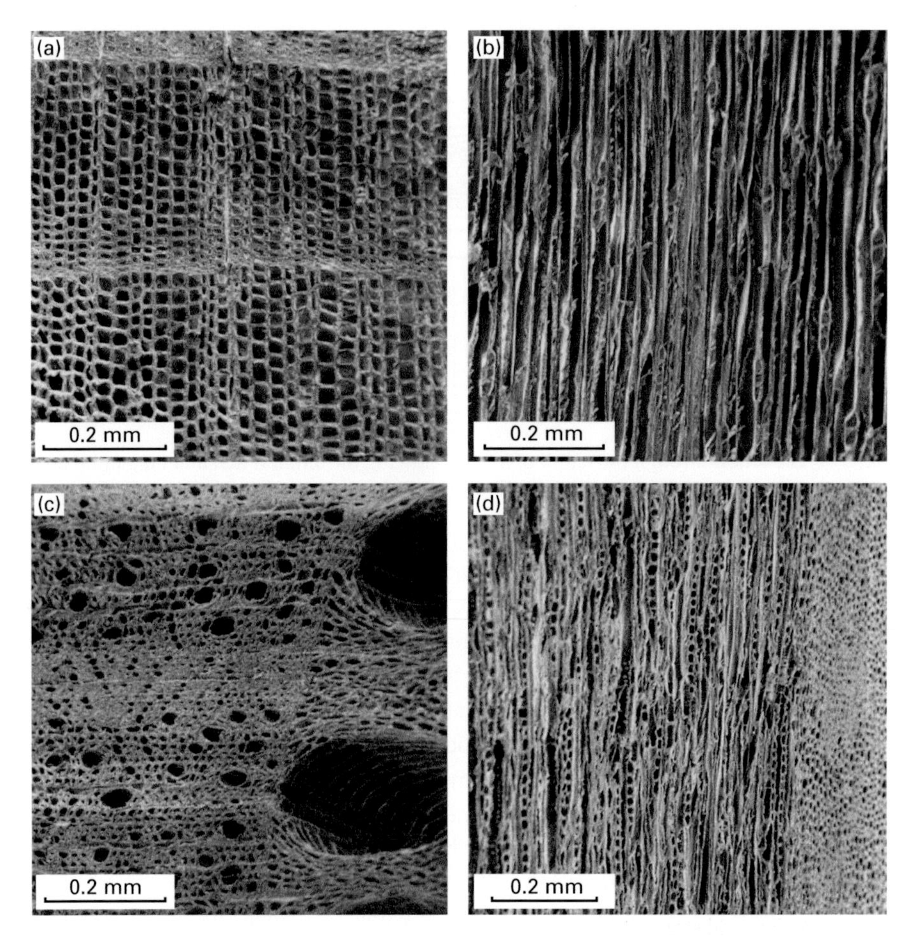

Fig. 4.3 Scanning electron micrographs of wood: (a) cedar, cross-section; (b) cedar, longitudinal section; (c) oak, cross-section; (d) oak, longitudinal section.

4.2.3 Young's modulus and compressive strength of wood

Compressive stress–strain curves for a number of woods with relative densities ranging from 0.05 to 0.5, loaded in the tangential and axial directions, are shown in Fig. 4.4. The general observations are as follows. At small strains (less than about 0.02) the behavior is linear-elastic in all three directions. The Young's modulus in the axial direction is much larger than those in the tangential and radial directions, which are roughly equal. Beyond the linear-elastic regime, the stress–strain curves for loading in all three directions show a stress plateau extending to strains between 0.2 and 0.8 depending on the density of the wood. At the end of the plateau the stress rises steeply. Compression in the tangential direction gives a smooth stress–strain curve which rises gently throughout the plateau. Compression along the radial direction is distinguished by a small yield drop at the end of the linear-elastic regime, followed by a slightly irregular or wavy stress plateau (Fig. 4.5b). The tangential and radial yield stresses are about equal. That in the axial direction is much higher and is followed by a sharply serrated plateau (Fig. 4.4b). As the density of the wood increases, the moduli

Table 4.1 Volume fractions and dimensions of wood cells

	Softwoods		Hardwoods		
	Tracheid	Ray cells	Fiber	Vessel	Ray cells
Volume fraction (%)	85–95	5–12	37–70	6–55	10–32
Axial dimension (mm)	2.5–7.0		0.6–2.3	0.2–1.3	
Tangential dimension (μm)	25–80		10–30	20–500	
Radial dimension (μm)	17–60		10–30	20–350	
Cell-wall thickness, t (μm)	2–7		1–11	–	

Source: Bodig and Jayne (1982), p. 14.

Table 4.2 Cell wall properties for wood

Property	Literature value	Value inferred from data plots
Density, ρ_s (kg/m^3)	1500[a]	–
Axial Young's modulus, E_s (GN/m^2)*	35[b] 28[c] 25[d]	35
Transverse Young's modulus (GN/m^2)	10[b]	19
Shear modulus, from A–R loading (GN/m^2)	–	2.6
Shear modulus, from R–T loading (GN/m^2)	–	2.6
Axial yield strength, σ_{ys} (MN/m^2)	350[e]	120
Transverse yield strength, σ_{ys} (MN/m^2)	135[e]	50
Shear yield strength	–	30
Toughness, peeling mode, $G_{CS}{}^{p}$(J/m^2)	–	350
Toughness, breaking mode, $G_{CS}{}^{b}$(J/m^2)	–	1650
Fracture toughness, peeling mode, $(K_{IC}{}^{p})_s$ (MN/m$^{3/2}$)	–	1.9
Fracture toughness, breaking mode, $(K_{IC}{}^{b})_s$ (MN/m$^{3/2}$)	–	4.1

Sources: [a] Dinwoodie (1981); [b] Cave (1968), from tensile tests on 2 mm × 2 mm × 60 mm specimens (10% moisture content); [c] Orso *et al.* (2006), from cell walls loaded as cantilever beams using an atomic force microscope tip mounted on a micromanipulator (vacuum dry); [d] Gierlinger *et al.* (2006) from tensile tests on single tracheids (dry); [e] Cave (1969), from tensile tests on 2 mm × 2 mm × 60 mm specimens (10% moisture content).

* *Note:* recent measurements of the axial Young's modulus, E_s, using nanoindentation give lower values, in the range 14–21 GPa, probably due to the anisotropic nature of the cell wall which is not accounted for in the analysis (e.g. see Wimmer *et al.*, 1997; Gindl and Schoberl, 2004; Tze *et al.*, 2007)

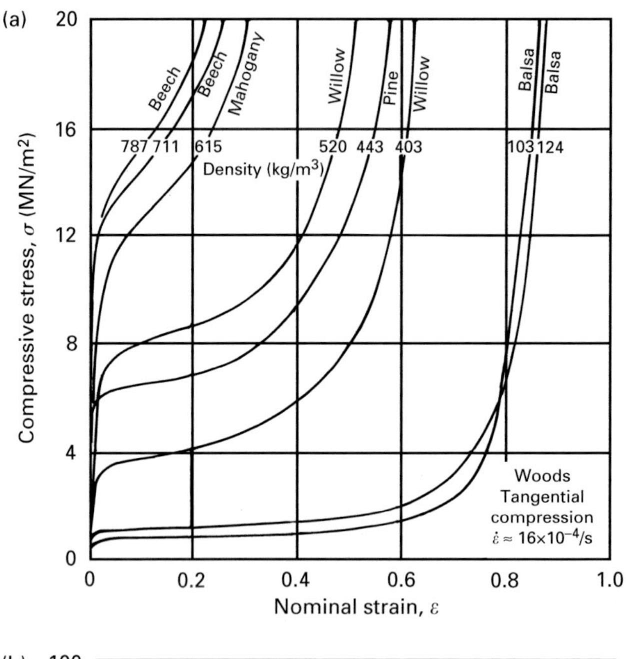

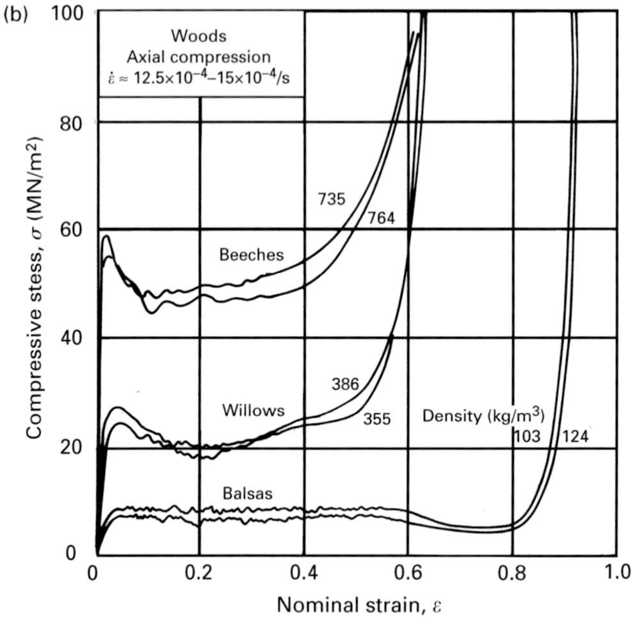

Fig. 4.4 Compressive stress–strain curves for several species of wood, loaded in the (a) tangential and (b) axial directions. The curves for radial loading are similar to those for tangential loading.

and plateau strengths increase. The moduli and strength of wood are also affected by the age and moisture content of the wood and the temperature and strain-rate at which testing is carried out: the data analysed in this chapter refer to well-seasoned wood, with a constant moisture content of about 12%, tested at 18°C ± 2°C and at a strain-rate of close to 10^{-3}/s.

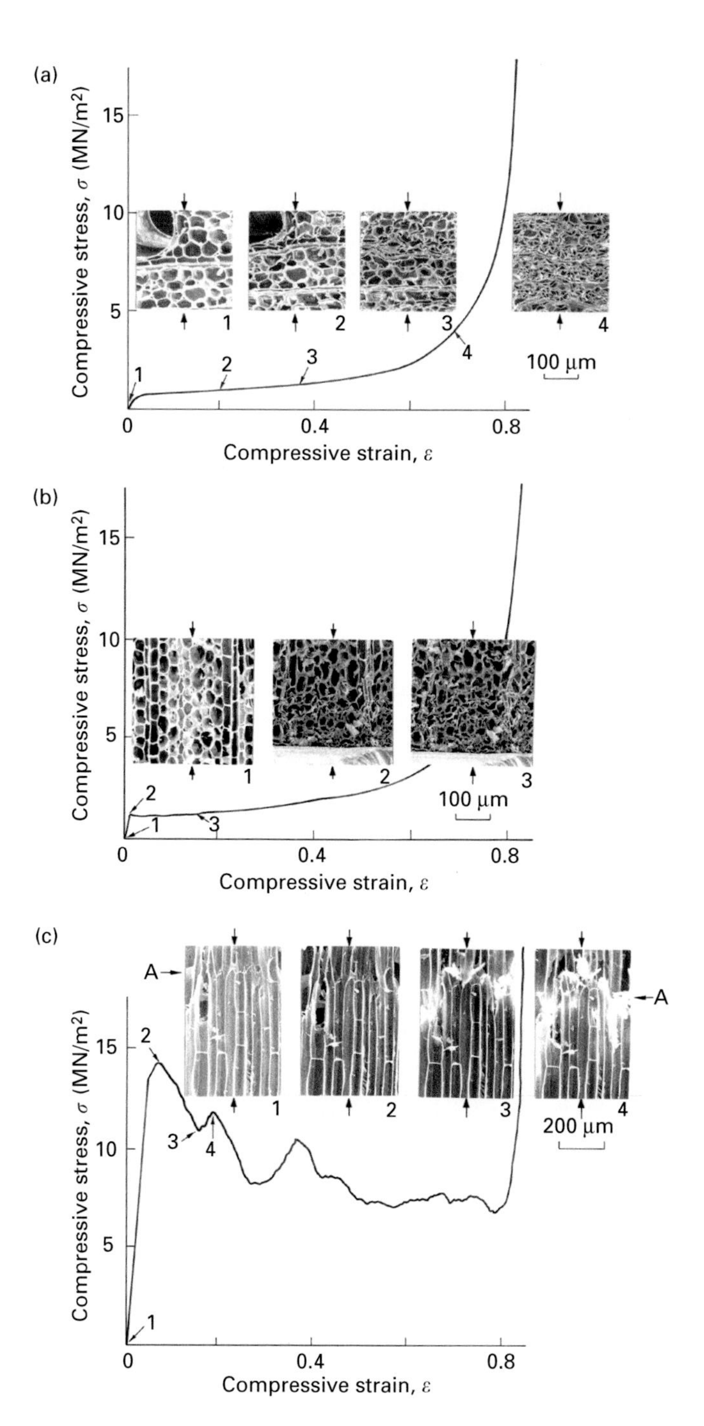

Fig. 4.5 Compressive stress–strain curves for balsa, along with micrographs showing the deformation of the cells with increasing strain, for loading in the (a) tangential, (b) radial and (c) axial directions. (Reprinted from Easterling *et al.*, 1982.)

The mechanisms by which wood cells deform and fail can be observed by loading balsa in a deformation stage mounted in a scanning electron microscope (Easterling *et al.*, 1982). Compression in the tangential direction causes bending and collapse by plastic yielding of the cell walls (Fig. 4.5a). As the strain increases, the deformation is uniform throughout the specimen and the stress–strain curve rises monotonically, similar to the rubber honeycombs described in Section 3.2. In species of wood with more pronounced growth rings than balsa, the growth rings act as stiff reinforcement; yielding corresponds to their plastic buckling on a macroscopic scale. Compression of balsa in the radial direction also initially causes bending of the cell walls (Fig. 4.5b). Plastic collapse of the cells is non-uniform, starting at the surface of the loading platen and propagating inward along the length of the specimen. The same feature has been observed in higher density woods, such as Douglas fir (Bodig and Jayne, 1982). As discussed in Section 3.2, the localization of cell collapse is associated with the small yield drop in the stress–strain curve, similar to the aluminum honeycombs shown in Fig. 3.6.

The mechanism of compressive deformation in the axial direction is fundamentally different: there is little evidence of cell-wall bending in the linear-elastic regime, instead, the cells compress uniaxially. At a sufficiently high strain the cells collapse by the yielding and fracturing of planes of material where the ends of adjacent layers of cells join together (section A–A in Fig. 4.5c). In woods of low density, the cell walls are forced between each other like the teeth of two combs meshing together. The stress drops until the next layer of cells is intercepted, one cell length away. The stress increases until this second layer of cell ends fails and then drops off again as the cells mesh. This process is repeated, giving the dramatically serrated plateau, until the wood has almost completely densified at a strain of about 0.8. In woods of higher density the mechanism can be different: cells undergo local, plastic buckling.

Data for the Young's moduli and compressive strengths of a number of species of wood are plotted in Fig. 4.6. The lower axis shows the density normalized by that of the cell wall (1500 kg/m^3). That on the left-hand side shows the modulus normalized by Cave's (1968) value for the axial modulus of the cell wall (35 GN/m^2) or the strength normalized by Cave's (1969) value for the axial strength of the cell wall in tension (350 MN/m^2). The unnormalized data are shown on the remaining two sides. Data for Poisson's ratios are listed in Table 4.3.

The cellular structure of wood is shown, somewhat idealized, in Fig. 4.7. The tracheids and fibers that make up the bulk of the cells in softwoods and hardwoods, respectively, can be thought of as a regular array of long, hexagonal prisms with occasional transverse membranes. They are traversed by rays: radial bands of shorter, more rectangular cells. Circular sap channels run up the axis of the tree. We seek to model the mechanical behavior of this structure. The problem is further complicated by the differences between softwoods, ring-porous hardwoods and diffuse-porous hardwoods. The rays and growth rings of softwoods and hardwoods differ: in softwoods the rays are narrow and only a few cells long, while in hardwoods they are much wider and longer; and the growth rings in softwoods are made up of

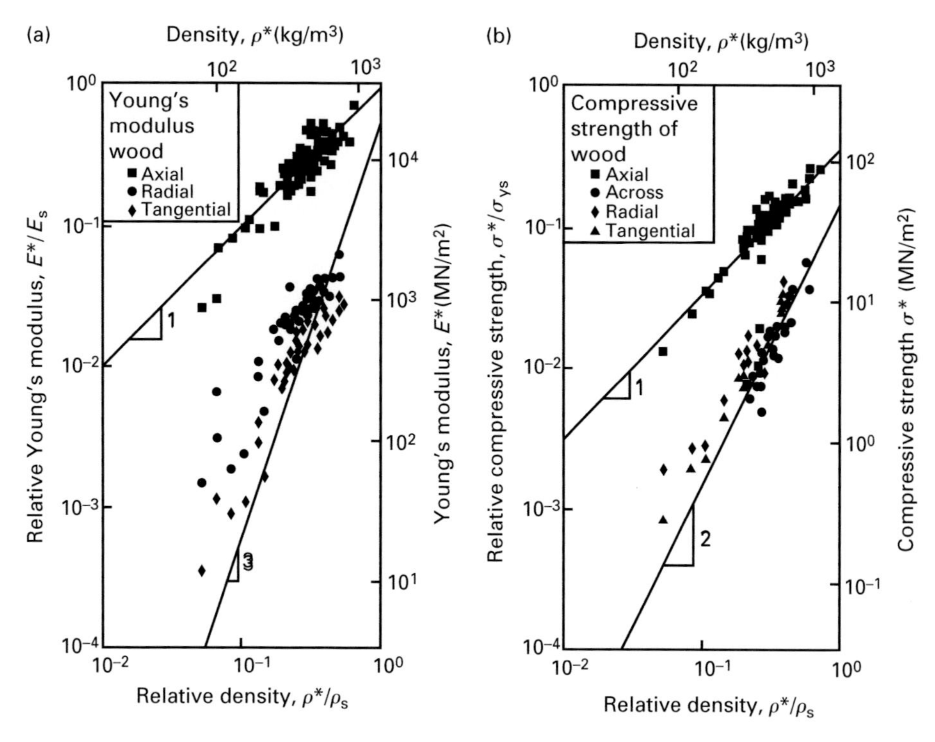

Fig. 4.6 (a) Young's modulus and (b) compressive strength of wood plotted against density. One pair of axes is normalized by the cell wall modulus or strength in the axial direction and by the cell wall density. The other pair corresponds to the raw data. Data from Goodman and Bodig (1970, 1971); Bodig and Goodman (1973); *Wood Handbook* (1974); Dinwoodie (1981); Bodig and Jayne (1982) and Easterling *et al.* (1982).

alternating circumferential bands of thick- and thin-walled tracheids, while those in ring-porous hardwoods have bands of large- and small-diameter sap channels, or vessels. The distribution of sap channels in diffuse-porous hardwoods is almost uniform, so that they do not have obvious growth rings. In devising mechanical models for wood these differences are introduced only in the most approximate way.

Attempts to model the elastic moduli of wood by analyzing the deformation of an idealized cell date back at least 80 years. Price (1928), in a remarkably detailed paper, modeled the elongated wood cells as a parallel array of cylindrical tubes. Loaded axially, the tubes are extended or compressed; loaded transversely, they distort from a circular to an oval cross-section, predominantly by bending. Price analyzed the distortion of a single tube and found that the transverse Young's modulus should vary as the cube of the density while the axial modulus should vary linearly with density – precisely the same as the results for a hexagonal honeycomb given in Chapter 3. Price suggested that the radial modulus should be somewhat greater than the tangential one and also calculated Poisson's ratios for each loading direction. As discussed below, his results are in broad agreement with the observations. The model was developed further by Srinavasan (1942) who added a second set of tubes at right angles to the first to simulate the rays.

Table 4.3 Poisson's ratios for wood

Species	Density(kg/m³)	ν_{RT}^*	ν_{RA}^*	ν_{TR}^*	ν_{TA}^*	ν_{AR}^*	ν_{AT}^*	Reference
Hardwoods								
Balsa	200	0.66	0.02	0.24	0.01	0.23	0.49	a
Aspen	300	–	–	0.50	–	0.49	0.37	b
Yellow poplar	380	0.70	0.03	0.33	0.02	0.32	0.39	c
Khaya	440	0.60	0.03	0.26	0.03	0.30	0.64	a
Sweetgum	530	0.68	0.04	0.30	0.02	0.33	0.40	c
Oak	580	–	–	0.26	–	0.29	0.48	b
Walnut	590	0.72	0.05	0.37	0.04	0.49	0.63	a
Birch	620	0.78	0.03	0.38	0.02	0.49	0.43	a
Ash	670	0.71	0.05	0.36	0.03	0.46	0.51	a
Beech	750	0.75	0.07	0.36	0.04	0.45	0.51	a
Softwoods								
Engleman spruce	350	–	–	0.22	–	0.44	0.50	b
Norway spruce	390	0.51	0.03	0.31	0.03	0.38	0.51	a
Sitka spruce	390	0.43	0.03	0.25	0.02	0.37	0.47	a
Scotch pine	550	0.68	0.04	0.31	0.02	0.42	0.51	a
Douglas fir	430	–	–	0.56	–	0.28	0.50	b
Douglas fir	590	0.63	0.03	0.40	0.02	0.43	0.37	a

$^*\nu_{ij} = -\varepsilon_j/\varepsilon_i$

Sources: [a] Dinwoodie (1981); [b] Goodman and Bodig (1970); [c] *Wood Handbook* (1974).

When wood is compressed in the tangential direction the cell walls bend, corresponding to in-plane loading of a honeycomb, so that (Easterling *et al.*, 1982)

$$\frac{E_T^*}{E_s} = C_1 \left(\frac{\rho^*}{\rho_s} \right)^3 \qquad (4.1)$$

where E_s is the Young's modulus of the cell wall in the axial direction (35 GN/m²). The anisotropy of the cell wall itself introduces complications, which we deal with by normalizing in every equation by the axial property of the cell wall (as here), and incorporating the difference between the axial and transverse cell-wall property in the constant C_1. The rays, the end-caps of the cells and the transverse membranes all act to increase the stiffness above the value given in (4.1). The rays are denser and stiffer than the bulk of the cells and the end-caps and transverse membranes deform by stretching; because they are stiffer, they also constrain the surrounding cells. It is perhaps for this reason that the data plotted in Fig. 4.6a lie somewhat above the line corresponding to (4.1). The transverse cell-wall modulus can be found from the intercept at a relative density of one; it is 19 GN/m², somewhat larger than the value found by Cave (1968). C_1 is equal to 0.54.

At first sight one might expect the moduli in the radial and the tangential directions to be equal. But the rays, which merely constrain lateral spreading in tangential

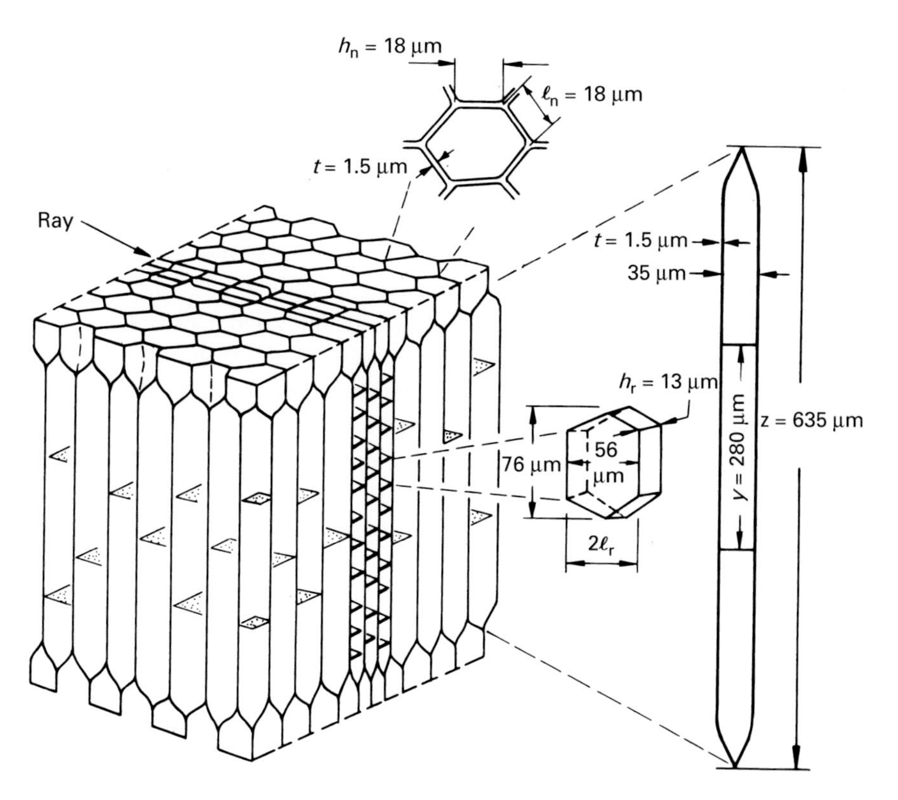

Fig. 4.7 Model for the cellular structure of wood. The cell dimensions are shown for balsa; typical cell dimensions for softwoods and hardwoods are given in Table 4.1. (Reprinted from Easterling *et al.*, 1982.)

loading, act as reinforcing plates in radial loading. Using a simple rule of mixtures approach, we find that

$$E_R^* \approx 1.5E_T^* \tag{4.2}$$

The axial modulus of wood is modeled by out-of-plane loading of a honeycomb, with uniaxial, rather than bending deformation of the cell walls, so that

$$\frac{E_A^*}{E_s} = C_2\left(\frac{\rho^*}{\rho_s}\right) \tag{4.3}$$

where $C_2 \sim 1$. This result is identical to that given by Price (1928).

Equations (4.1) and (4.3) for the Young's moduli of wood in the tangential and axial directions are shown by the bold lines on Fig. 4.6a. Equation (4.3) gives a good description of the axial modulus. The data for the tangential and radial moduli lie somewhat above the slope of 3 given by (4.1), because of the axial stiffening of the end-caps and transverse membranes, as described above. Equations (4.1) and (4.3) indicate that the different mechanisms of deformation (bending vs. axial compression) give rise to a density-dependent anisotropy in wood: the ratio of the axial to tangential moduli for low-density woods like balsa is almost 100 while that for very dense woods is about 4. The shear moduli and Poisson's ratios of wood can be treated in the same manner; they are described in detail in Gibson and Ashby (1997).

The anisotropy of wood also depends, but to a lesser extent, on the anisotropic, fiber composite nature of the cell walls. Figure 4.8 shows the interplay between influence of the fiber composite cell wall and the cellular structure of wood on modulus. The figure shows envelopes of Young's modulus and density for the constituents of wood (cellulose, lignin and hemicellulose), the cell wall of wood, and a number of different species of woods. The properties of the constituents of the wood cell wall are given in Table 2.2. The upper and lower bounds for the moduli of composites made from them can be estimated by simple composites theory; they are plotted with the envelope bounded by the data for cellulose and lignin in Fig. 4.8. The envelope of the Young's moduli of the wood cell walls is close to that of cellulose, as a result of the small angle of the cellulose microfibrils in the lignin/hemicellulose matrix in the thick S_2 layer of the cell wall (Fig. 2.7). The moduli of the woods loaded along the grain varies with linearly with density (4.3); extrapolating to the solid cell wall properties gives the value of E_s = 35 GPa. The moduli of woods loaded across the grain varies roughly with the cube of density (4.1); extrapolating to the solid cell wall properties gives the lower value of E_s = 19 GPa.

The compressive strength of wood can also be modeled using the results for honeycombs from Chapter 3. For loading in the tangential direction, the compressive strength is reached when the cell walls fail by plastic bending, or by plastic buckling preceded by plastic bending. Plastic collapse occurs at a stress of

$$\frac{\sigma_T^*}{\sigma_{ys}} = C_3 \left(\frac{\rho^*}{\rho_s}\right)^2 \tag{4.4}$$

where C_3 is a constant that incorporates the ratio of the transverse to the axial strength of the wall. For loading in the radial direction, the rays act as reinforcement. As for the modulus, the compressive strength in the radial direction can be obtained using a rule of mixtures approach. For low-density woods, we find that the radial compressive strength is about 1.4 times that in the tangential direction and that the difference between the two decreases as the density increases (Easterling *et al.*, 1982). The radial strength, then, is a little larger than that in the tangential direction, and both vary approximately as the second power of the density.

Axial crushing occurs by either yield and fracture of the pyramidal end caps in low-density woods like balsa, or by yielding followed by local plastic buckling in higher density woods such as spruce. In either case, the failure is initiated by axial plastic yielding in compression, giving

$$\frac{\sigma_A^*}{\sigma_{ys}} = C_4 \left(\frac{\rho^*}{\rho_s}\right) \tag{4.5}$$

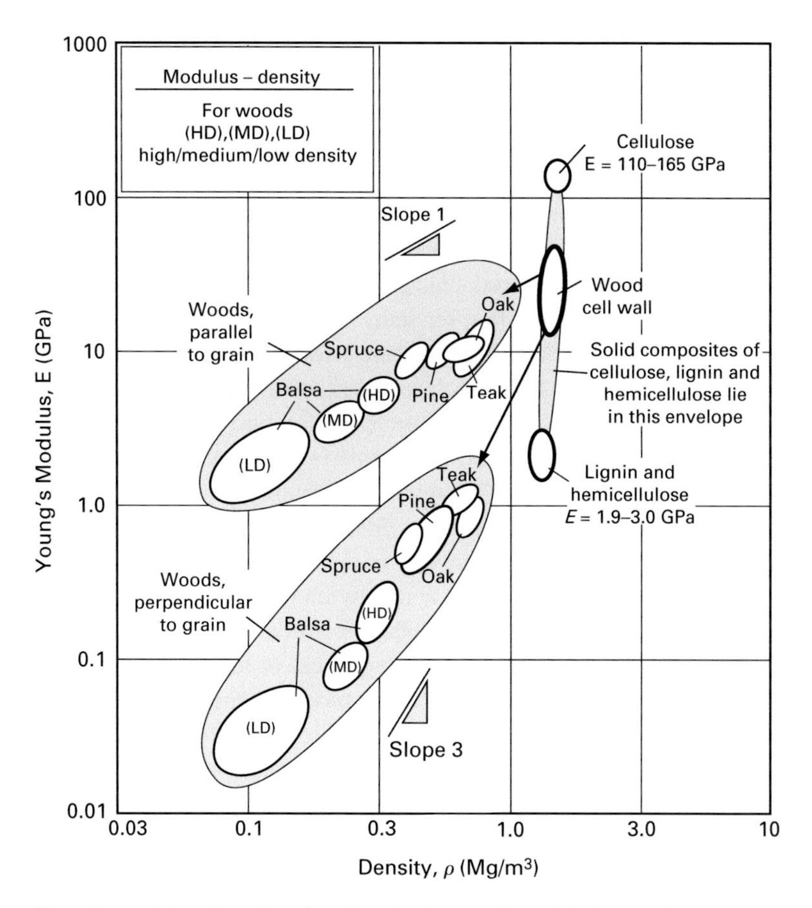

Fig. 4.8 Young's modulus plotted against density for woods and their constituents.

Data for the compressive strengths of wood in the tangential, radial and axial directions are plotted against density in Fig. 4.6b. The data are well described by (4.4) and (4.5). Equations (4.4) and (4.5) indicate that the different mechanisms of deformation (bending vs. axial) give rise to a density-dependent anisotropy in the compressive strength of wood, too: the ratio of the axial to tangential strengths for low-density woods like balsa is over 10 while that for very dense woods is about 2. The cell-wall yield strengths in transverse and axial compression, found from the intercepts at $\rho^* / \rho_s = 1$, are 50 and 120 MN/m^2, giving $C_3 = 0.14$ and $C_4 = 0.34$. These are roughly one-third the strengths given by Cave (1969). The shear strengths are treated in a similar way in Gibson and Ashby (1997).

4.2.4 Fracture and toughness

Fracture mechanics has been applied to wood for over 40 years (Wu, 1963; Schniewind and Pozniak, 1971; Walsh, 1971; Johnson, 1973; Schniewind and Centano, 1973; Mindess *et al.*, 1975; Williams and Birch, 1976; Jeronimidis, 1980; Barrett *et al.*, 1981; Nadeau *et al.*, 1982; Ashby *et al.*, 1985; Bentur and Mindess, 1986). The mechanics is complicated by a number of factors. Wood is not, strictly speaking, a linear-elastic solid: even at room temperature, and loaded rapidly, it is viscoelastic. As we saw in the last section, wood is orthotropic, so that standard solutions for isotropic materials do not apply. And there is great variability in the microstructure: the types and arrangement of the cells differs from softwoods to hardwoods and the relative density of earlywood is lower than that of latewood. Because of all of these factors, the standard tests of linear-elastic fracture mechanics do not always give consistent results. Notched three-point bend and double edge-notch specimens, for instance, give values of fracture toughness that differ, for some woods, by up to 30% (Ashby *et al.*, 1985). None-the-less, the results of a given test are broadly reproducible and the conditions for valid fracture toughness tests are approached nearly enough for the data to be useful for mechanical design. Here, we examine the fracture mechanics of wood assuming linear-elastic fracture mechanics and considering fracture toughnesses for initial crack propagation along and across the grain. More detailed treatments that use finite element analysis to account for orthotropy (Stanzl-Tschegg *et al.*, 1995; Schachner *et al.*, 2000) and that consider the differences in the microstructure between softwoods and hardwoods (Stanzl-Tschegg, 2006) and between earlywood and latewood (Thuvander and Berglund, 2000; Reiterer *et al.*, 2002) are available.

In wood, eight systems of crack propagation can be identified. They are illustrated in Fig. 4.9. Each system is identified by a pair of letters, the first indicating the direction normal to the crack plane, the second describing the direction of crack propagation. The LT and LR systems correspond to crack propagation across the grain, the others to crack propagation along the grain. The growth rings, which contain a density gradient, introduce further loss of microscopic symmetry. The + sign means that the crack propagates away from the center of the tree; the – sign, that it propagates towards the center.

Data for the fracture toughness of woods are plotted against relative density in Fig. 4.10. Values for K_{IC}^* for crack propagation normal to the grain (K_{IC}^{*n}) are roughly ten times greater than those for crack propagation along the grain (K_{IC}^{*a}), regardless of the relative density of the wood. Some researchers detected significant differences

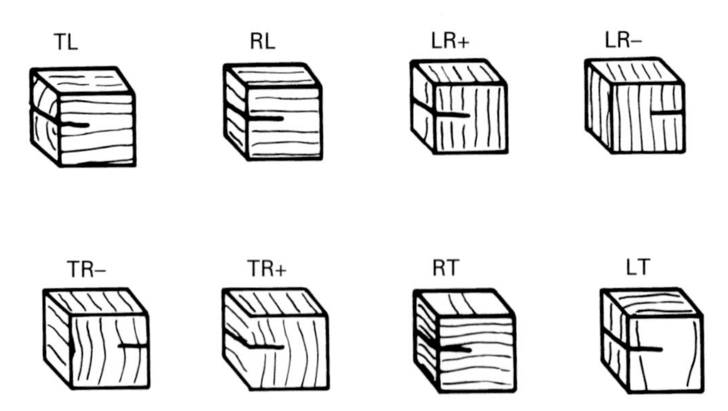

Fig. 4.9 The eight modes of crack propagation in wood. The distinction between the + and – directions arises because of the asymmetric structure of the growth rings. (Reprinted from Ashby *et al.*, 1985.)

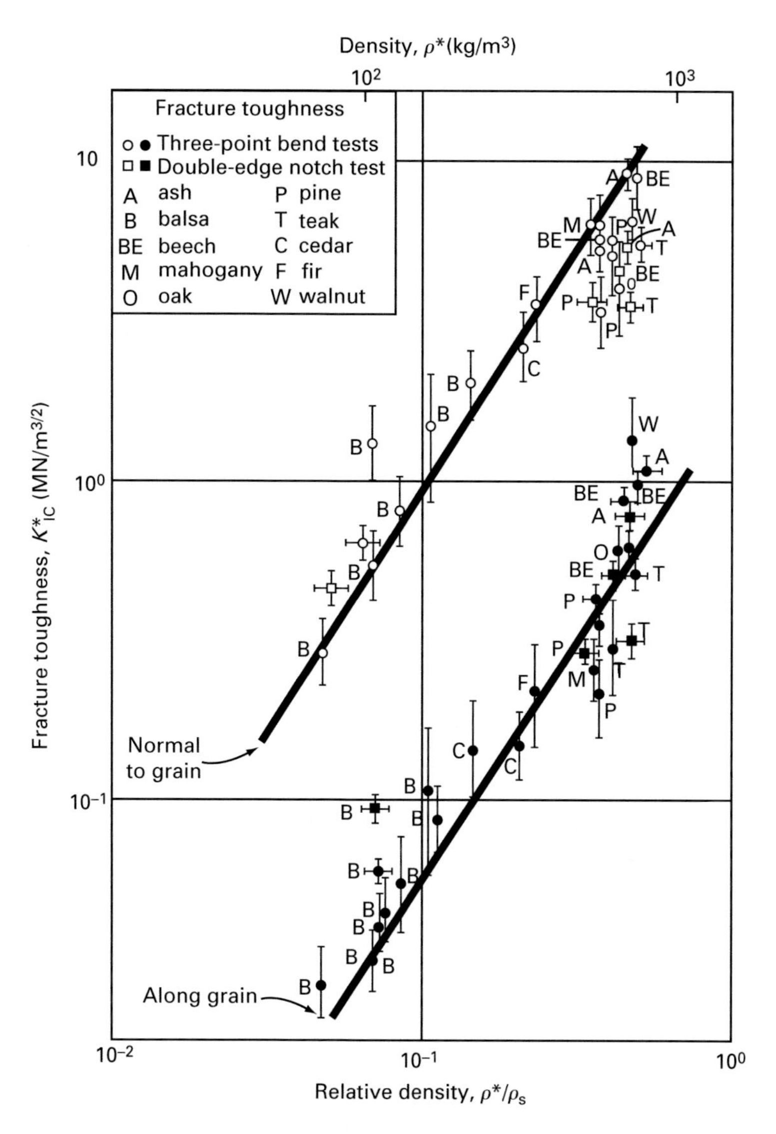

Fig. 4.10 The fracture toughness of wood plotted against the relative density. Data from Wu (1963); Schniewind and Pozniak (1971); Walsh (1971); Johnson (1973); Schniewind and Centano (1973); Williams and Birch (1976); Jeronimides (1980) and Ashby *et al.* (1985). (Reprinted from Ashby *et al.*, 1985.)

between the different directions which lie parallel to the grain – the TL or RL orientations, for instance (Schniewind and Pozniak, 1971; Williams and Birch, 1976) – but these are small (Schniewind and Centano, 1973; Ashby *et al.*, 1985), and largely masked by the sample-to-sample variations inevitable in a natural material like wood. Large variations in temperature (from –195 to 20 °C), in moisture content (from 1 to 72%) and in the rate of deformation (from 10^{-4} to 10^{2} mm/s) have a significant effect on the modulus, strength and fracture toughness of woods (Jeronimides, 1980). The small differences between the conditions of the tests of Fig. 4.10 (18–22 °C, 7–12% moisture content, deformation rates around 10^{-2} mm/s) do not introduce significant variation into the data (Silvester, 1967; Dinwoodie, 1981). The obvious and important implication of Fig. 4.10 is that, for wood in a standard state, the fracture toughness depends principally on its density.

Crack propagation through wood can be studied by loading suitable specimens in the scanning electron microscope. The observations from an extensive study of knot-free samples of six different woods, loaded in tension normal to the crack plane, were as follows (Ashby *et al.*, 1985):

(1) The crack is stable until a critical load is reached. As this load is approached, the crack first advances stably by one or a few cell diameters and then becomes unstable and propagates rapidly over many hundred cell diameters. The initial crack extension is almost always parallel to the grain, even when the starter crack lies across the grain (i.e. in the LT or LR orientations).

(2) Crack advance in low-density woods ($\rho^*/\rho_s < 0.2$) and in the thin-walled early wood of higher-density woods is commonly by cell-wall breaking (Koran, 1966; Bodig and Jayne, 1982; Ashby *et al.*, 1985). It is shown in the micrograph of Fig. 4.11. When cracks propagate in the RT orientation, the crack advance is almost entirely by cell-wall breaking (Bodig and Jayne, 1982; Ashby *et al.*, 1985). Johnson (1973) reports the same thing for the RL orientation.

(3) Crack advance in woods of higher density ($\rho^*/\rho_s > 0.2$) involves both cell-wall breaking and cell-wall peeling: the pulling apart of two halves of the cell wall which debond along the central lamella (Bodig and Jayne, 1982; Ashby *et al.*, 1985). It is shown in the micrograph of Fig. 4.12. When cracks propagate in the TR orientation, cell-wall peeling predominates. A crack running at an angle between the RT and TR orientations tends to deviate towards TR and adopt a peeling mode, suggesting that the toughness in this mode is lower than that for cell-wall breaking. Cracks in the TR orientation seek rays and propagate along them (Koran, 1966; Schniewind

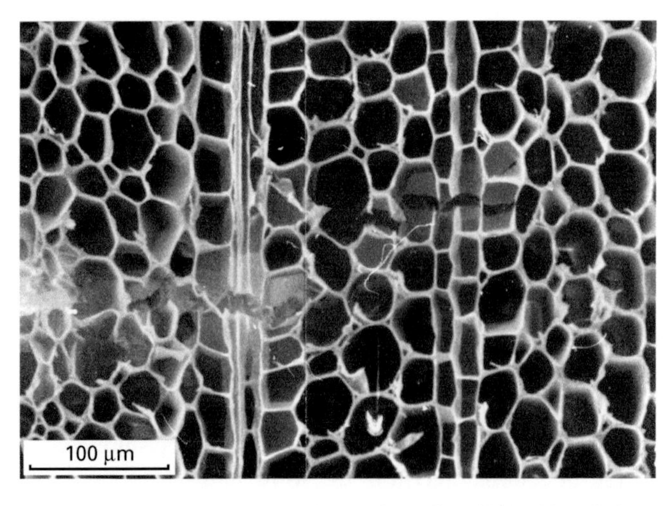

Fig. 4.11 A micrograph showing crack advance by cell-wall breaking (balsa, RL loading). (Reprinted from Ashby *et al.*, 1985.)

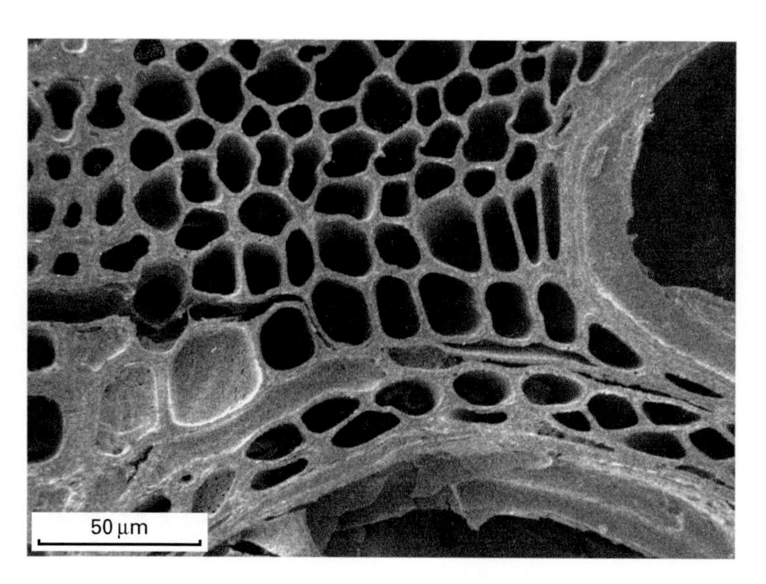

Fig. 4.12 A micrograph showing crack advance by cell-wall peeling (ash, TR loading). (Reprinted from Ashby *et al.*, 1985.)

and Pozniak, 1971; Johnson, 1973). When they do, they advance mainly by cell-wall peeling even in low-density woods. Figures 4.13(a) and (b) show, for the same balsa, cell-wall peeling (along a ray) for TR propagation and cell-wall breaking for RT propagation. Figures 4.13(c) and (d) show the same two fracture mechanisms in a softwood.

(4) Clusters of sap channels and, less commonly, single sap channels, can act as crack arrestors as shown in the micrograph of Fig. 4.14. The crack tends to deviate towards a sap channel, and either enter it or run partly around its periphery and then stop. Ashby *et al.* (1985) observed that a crack often jumped from one layer of sap channels to the next, arresting at each layer, confirming an inference made by Schniewind and Pozniak (1971) when they saw discontinuous crack growth on a macroscopic scale in their wood specimens loaded in the TR direction.

Additional observations of the mechanisms of cell-wall breaking for loading across the grain (LT and LR systems of crack propagation) have been made by Jeronimidis (1980). He finds that the cell-wall fracture is associated with "pseudo-buckling" caused by the shear stresses induced in a helically wound fiber composite loaded in axial tension.

When wood containing a crack which lies in the plane of the grain is subjected to mode I, or crack-opening, loads, the crack advances in its own plane. The load deflection curve is linear to failure, which occurs by fast, unstable fracture. Data for the fracture toughness,, for this sort of crack propagation are plotted as a function of relative density in Fig. 4.10.

If, instead, the crack lies normal to the grain, the problem is more complicated. The load-deflection curve is linear up to the load at which the crack first extends. This

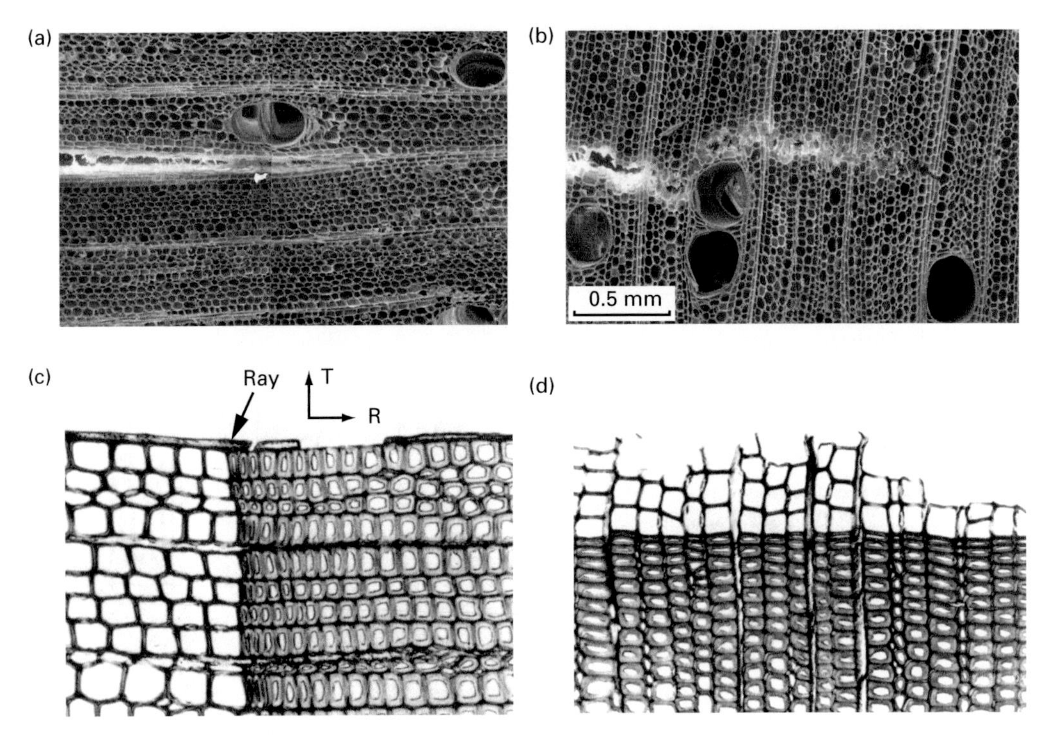

Fig. 4.13 Crack propagation in balsa: (a) cell-wall peeling in TR mode; (b) cell-wall breaking in RT mode. Crack propagation in a softwood: (c) cell-wall peeling in TR mode; (d) cell-wall breaking in RT mode. (a,b, Reprinted from Ashby *et al.*, 1985; c,d, reprinted from Koran, 1966, with kind permission of Paprican.)

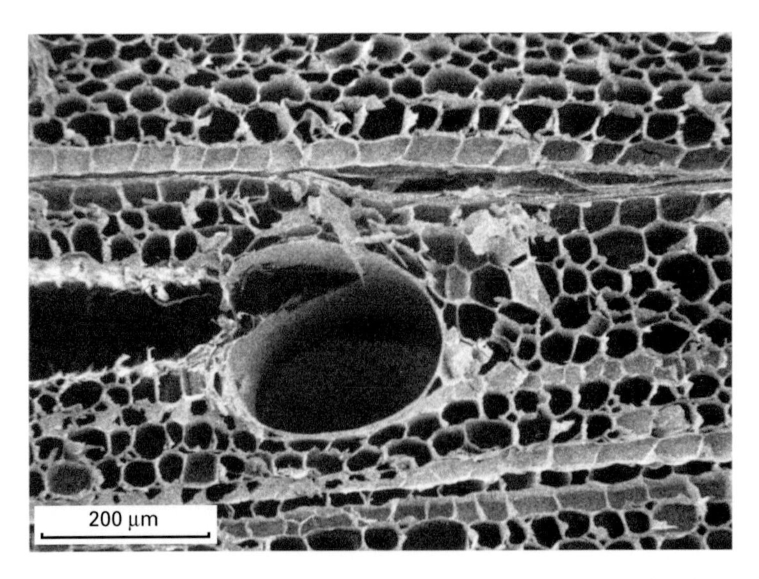

Fig. 4.14 A micrograph showing a crack breaking into a sap channel, arresting the crack. (Reprinted from Ashby *et al.*, 1985.)

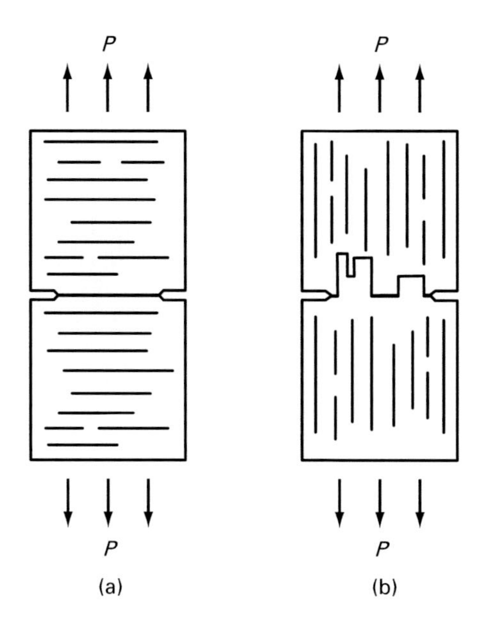

Fig. 4.15 Crack propagation in wood: (a) initial crack parallel to the grain; (b) initial crack perpendicular to the grain. (Reprinted from Ashby *et al.*, 1985.)

initial extension is stable and, almost always, parallel to the grain (and thus perpendicular to the starter crack). Thereafter the crack extends in a step-wise manner, partly along the grain and partly across it (Fig. 4.15), giving a load–deflection curve which passes through a maximum and then falls. It is then necessary to distinguish two values of K_{IC}^*: that for initial crack extension, and that for failure. Data for the smaller of the two – that for initial extension, K_{IC}^{*n} – are plotted as a function of relative density in Fig. 4.10.

The figure shows that the fracture toughness for cracking normal to the grain, K_{IC}^{*n}, is about ten times greater than that for cracking along the grain, K_{IC}^{*a}. To understand this we must first examine the stress state around the tip of a mode I crack in an orthotropic material (Sih *et al.*, 1965). When the crack is loaded (Fig. 4.16), tensile stresses appear on the plane TB ahead of its tip. But there are also tensile stresses on the plane TA: at a given distance from the tip they are less than those on the plane TB by a factor, *F*, which depends on the degree of anisotropy in the material. For woods this factor is between 5 and 12 (Ashby *et al.*, 1985). But the fracture toughness on plane TA is much lower than that on plane TB, so that the crack starts to propagate on TA when the load is larger, by the factor, F, than that required to propagate a simple mode I crack parallel to the grain: that is why K_{IC}^{*n} is a constant factor roughly ten times larger than K_{IC}^{*a}, independent of the density of the wood. As the crack propagates along the grain it seeks out weak fibers, or defects in the wood (knots for instance) and at these points it breaks across the grain, giving the zig-zag path sketched in Fig. 4.15.

Figure 4.10 shows that the fracture toughness depends principally on the relative density of the wood. For crack propagation along the grain the data is described by

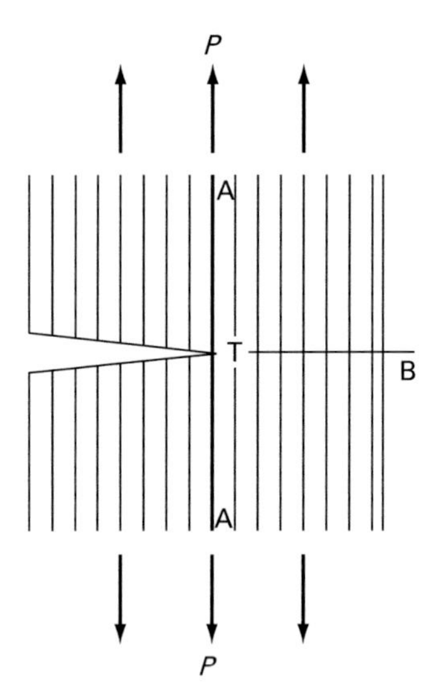

Fig. 4.16 Crack propagation across the grain in wood.

$$K_{IC}^{*a} = 1.8\left(\frac{\rho^*}{\rho_s}\right)^{3/2} \text{ MN/m}^{3/2} \tag{4.6}$$

This can be understood as follows. When cell walls peel apart along the central lamella, as in Fig. 4.12, the fracture process is like that of the peeling apart of an adhesion joint. Let the energy absorbed per unit area of peeling be G_{cs}^{p}; we expect this to be about constant for all woods (at a given moisture content) since the composition and structure of the cell walls varies very little between species. During fracture this energy is supplied by the release of elastic energy from the surrounding wood, plus any work done by the applied loads. Using standard results the energy release rate, G_I is, to a sufficient approximation, given by

$$G_I = \frac{K_I^2}{E_R^*} \tag{4.7}$$

where E^*_R is the Young's modulus of the wood across the grain and K_I is the stress intensity. Equating this to G_{cs}^{p} and using the result that the transverse modulus is related to the relative density and Young's modulus, E_s, for the cell-wall material by (4.1) and (4.2) gives, for the peeling mode,

$$K_{IC}^{*a} = \left(C_1 E_s G_{cs}^{p}\right)^{1/2}\left(\frac{\rho^*}{\rho_s}\right)^{3/2} \tag{4.8}$$

The quantity $(C_1\, E_s\, G_{cs}^{p})^{1/2}$ is simply the fracture toughness of the cell wall in the peeling mode, (K_{Ic}^{p}), so that

$$K_{IC}^{*a} = \left(K_{IC}^{p}\right)_s \left(\frac{\rho^*}{\rho_s}\right)^{3/2} \qquad \text{(peeling mode)} \quad (4.9)$$

When the cell walls break, as in Figs. 4.11 and 4.13, the energy is absorbed, G_{cs}^{b} per unit area, in breaking a cell wall. Fibers of cellulose in the cell wall must be broken or pulled out when the cell wall breaks, so we expect this energy to be larger than that for the peeling mode, G_{cs}^{p}, which does not involve pull-out. The area fraction occupied by cell walls in the crack plane of Fig. 4.11 is approximately t/l, or roughly ρ^*/ρ_s. Thus, the energy absorbed per unit area of crack is $(\rho^*/\rho_s)G_{cs}^{b}$. Equating this to the energy-release rate (4.7), and using (4.1) and (4.2) for E_R^*, gives, for the fracture toughness in the breaking mode,

$$K_{IC}^{*a} = \left(C_1 E_s G_{cs}^{b}\right)^{1/2} \left(\frac{\rho^*}{\rho_s}\right)^{2} \qquad (4.10)$$

or

$$K_{IC}^{*a} = \left(K_{IC}^{*b}\right)_s \left(\frac{\rho^*}{\rho_s}\right)^{2} \qquad \text{(breaking mode)} \quad (4.11)$$

Thus the fracture toughness for a crack propagating along the grain in the breaking mode varies as a higher power of the density than that for one propagating in the peeling mode. Remembering that $G_{cs}^{b} > G_{cs}^{p}$ and that $\rho^*/\rho_s < 1$, we find an interesting result: cracks propagating along the grain should do so by cell-wall breaking when the relative density is low, and by cell-wall peeling when it is high. There is experimental evidence suggesting that this is the case; the transition occurs at a relative density of about 0.2. Equating (4.6) and (4.8) gives a value for the peeling toughness, G_{CS}^{p}:

$$G_{cs}^{p} \approx 350 \ \text{J/m}^2 \qquad (4.12)$$

A value for the breaking toughness G_{cs}^{b} can be inferred from the transition from breaking to peeling at a relative density of about 0.2. Equating (4.8) and (4.10) at this density gives

$$G_{cs}^{b} \approx 1650 \ \text{J/m}^2 \qquad (4.13)$$

It is easy, in focusing on the cellular structure of wood, to forget that it may contain defects on a larger scale that affect its properties. Any wood sample with a volume of more than a few cubic centimeters contains knots where branches were accommodated into the trunk of the tree. The grain around a knot is distorted, and the knot itself may be poorly bonded to the rest of the wood.

Knots reduce both the stiffness and the strength of wood (Dinwoodie, 1981). More important, they reduce the fracture strength. The knot and the distorted grain around it are a center of weakness, and can behave like an incipient crack so that, when the wood is loaded, failure starts from the knot. The behavior is sometimes characterized in terms of the knot-area ratio R_k (defined as the total area of knots divided by that of the cross-section); then, very roughly, the fracture strength σ_f^* falls as

$$\sigma_f^* = \sigma_f^{*0}(1 - 1.2R_k) \tag{4.14}$$

where $\sigma_f^{*0}$is the strength of the knot-free wood (Dinwoodie, 1981). But this is not the best approach. A more complete understanding awaits the application of fracture mechanics to the problem. We shall not attempt it here.

4.2.5 Creep

The bowing of roof timbers in old barns is evidence that wood creeps: under a constant load, the strain increases with time. Typically, for wood, the strain, ε, at any time, t, is the sum of the initial elastic deformation, ε_o, and the time-dependent creep strain:

$$\varepsilon(t) = \varepsilon_o + at^m \tag{4.15}$$

where a and m are constants that depend on the species of wood and environmental conditions such as temperature and moisture content (Dinwoodie, 1981; Holzer *et al.*, 1989). The creep strain has both recoverable and irrecoverable components. At sufficiently long times, creep leads to failure; (4.15) does not describe creep failure.

As we saw in Chapter 3, in a linear viscoelastic material, the strain at a given time is linearly related to the applied stress, so that the creep compliance $J(t) = \varepsilon(t)/\sigma$ is a constant for a given time, independent of stress. While there is some evidence that wood is non-linear viscoelastic at all stress levels, the creep response of wood is generally approximated as linear viscoelastic at stresses less than about 40% of the ultimate strength (Dinwoodie, 1981; Morlier, 1994). Wood is anisotropic: as for the elastic constants, there are nine independent creep compliances. Linear viscoelasticity can be described by spring–dashpot models: the creep behavior of wood at low stresses has been successfully described using a four-element model (Dinwoodie, 1981; Fridley *et al.*, 1992) (Fig. 4.17) for which

$$\varepsilon(t) = \frac{\sigma}{E_1} + \frac{\sigma t}{\eta_1} + \frac{\sigma}{E_2}\left(1 - e^{-tE_2/\eta_2}\right) \tag{4.16}$$

where E_1 and E_2 are the spring stiffnesses and η_1 and η_2 are the dashpot viscosities. There are no systematic data for the creep response of woods of varying relative density, so that a four-element spring–dashpot model for honeycombs, similar to the standard linear solid model for honeycombs introduced in Chapter 3, cannot be tested.

The creep response of wood is sensitive to temperature and moisture content. The usual WLF equations for time–temperature superposition for creep of linear

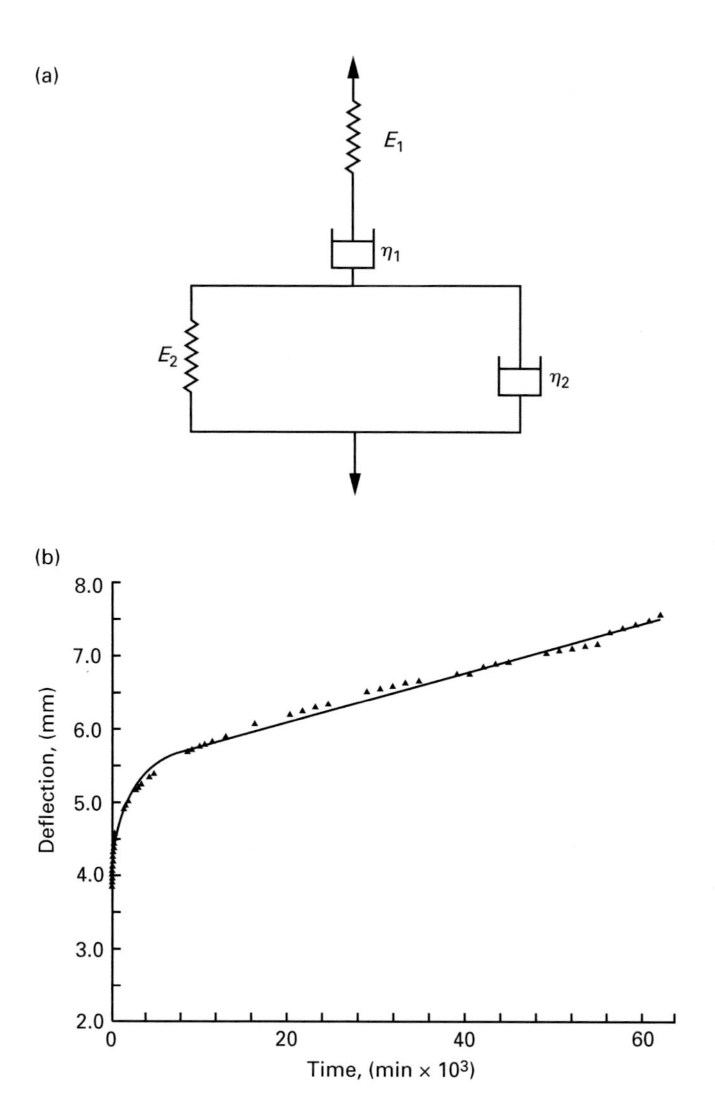

Fig. 4.17 (a) A four-element spring–dashpot model for the creep of wood; (b) the creep deformation of urea-formaldehyde bonded chipboard. The solid curve is fitted to (4.16) for the four-element spring dashpot. (Reprinted from Dinwoodie (1981), copyright Building Research Establishment, UK, reproduced with permission.)

viscoelastic polymers are not valid for wood, however (Dinwoodie, 1981; Holzer *et al.*, 1989; Morlier, 1994). Wood is especially sensitive to cyclic increases and decreases in moisture content. In one study, one specimen was loaded to three-eighths of the ultimate short-term strength at 93% relative humidity, while a second, similar specimen at the same load was subjected to cycles of 0% to 93% relative humidity. After four weeks, the deflection of the second specimen was over 12 times that of the first, at which point it failed, while the first specimen was still intact (Hearmon and Paton, 1964).

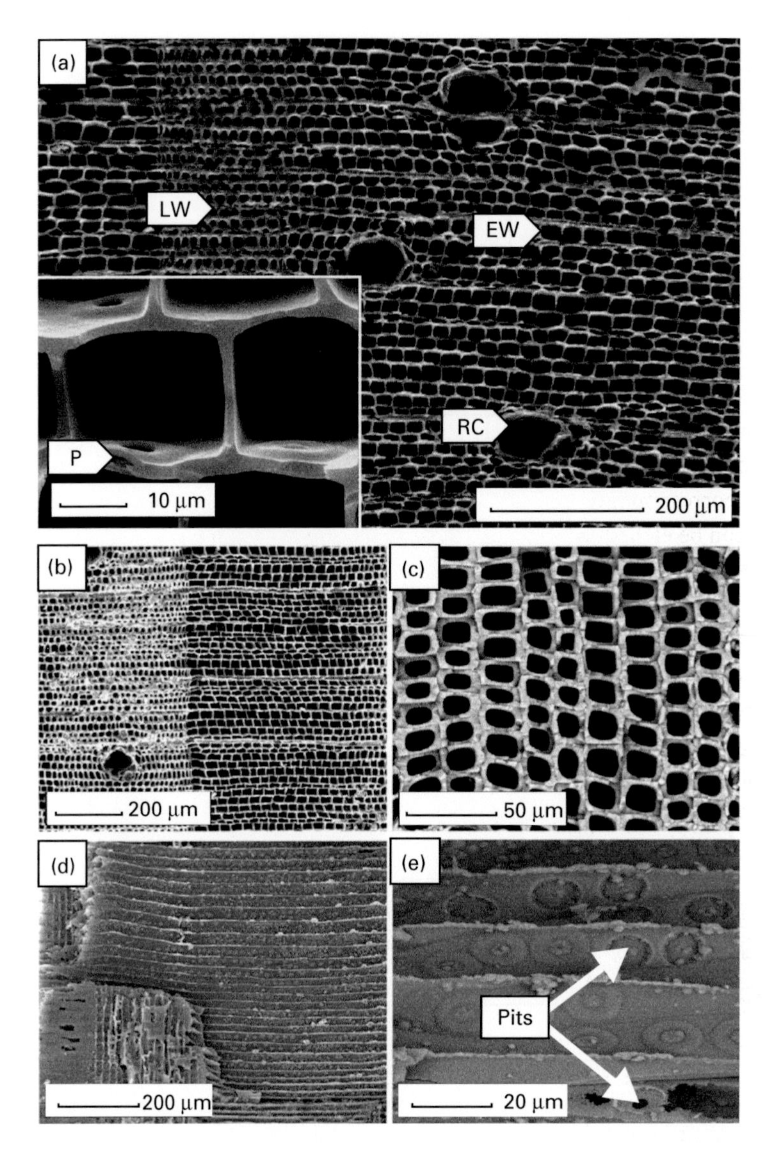

Fig. 4.18 Scanning electron micrographs of (a) pyrolyzed pine wood (EW = earlywood, LW = latewood, RC = sap channel, P = pit) and (b,c,d,e) SiC ceramic made by Si-vapor infiltration of pyrolyzed pine; (b,c) show axial sections, (d,e) show tangential sections. (Reprinted from Vogli *et al.*, 2002, with permission from Elsevier.)

4.2.6 Biomimicking of wood

Engineers have recently developed techniques for replicating the cellular structure of wood in ceramics with remarkable fidelity (Fig. 4.18). Replication of the wood structure allows the fabrication of honeycomb-like ceramics with cell sizes of the order of 50 μm, much smaller than conventional ceramic honeycombs, making them attractive for high-temperature filters, catalyst carriers, and heat exchangers. Typically, wood is first pyrolyzed at 800 °C in an inert atmosphere to form a biocarbon template for further processing steps (Sieber, 2005). Although there is significant shrinkage during pyrolysis (20–40%), the cellular structure is maintained (Paris *et al.*, 2005). Silicon carbide replicas can be made either by infiltration of the biocarbon template with gaseous Si (Vogli *et al.*, 2001, 2002) or by infiltration with a sol, such as tetraethyl-orthosilicate (TEOS) ($Si(OC_2H_5)_4$), drying in air to form a gel, annealing to decompose the gel into an oxide and reducing the oxide to SiC (Rambo *et al.*, 2005). Variations of the latter process have been used to produce TiC, ZrC, Al_2O_3 and TiO_2 replicas of the cellular structure of wood (Cao *et al.*, 2004a,b; Rambo *et al.*, 2004, 2005).

Nearly fully dense composites of Si-SiC have been made by liquid silicon infiltration of biocarbon templates of wood (Griel *et al.*, 1998; Zollfrank and Sieber, 2004, 2005). In the lower density earlywood, the cell walls are fully converted to SiC, while in the higher density latewood some unconverted carbon remains in the walls. The lumens, or voids within the wood cells, are nearly fully dense Si. The final composite has SiC fibers, corresponding to the wood cell wall, in a matrix of Si (Fig. 4.19). Alternatively, the Si can be etched out with acid treatment and the SiC wood replica reinfiltrated with Al to form an Al-SiC composite (Wilkes *et al.*, 2006).

The Young's moduli of the nearly fully dense Si-SiC composites, in the direction along the grain of the precursor wood, increase linearly with density, reflecting the uniaxial orientation of the SiC replicating the wood cell walls and of the Si filling the cell lumens (Singh and Salem, 2002). The bending strengths of the biocarbon template are higher when stressed along the grain of the precursor wood than across it, reflecting the anisotropy of the underlying wood microstructure (Sieber *et al.*, 2000). The bending strength of commercially available Si-SiC lies in between the values for the

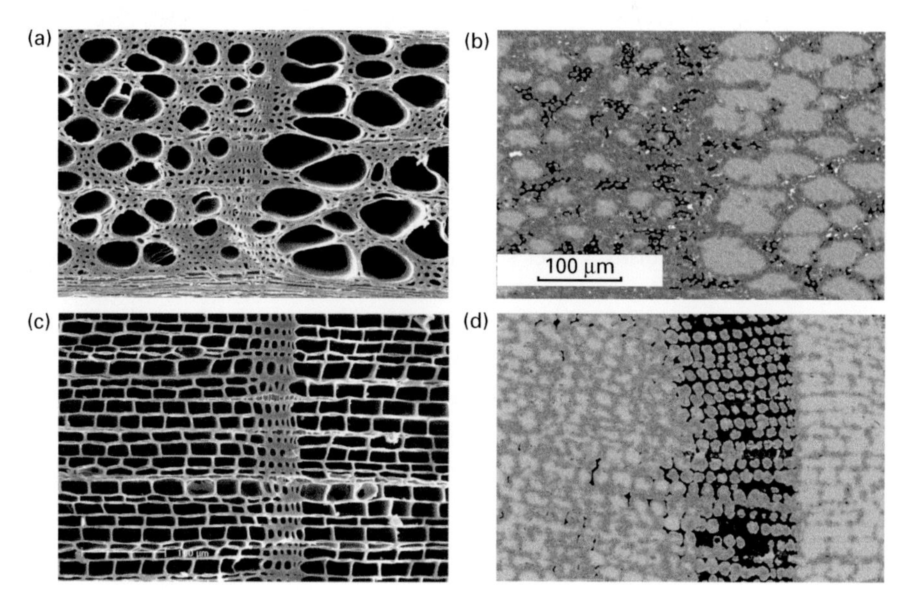

Fig. 4.19 Scanning electron micrographs of (a) biocarbon template of beech, (b) Si-SiC composite made by liquid Si infiltration of the beech template, (c) biocarbon template of pine and (d) Si-SiC composite made by liquid Si infiltration of the pine template. All images at same scale. (Reprinted from Zollfrank and Sieber, 2004, with permission from Elsevier.)

nearly fully dense Si-SiC composite along and across the grain, presumably as a result of its more random microstructure.

4.3 Cork

4.3.1 Introduction

Cork has a remarkable combination of properties. It is light yet resilient; it is an outstanding insulator for heat and sound; it has a high coefficient of friction; and it is impervious to liquids, chemically stable and fire-resistant. Such is the demand that production now exceeds 350 000 tonnes a year (and 1 tonne of cork has the volume of 56 tonnes of steel) (Pereira, 2007).

In pre-Christian times cork was used (as we still use it today) for fishing floats and the soles of shoes. When Rome was besieged by the Gauls in 400 BC, messengers crossing the Tiber clung to cork for buoyancy (Plutarch, AD 100). And ever since people have cared about wine, cork has been used to keep it sealed in flasks and bottles. "Corticum abstrictum pice demovebit amphorae" (Pull the cork, set in pitch, from the bottle) sang Horace (27 BC) to celebrate his miraculous escape from death from a falling tree. But it was in the Benedictine Abbey at Hautvilliers where, in the seventeenth century, the technology of stopping wine bottles with clean, unsealed cork was perfected. Its elasticity and chemical stability mean that it seals the bottle without contaminating the wine, even when it must mature for many years.

Commercial cork is the bark of an oak (*Quercus suber*) that grows in Portugal, Spain, Algeria and California. Pliny describes it thus (Pliny, AD 77): "The Cork Oak is a small tree; its only useful product is its bark which is extremely thick and which, when cut, grows again." Modern botanists add that the cork cells (phellem) grow from the cortex cells via an intermediate structure known as cork cambium (phellogen). Their walls are covered with thin layers of an unsaturated fatty acid (suberin) and waxes which make them impervious to air and water, and resistant to attack by many acids (Eames and MacDaniels, 1951; Esau, 1965; Zimmerman and Brown, 1971). All trees have a thin layer of cork in their bark. *Quercus suber* is unique in that, at maturity, the cork forms a layer several centimeters thick around the trunk of the tree. Its function in nature is to insulate the tree from heat and loss of moisture, and perhaps to protect it from damage by animals (suberin tastes unpleasant). We use it today for thermal insulation in refrigerators and rocket boosters, acoustic insulation in submarines and recording studios, as a seal between mating surfaces in woodwind instruments and internal combustion engines, and as an energy-absorbing medium in flooring, shoes and packaging (Pereira, 2007). Its use has widened further since 1892, when a Mr John Smith of New York patented a process for making cork aggregate by the simple hot-pressing of cork particles: the suberin provides the necessary bonding.

In this section we describe the structure of cork, review data for the moduli and collapse stresses, and examine the way in which the theory for the mechanics of honeycombs in Chapter 3 can be used to explain them. We conclude with a survey of the ways in which the special properties of cork are exploited in several applications.

4.3.2 Structure of cork

Cork occupies a special place in the histories of microscopy and of plant anatomy. When Robert Hooke perfected his microscope, around 1660, one of the first materials he examined was cork. What he saw led him to identify the basic unit of plant and biological structure, which he termed "the cell" (Hooke, 1665). His careful drawings of cork cells (Fig. 4.20) show their roughly hexagonal shape in one section and their box-like shape in the other. Hooke noted that the cells were stacked in long rows, with very thin walls, like the wax cells of the honeycomb. Subsequent descriptions of cork-cell geometry add very little to this. Esau (1965), for example, describes cork cells as "approximately prismatic in shape – often somewhat elongated parallel to the long axis of the stem." Lewis (1928) concluded that their shape lay "somewhere between orthic and prismatic tetrakaidecahedrons," Eames and MacDaniels (1951) simply described them as "polygonal," but their drawing shows the same shape that Hooke described. These descriptions conflict, and none is quite correct. Figure 4.21 shows the three faces of a cube of cork (Gibson *et al.*, 1981). In one section the cells are roughly hexagonal; in the other two they are shaped like little bricks, stacked as one would stack them in building a wall. The similarity with Hooke's drawing is obvious.

From micrographs such as these the cell shape can be deduced. Roughly speaking, the cells are closed hexagonal prisms (Fig. 4.22) stacked in rows so that the hexagonal faces register and are shared by two cells; but the rows are staggered so that the membranes forming the hexagonal faces are not continuous across rows. Figure 4.23 shows how the cells lie with respect to the trunk of the tree. The axes of the hexagons lie parallel to the radial (X_3) direction. A cut normal to the radial direction shows the hexagonal cross-section; any cut containing the radial direction shows the rectangular section, stacked like bricks in a wall because of the staggering of the rows. At higher magnifications the scanning microscope reveals details that Hooke could not see, because their scale is comparable with the

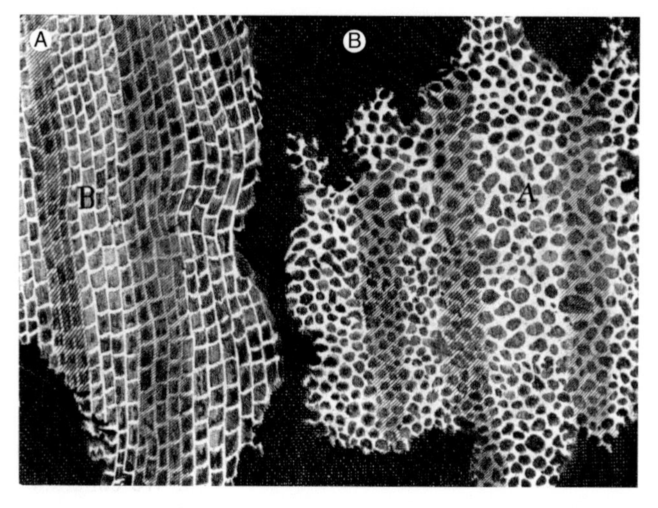

Fig. 4.20 (a) Radial and (b) tangential sections of cork, as seen by Robert Hooke through his microscope in 1665.

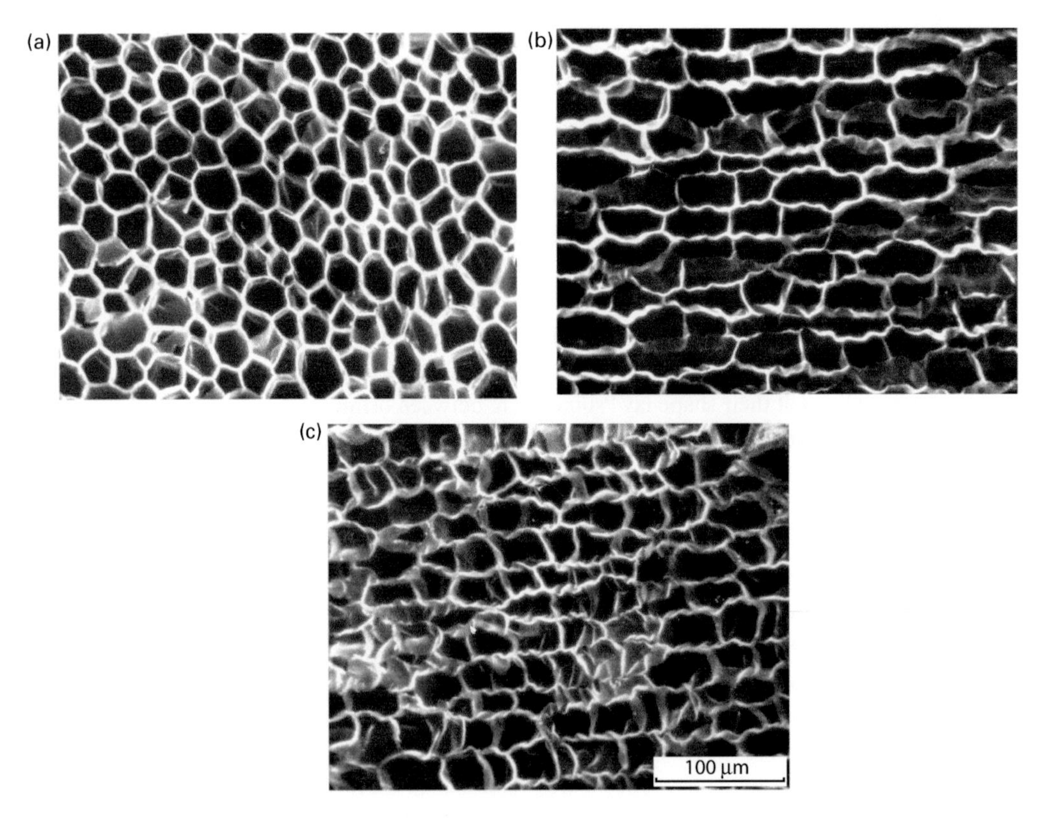

Fig. 4.21 Scanning electron micrographs of (a) radial, (b) axial and (c) tangential sections of cork. All images at same scale. (Reprinted from Gibson *et al.*, 1981.)

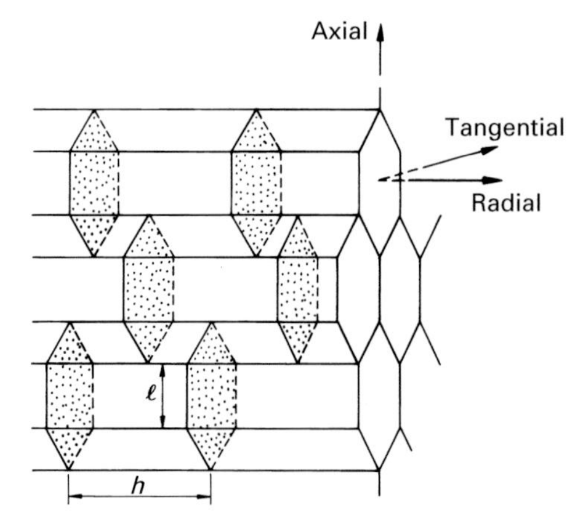

Fig. 4.22 The shape of cork cells, deduced from the micrographs of Fig. 4.21. They are hexagonal prisms. The cell walls are not straight, as shown here, but corrugated. (Reprinted from Gibson *et al.*, 1981.)

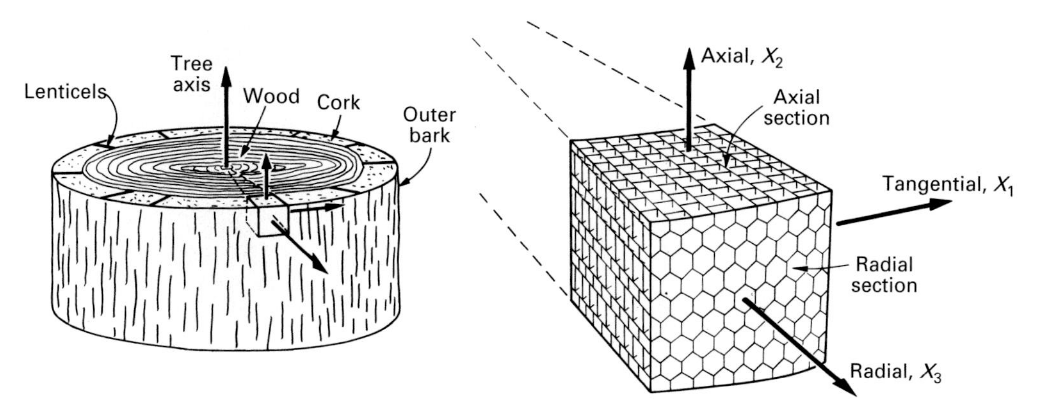

Fig. 4.23 Diagram of a cork tree and cork, showing axis system and cells. (Reprinted from Gibson *et al.*, 1981.)

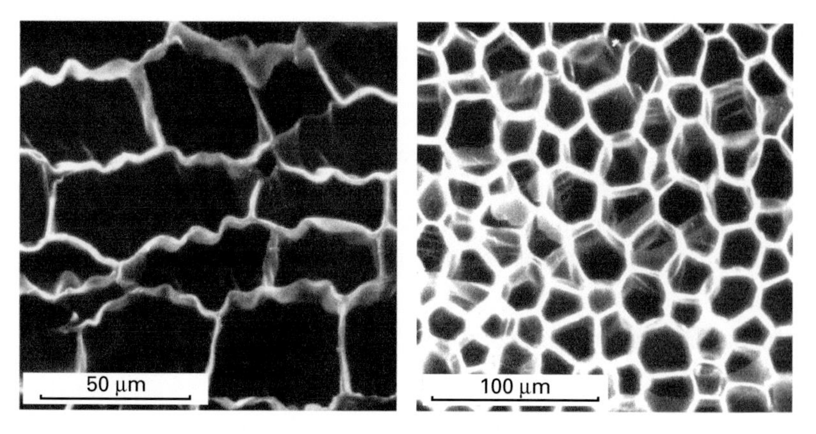

Fig. 4.24 Scanning electron micrographs of cork cells, showing corrugations. (Reprinted from Gibson *et al.*, 1981.)

wavelength of light. Six of the eight walls of each cell are corrugated (Fig. 4.24). Each cell has two or three complete corrugations, so that it is shaped like a concertina, or bellows.

Figure 4.25 summarizes the observations, and catalogues the dimensions of the cork cells. The cell walls have a uniform thickness (about 1 μm) of height h (about 40 μm) and hexagonal face-edge l (about 20 μm). The density ρ^* of the cork is related to that of the cell-wall material ρ_s and the cell-wall dimensions by

$$\frac{\rho^*}{\rho_s} = \frac{t}{l}\left(\frac{l}{h} + \frac{2}{\sqrt{3}}\right) \tag{4.17}$$

The density of the cell-wall material is close to 1150 kg/m^3 (Gibson *et al.*, 1981). The mean density of cork is roughly 170 kg/m^3, giving a relative density of 0.15.

The aspect ratio of the cells, h/l, is about 2; this is larger than the value (1.7) that minimizes the surface area of a close-packed array of hexagons. The radial section of the

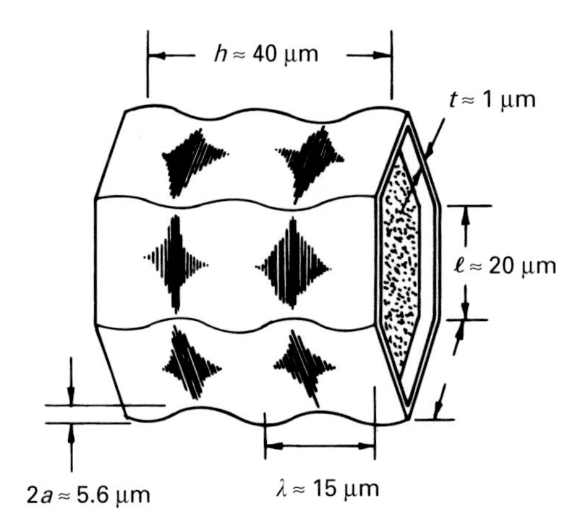

Fig. 4.25 A corrugated cell, showing dimensions. (Reprinted from Gibson *et al.*, 1981.)

structure does not always show hexagonal sections: five-, six-, seven- and eight-sided figures are all observed. But the average number of sides per cell in the radial section is six (Lewis, 1928, finds 5.978). This, of course, is an example of the operation of Euler's law, which asserts, when applied to a three-connected net, that the average number of sides per cell is six (Gibson and Ashby, 1997). The cells themselves are very small; there are about 20 000 of them in a cubic millimeter. They are much smaller than those in common foamed plastics, and comparable with those in "microporous" foams.

4.3.3 Mechanical properties of cork

Figure 4.26 is a complete compressive stress–strain curve for cork. It has all the characteristics we expect of a cellular solid (Chapter 3). It is linear-elastic up to about 7% strain, at which point elastic collapse gives an almost horizontal plateau which extends to about 70% strain when complete collapse of the cells causes the curve to rise steeply. Typical mean values for the Young's moduli, shear moduli and Poisson's ratios are recorded in Table 4.4. The Young's modulus along the prism axis is roughly one and a half times that along the other two directions; additional tests by Fortes and co-workers confirm this result (Fortes and Nogueira, 1989; Rosa and Fortes, 1991). The moduli (and the other properties) have circular symmetry about the prism axis. In the plane normal to this axis cork is roughly isotropic, as might be expected from its structure. The table lists the stress (σ_{el}^*) and the strain at the start of the plateau in compression, and the fracture stress (σ_f^*) and fracture strain in tension. Tensile fracture along the prism axis occurs at 5% strain, but in the other two directions the strain is larger – about 9%. The fracture toughness, K_{ICij}^* depends on both the direction of the normal to the crack plane, *i*, and the direction of crack propagation, *j*. It is roughly 115–130 $\text{kPa}\sqrt{\text{m}}$ for systems with the radial direction normal to the crack plane and 60–100 $\text{kPa}\sqrt{\text{m}}$ for all other systems (Rosa and Fortes, 1991). The final item in the table is the loss coefficient:

$$\eta^* = \frac{D}{2\pi U} \tag{4.18}$$

Table 4.4 The mechanical properties of cork

Young's modulus	
Tangential	$E_1^* = 13 \pm 5$ MN/m^2
Axial	$E_2^* = 13 \pm 5$ MN/m^2
Radial	$E_3^* = 20 \pm 7$ MN/m^2
Shear modulus	
In 1–2 planes	$G_{12}^* = 4.3 \pm 1.5$ MN/m^2
In 1–3 planes	$G_{13}^* = 2.5 \pm 1.0$ MN/m^2
In 2–3 planes	$G_{23}^* = 2.5 \pm 1.0$ MN/m^2
Poisson's ratio	$\nu_{12}^* = \nu_{21}^* = 0.25^a - 0.50$
	$\nu_{13}^* = \nu_{31}^* = \nu_{23}^* = \nu_{32}^* = 0 - 0.10^a$
Collapse stress (and strain)	
Tangential	$(\sigma_{el}^*)_1 = 0.7 \pm 0.2$ MN/m^2, 6% strain
Axial	$(\sigma_{el}^*)_2 = 0.7 \pm 0.2$ MN/m^2, 6% strain
Radial	$(\sigma_{el}^*)_3 = 0.8 + 0.29$ MN/m^2, 4% strain
Fracture stress (and strain)	
Tangential	$(\sigma_f^*)_1 = 1.1 \pm 0.2$ MN/m^2, 9% strain
Axial	$(\sigma_f^*)_2 = 1.1 \pm 0.2$ MN/m^2, 9% strain
Radial	$(\sigma_f^*)_3 = 1.0 \pm 0.2$ MN/m^2, 5% strain
Fracture toughness	$K_{Ic}^* = 60–130$ MPa $\sqrt{m}^b$
Loss coefficient	
Tangential	$\eta_1^* = 0.3$ at 20% strain
Axial	$\eta_2^* = 0.3$ at 20% strain
Radial	$\eta_1^* = 0.1$ at 1% strain

Sources: Data from Gibson *et al.* (1981), except [a] Fortes and Nogueira (1989) and [b] Rosa and Fortes (1991).

where D is the energy dissipated in a complete tension–compression cycle and U is the maximum energy stored during the cycle. It is roughly constant from 0.01 to 4 kHz, with a broad peak at 2 kHz (Fernandez, 1978) and increases from 0.1 at low strain amplitudes to 0.3 at high. It is this which gives cork good damping and sound-absorbing properties, and a high coefficient of friction. Additional data for creep and rate effects have been reported by Rosa and Fortes (1988a,b).

When cork deforms, the cell walls bend and buckle. The behavior when the axis of deformation lies along the prism axis differs from that when it lies across the prisms. Both tensile and compressive deformation across the prism axis first bend the cell walls. In compression, at higher strains, the cell walls buckle (Fig. 4.27) giving large recoverable strains of order 1.

Tensile deformation along the prism axis unfolds the corrugations (Fig. 4.28), straightening the prism walls. About 5% extension is possible in this way; by then the walls have become straight, and further tension at first stretches and then breaks them, causing the cork to fail. Compressive deformation, on the other hand, folds the corrugations (Fig. 4.29). The folding is unstable; once it reaches about 10% a layer of cells collapses completely, suffering a large compressive strain. Further compression makes the boundary of this layer propagate; cells collapse at the boundary, which moves through the cork like a Luders band through steel or a drawing band through polyethylene.

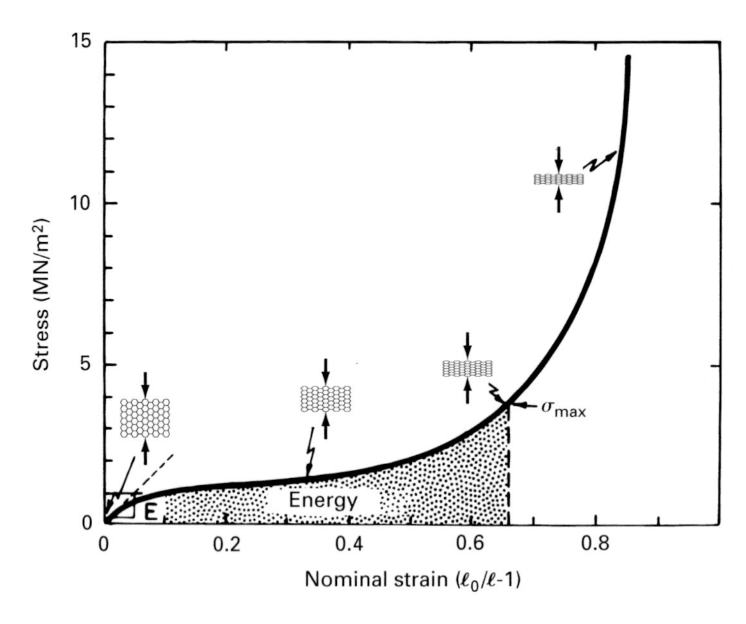

Fig. 4.26 Stress–strain curve for cork. (Reprinted from Gibson *et al.*, 1981.)

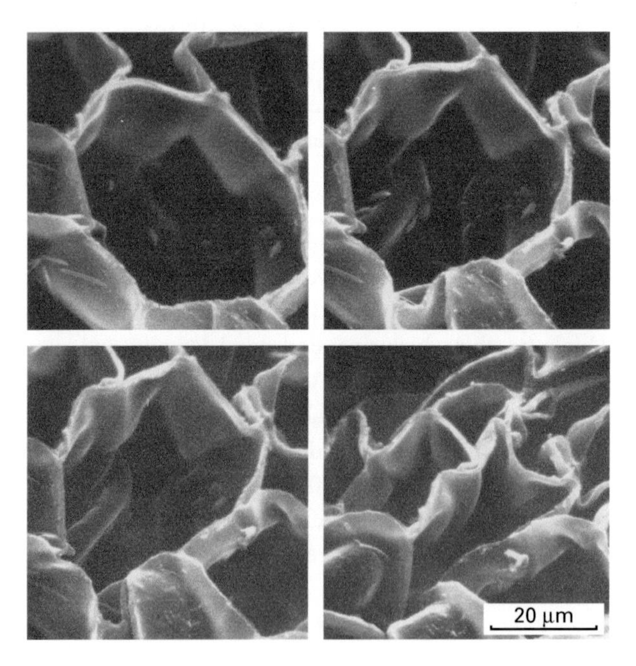

Fig. 4.27 Micrographs showing the bending and buckling of cell walls as cork is compressed across the prism axis. The load is applied from the top left to bottom right. (Reprinted from Gibson *et al.*, 1981.)

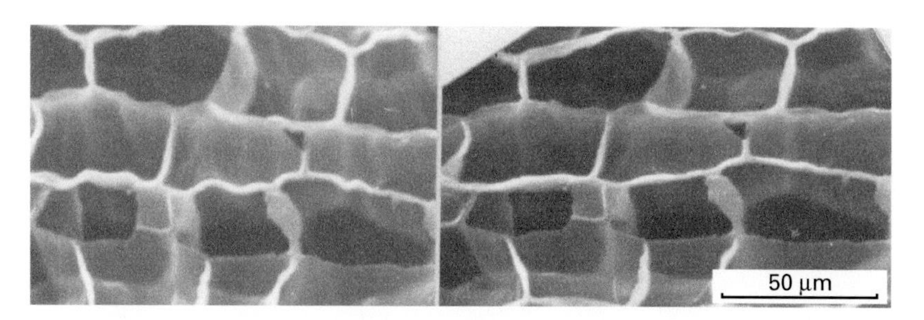

Fig. 4.28 Micrographs showing the progressive straightening of cell walls as cork is pulled in the radial direction (tensile axis parallel to the prism axis). (Reprinted from Gibson *et al.*, 1981.)

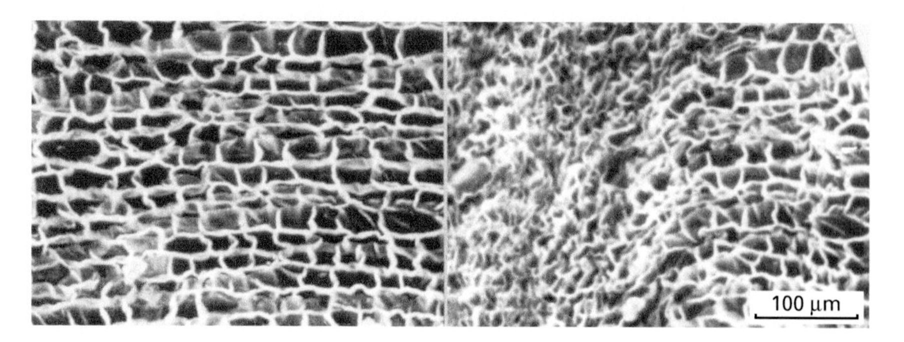

Fig. 4.29 The catastrophic collapse of cork cells compressed in the radial direction (along the prism axis). (Reprinted from Gibson *et al.*, 1981.)

The in-plane mechanical properties for loading normal to the prism axis can be understood in terms of the models developed in Chapter 3 for a two-dimensional array of honeycomb-like hexagonal cells. Results developed there for the in-plane properties explain the values of the moduli we have called E_1^*, E_2^*, G_{12}^*, ν_{12}^* and ν_{21}^* the elastic collapse stresses $(\sigma_{el}^*)_1$ and $(\sigma_{el}^*)_2$. It is convenient to recall the results, obtained by substituting (4.17) into the appropriate equations in Table 3.1:

$$E_1^* = E_2^* = 0.5E_s\left(\frac{\rho^*}{\rho_s}\right)^3 \tag{4.19a}$$

$$G_{12}^* = G_{21}^* = 0.13E_s\left(\frac{\rho^*}{\rho_s}\right)^3 \tag{4.19b}$$

$$\nu_{12}^* = \nu_{21}^* = 1.0 \tag{4.19c}$$

$$\left(\sigma_{el}^*\right)_2 = 0.05E_s\left(\frac{\rho^*}{\rho_s}\right)^3 \tag{4.19d}$$

Here E_s and ρ_s are the modulus and density of the solid of which the cell walls are made and ρ^* is the overall density of the cork. The collapse strain associated with elastic buckling of the cell walls is given by $\varepsilon^* = \sigma_{el}^*/E^* = 0.1$. Below this strain the structure is linear-elastic; above it is non-linear but still elastic. Buckling allows deformation to continue until the cell walls touch at a nominal strain ε_D of about

$$\varepsilon_D = 1 - 1.4\left(\frac{\rho^*}{\rho_s}\right) \tag{4.20}$$

If a honeycomb of regular prismatic cells like that of Fig. 4.22 is compressed parallel to the prism axis, the modulus is determined by the axial compression of the material in the cell walls. This leads to the obvious result (Table 3.1)

$$\frac{E_3^*}{E_s} = \frac{\rho^*}{\rho_s} \tag{4.21}$$

This equation properly predicts the axial modulus of wood (Section 4.2) but it over-estimates, by a factor of 50 or more, that of cork. The discrepancy arises because the cell walls have corrugations which fold or unfold like the bellows of a concertina when the cork is deformed (Fig. 4.28). The axial stiffness of a corrugated cell wall of thickness, t, and corrugation amplitude, a, is derived by Gibson *et al.* (1981). It is

$$\frac{E_3^*}{E_s} = 0.7\left(\frac{\rho^*}{\rho_s}\right)\left\{\frac{1}{1+6\left[\frac{a}{t}\right]^2}\right\} \tag{4.22}$$

This deformation has another interesting feature. Axial compression produces no lateral expansion, because the cells simply fold up. We therefore expect

$$\nu_{13}^* = \nu_{31}^* = \nu_{23}^* = \nu_{32}^* = 0 \tag{4.23}$$

The properties ρ_s and E_s of the cell walls of cork are discussed by Gibson *et al.* (1981). The best estimates of their values are:

$$\rho_s = 1150 \text{ kg/m}^3$$
$$E_s = 9 \text{ GN/m}^2$$

This information, together with the dimensions of the cells given earlier, give the moduli and collapse stresses for cork given in Table 4.5. Agreement is remarkably good. In particular, the understanding of the cork structure explains the isotropy in the plane normal to the radial (X_3) direction and the factor of 1.5 difference between the Young's modulus in the radial direction and in the other two; and it explains the

Table 4.5 Comparison between calculated and measured properties of cork

	Calculated	Measured
Moduli		
E_1^*, E_2^* (MN/m^2)	15	13 ± 5
E_3^* (MN/m^2)	20	20 ± 7
G_{12}^*, G_{21}^* (MN/m^2)	4	4.3 ± 1.5
$G_{13}^*, G_{31}^*, G_{23}^*, G_{32}^*$ (MN/m^2)	–	2.5 ± 1
$\nu_{12}^* = \nu_{21}^*$	1.0	0.25[a]–0.50
$\nu_{13}^* = \nu_{31}^* = \nu_{23}^* = \nu_{32}^*$	0	0–0.10[a]
Compressive collapse stress		
$(\sigma_{el}^*)_1, (\sigma_{el}^*)_2$ (MN/m^2)	1.5	0.7 ± 0.2
$(\sigma_{el}^*)_3$ (MN/m^2)	1.5	0.8 ± 0.2

Source: Data from Gibson *et al.* (1981), except for [a] Fortes and Nogueira (1989).

striking anisotropy in the values of Poisson's ratios and the magnitude of the elastic collapse loads. The biggest discrepancy is in the value of Poisson's ratio, ν_{12}^*, and is probably due to a variation in cell shape and orientation, and to the constraining effect of the membranes which form the hexagonal end-faces of the cells.

4.3.4 Uses of cork

For at least 2000 years cork has been used for "floats for fishing nets, and bungs for bottles, and also to make the soles of woman's winter shoes" (Pliny, AD 77). Few materials have such a long history, or have survived so well the competition from man-made substitutes (Pereira, 2007). We now examine briefly how the special structure of cork has suited it so well to its uses.

Bungs for bottles and gaskets for woodwind instruments

Connoisseurs of wine agree that there is no substitute for cork. Plastic corks are hard to insert and remove, they do not always make a very good seal, and they may contaminate the wine. Cork corks have none of these problems. The excellence of the seal is a result of the elastic properties of the cork. It has a low Young's modulus (E^*) but, much more important, it also has a low bulk modulus ($K^* = (1/3)E^*$, Fortes *et al.*, 1989). Solid rubber and solid polymers above their glass transition temperature have a low E but a large K, and it is this which makes them hard to force into a bottle, and gives a poor seal when they are inserted.

One might expect that the best seal would be obtained by cutting the axis of the cork parallel to the prism axis of the cork cells, so that the circular symmetry of the cork and of its properties could be used to best advantage. This idea is correct: the best seal is obtained by cork cut in this way. But natural cork contains lenticels, tubular channels that connect the outer surface of the bark to the inner surface, allowing oxygen into, and carbon dioxide out from, the new cells that grow there. A glance at Fig. 4.23 shows that the lenticels lie parallel to the prism axis, and that a cork cut parallel to this axis

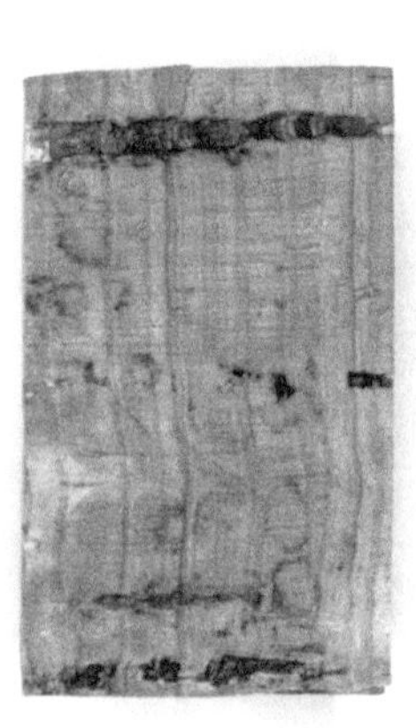

Fig. 4.30 Sections through a champagne cork and a still wine cork. (Reprinted from Gibson *et al.*, 1981.)

will therefore leak. That is why most commercial corks are cut with the prism axis (and the lenticels) at right angles to the axis of the bung.

A way out of this problem is shown in Fig. 4.30. The base of the cork, where sealing is most critical, is made of two discs cut with the prism axis (and lenticels) parallel to the axis of the bung itself. The leakage problem is overcome by laminating the two discs together so that the lenticels do not connect. Then the cork, when forced into the bottle, is compressed (radially) in the plane in which it is isotropic, and it therefore exerts a uniform pressure on the inside of the neck; the axial load needed to push the cork into the bottle produces no radial expansion (which would hinder insertion) because this Poisson's ratio (ν_{31}^*) is zero.

Cork makes good gaskets for the same reason that it makes good bungs: it accommodates large elastic distortion and volume change, and its closed cells are impervious to water and to oil. Thin sheets of cork are used, for instance, for the joints of woodwind and brass instruments. The sheet is always cut with prism axis (and lenticels) normal to its plane. The sheet is then isotropic in its plane, and this may be the reason for cutting it so. But it seems more likely that it is cut like this because the Poisson ratio for compression down the prism axis is zero. Then, when the joints of the instrument are mated, there is no tendency for the sheet to spread in its plane, and wrinkle.

Friction for shoes and floor coverings

Manufacturers who sell cork flooring sometimes make remarkable claims that it retains its friction even when polished or wet. Friction between a shoe and a cork floor has two origins. One is adhesion: atomic bonds form between the two contacting surfaces, and work must be done to break and re-form them if the shoe slides. Between a hard shoe and a tiled or stone floor this is the only source of friction and, since it is a surface effect, it is completely destroyed by a film of polish or soap. The other source of friction is due to anelastic loss. When a rough shoe slides on a cork floor the bumps on the shoe deform the cork. If the cork were perfectly elastic no net work would be done: the work done in deforming the cork ahead of a bump is recovered as the bump moves on. But if the cork has a high loss coefficient (as it does) then it is like riding a bicycle through sand: the work done in deforming the material ahead of the bump is not recovered, and a large coefficient of friction appears. This anelastic loss is the main source of friction when rough surfaces slide on cork; and since it depends on processes taking place below the surface, it is not affected by films of polish or soap. Exactly the same thing happens when a cylinder or sphere rolls on cork, which therefore shows a high coefficient of rolling friction.

Energy absorption and packaging

Many of the uses of cork depend on its capacity to absorb energy. Cork is attractive for soles of shoes and flooring because, as well as having frictional properties, it is resilient under foot, absorbing the shocks of walking. It makes good packaging because it compresses on impact, limiting the stresses to which the contents of the package are exposed. It is used as handles of tools to insulate the hand from the impact loads applied to the tool. In each of these applications it is essential that the stresses generated by the impact are kept low, but that considerable energy is absorbed.

Cellular materials are particularly good at this. The stress–strain curve for cork (Fig. 4.26) shows that the collapse stress of the cells is low, so that the peak stress during impact is limited. But large compressive strains are possible, absorbing a great deal of energy as the cell walls progressively collapse. In this regard its structure and properties resemble polystyrene foam, which has replaced cork (because it is cheap) in many packaging applications.

Insulation

The cork tree, it is thought, surrounds itself with cork to prevent loss of water in hotter climes. The properties involved – low thermal conductivity and low permeability to water – make it an excellent material for the insulation of cold, damp habitations. Caves fall into this category: the hermit caves of southern Portugal, for example, are liberally lined with cork. For the same reasons, crates and boxes are sometimes lined with cork. And the original cork tips of cigarettes may have appealed to smokers because they insulated (a little) and prevented the tobacco from becoming moist.

Heat flow through cellular materials is discussed in Gibson and Ashby (1997). Flow by conduction depends only on the amount of solid in the foam (ρ^*/ρ_s) and therefore does not depend on the cell size. Flow by convection does depend on cell size because convection currents in large cells carry heat from one side of the cell to the other. But when cells are less than about 10mm in size, convection does not contribute significantly. Flow by radiation, too, depends on the cell size: the smaller the cells, the more times the heat has to be absorbed and re-radiated, and the lower is the rate of flow. So the small cells are an important feature of cork. They are very much smaller than those in ordinary foamed plastics, and it is this, apparently, that imparts exceptional insulating properties to the material.

Indentation and bulletin boards

Cellular materials densify when they are indented: the requirement that volume is conserved, so important in understanding indentation problems for fully dense solids, no longer applies. So when a sharp object like a drawing pin is stuck into cork, the deformation is highly localized. A layer of cork cells, occupying a thickness of only about one quarter of the diameter of the indenter, collapses, suffering a large strain. The volume of the indenter is taken up by the collapse of the cells so that no long-range deformation is necessary. For this reason the force needed to push the indenter in is small. And since the deformation is (non-linear)-elastic, the hole closes up when the pin is removed.

Cork is the epitome of an elastomeric cellular solid. Its low density ($\rho^*/\rho_s = 0.15$) and closed cells, and its chemical stability and resilience, give it special properties which have been exploited by man for at least 2000 years. These properties derive from its cellular structure, and can be understood in terms of the models of Chapter 3, with modifications to include the curious shape of the cells in cork. This shape gives the cork anisotropic elastic properties, and even these can be exploited to advantage in its applications.

4.4 Summary

Woods are among the oldest construction materials, and remain, today, one of those used in the largest quantities. The number of species is very large, and the corresponding range of mechanical properties is also great. Cork, like wood, has a long history of use and although, today, its use to cork wine is diminishing slightly, its application in interior design, for footwear, and as a decorative finish increases.

Structurally, both wood and cork are complex, but both have in common that they are made up of aligned, prismatic cells. Their mechanical properties – their moduli and strength, for instance – are well studied. They can be explained by treating both materials as hexagonal honeycombs. The axial properties are well modeled as those of prismatic hexagonal cells loaded along the prism axis, and the radial and tangential properties (which are similar) by those of an assembly of such cells loaded transversely to the prism axis. The models explain both the magnitude of the properties as a function of density and their dependence on the direction of loading.

References

Wood

Ashby MF, Easterling KE, Harrysson R and Maiti SK (1985) Fracture and toughness of woods. *Proc. Roy. Soc.* **A398**, 261–80.

Barrett JD, Haigh IP and Lovegrove HM (1981) Fracture mechanics and the design of wood structures. *Phil. Trans. Roy. Soc.* **A299**, 217–26.

Bentur A and Mindess S (1986) Characterization of load-induced cracks in balsa wood. *J. Mat. Sci.* **21**, 559–65.

Bodig J and Goodman JR (1973) Prediction of elastic parameters for wood. *Wood Sci.* **5**, 249–64.

Bodig J and Jayne BA (1982) *Mechanics of Wood and Wood Composites*. New York: Van Nostrand Reinhold.

Cao J, Rambo CR and Sieber H (2004a) Preparation of porous Al_2O_3 ceramics by biotemplating of wood. *J. Porous Mat.* **11**, 163–72.

Cao J, Rusina O and Sieber H (2004b) Processing of porous TiO_2-ceramics from biological performs. *Ceramics Int.* **30**, 1971–4.

Cave ID (1968) Anisotropic elasticity of the plant cell wall. *Wood Sci. Tech.* **2**, 268–78.

Cave ID (1969) Longitudinal Young's modulus of Pinus radiata. *Wood Sci. Tech.*, **3**, 40–8.

Dinwoodie JM (1981) *Timber, its Nature and Behaviour*. New York: Van Nostrand Reinhold.

Easterling KE, Harrysson R, Gibson LJ and Ashby MF (1982) On the mechanics of balsa and other woods. *Proc. Roy. Soc.*, **A383**, 31–41.

Fridley KJ, Tang RC and Soltis LA (1992) Creep behaviour model for structural lumber. *J. Struct. Eng. ASCE* **118**, 2261–79.

Gibson LJ and Ashby MF (1997) *Cellular Solids: Structure and Properties*, 2nd edn. Cambridge: Cambridge University Press.

Gierlinger N, Schwanninger M, Reinecke A and Burgert I (2006) Molecular changes during tensile deformation of single wood fibers followed by Raman microscopy. *Biomacromolecules* **7**, 2077–81.

Gindl W and Schoberl T (2004) The significance of the elastic modulus of wood cell walls obtained by nanoindentation measurements. *Composites Part A: Appl. Sci. Manuf.* **35**, 1345–9.

Goodman JR and Bodig J (1970) Orthotropic elastic properties of wood. *J. Struct. Div., ASCE* **ST11**, 2301–19.

Goodman JR and Bodig J (1971) Orthotropic strength of wood in compression. *Wood Sci.*, **4**, 83–94.

Griel P, Lifka T and Kaindl A (1998) Biomorphic cellular silicon carbide ceramics from wood: I. Processing and microstructure. *J. Euro. Ceram. Soc.* **18**, 1961–73.

Hearmon RFS and Paton JM (1964) Moisture content changes and creep in wood. *Forest Products J.* **14**, 357–9.

Holzer SM, Loferski JR and Dillard DA (1989) A review of creep in wood: concepts relevant to develop long-term behavior predictions for wood structures. *Wood Fiber Sci.* **21**, 376–92.

International Iron and Steel Institute (2008) June 2008 crude steel production, available at: www.worldsteel.org/?action=newsdetail&id=244.

Jeronimidis G (1980) Fracture behaviour of wood and the relations between toughness and morphology. *Proc. Roy. Soc.* **B208**, 447–60.

Johnson JA (1973) Crack initiation in wood plates. *Wood Sci.* **6**, 151–8.

Kettunen PO (2006) *Wood: Structure and Properties*.Enfield, NH: Trans Tech Publications.

Koran Z (1966) Pulp and Paper Research Institute of Canada Technical Report No. 472.

Mindess S, Nadeau JS and Barrett JD (1975) Slow crack growth in Douglas Fir. *Wood Sci.* **8**, 389–96.

Morlier P (editor) (1994) *Creep in Timber Structures*. London: Chapman and Hall.

Nadeau JS, Bennett R and Fuller ER (1982) Explanation for the rate-of-loading and the duration-of-load effects in wood in terms of fracture mechanics. *J. Mat. Sci.* **17**, 2831–40.

Orso A, Wegst UGK and Arzt E (2006) The elastic modulus of spruce wood cell wall material measured by an in situ bending technique. *J. Mat. Sci.* **41**, 5122–6.

Paris O, Zollfrank C, Zickler GA (2005) Decomposition and carbonization of wood biopolymers – a microstructural study of softwood pyrolysis. *Carbon* **43**, 53–66.

Price AT (1928) Mathematical discussion on the structure of wood in relation to its elastic properties. *Phil. Trans.* **A228**, 1–62.

Rambo CR, Cao J, and Sieber H (2004) Preparation and properties of highly porous, biomorphic YSZ ceramics. *Mater. Chem. Phys.* **87**, 345–52.

Rambo CR, Cao J, Rusina O and Sieber H (2005) Manufacturing of biomorphic (Si, Ti, Zr)-carbide ceramics by sol-gel processing. *Carbon* **43**, 1174–83.

Reiterer A, Sinn G and Stanzl-Tschegg SE (2002) Fracture characteristics of different wood species under mode I loading perpendicular to the grain. *Mat. Sci. Eng.* **A332**, 29–36.

Schachner H, Reiterer A and Stanzl-Tschegg SE (2000) Orthotropic fracture toughness of wood. *J. Mat. Sci. Lett.* **19**, 1783–5.

Schniewind AP and Pozniak RA (1971) On the fracture toughness of Douglas fir wood. *Eng. Fract. Mech.* **2**, 223–33.

Schniewind AP and Centano JC (1973) Fracture toughness and duration of load factor I. Six principal systems of crack propagation and the duration factor for cracks propagating parallel to grain. *Wood Fiber* **5**, 152–9.

Sieber H, Hoffmann C, Kaindl A and Greil P (2000) Biomorphic cellular ceramics. *Adv. Eng. Mat.* **2**, 105–9.

Sieber H (2005) Biomimetic synthesis of ceramics and ceramic composites. *Mat. Sci. Eng.* **A412**, 43–7.

Sih GC, Paris PC and Irwin GR (1965) On cracks in rectilinearly anisotropic bodies. *Int. J. Fract. Mech.*, **1**, 189–203.

Silvester FD (1967) *Mechanical Properties of Timber.* Oxford: Pergamon Press.

Singh M and Salem JA (2002) Mechanical properties and microstructure of biomorphic silicon carbide ceramics fabricated from wood precursors. *J. Euro. Ceram. Soc.* **22**, 2709–17.

Srinavasan PS (1942) *J. Indian Inst. Sci.*, **23B**, 222.

Stanzl-Tschegg SE, Tan DM and Tschegg EK (1995) New splitting method for wood fracture characterization. *Wood Sci. Technol.* **29**, 31–50.

Stanzl-Tschegg SE (2006) Microstructure and fracture mechanical response of wood. *Int. J. Fracture* **139**, 495–508.

Thompson DW (1961) *On Growth and Form*, abridged edition, edited by JT Bonner. Cambridge: Cambridge University Press.

Thuvander F and Berglund LA (2000) In situ observations of fracture mechanisms for radial cracks in wood. *J. Mat. Sci.* **35**, 6277–83.

Tze WTY, Wang S, Rials TG, Pharr GM and Kelley SS (2007) Nanoindentation of wood cell walls: Continuous stiffness and hardness measurements. *Composites Part A: Appl. Sci. Manuf.* **38**, 945–53.

United Nations Food and Agricultural Organization (2005) *Forest Products Yearbook 1999–2003.* Rome: United Nations.

Vogli E, Mukerji J, Hoffman C, Kladny R, Sieber H and Greil P (2001) Conversion of oak to cellular silicon carbide ceramic by gas-phase reaction with silicon monoxide. *J. Amer. Ceram. Soc.* **84**, 1236–40.

Vogli E, Sieber H and Greil P (2002) Biomorphic SiC-ceramic prepared by Si-vapor phase infiltration of wood. *J. Euro. Ceram. Soc.* **22**, 2663–8.

Walsh PF (1971) *Cleavage Fracture in Timber.* Div. Forest Prod., Tech. Paper No. 65. Melbourne, Australia: CSIRO.

Wilkes TE, Young ML, Sepulveda RE, Dunand DC and Faber KT (2006) Composites by aluminum infiltration of porous silicon carbide derived from wood precursors. *Scripta Materialia* **55**, 1083–6.

Williams JG and Birch MW (1976) Mixed mode fracture in anisotropic media, *in Cracks and Fracture*, ASTM Stand. No. 601, p. 125.

Wimmer R, Lucas BN, Tsui TY and Oliver WC (1997) Longitudinal hardness and Young's modulus of spruce tracheid secondary walls using nanoindentation technique. *Wood Sci. Technol.* **31**, 131–41.

Wood Handbook (1974) US Dept. of Agriculture, Agricultural Handbook No. 72, Forest Products Lab., Madison, WI.

Wu EM (1963) *Applications of Fracture Mechanics to Orthotropic Plates.* Theor. Appl. Mech. Report No. 1418. Urbana, IL: University of Illinois.

Zollfrank C and Sieber H (2004) Microstructure and phase morphology of wood derived biomorphous SiSiC-ceramics. *J. Euro. Ceram. Soc.* **24**, 495–506.

Zollfrank C and Sieber H (2005) Microstructure evolution and reaction mechanism of biomorphous SiSiC ceramics. *J. Amer. Ceram. Soc.* **88**, 51–8.

Cork

Eames AJ and Macdaniels LH (1951) *An Introduction to Plant Anatomy.* London: McGraw-Hill.

Esau KE (1965) *Plant Anatomy.* New York: Wiley, p. 340.

Fernandez LV (1978) *Inst. Nac. Invest. Agrar.* (Spain). Cuad. No. 6, p. 7.

Fortes MA, Fernandes JJ, Serralheiro I and Rosa ME (1989) Experimental determination of hydrostatic compression versus volume change curves for cellular solids. *J. Testing Eval.* **17**, 67–71.

Fortes MA and Nogueira MT (1989) The Poisson effect in cork. *Mat. Sci. Eng.* **A122**, 227–32.

Gibson LJ, Easterling KE and Ashby MF (1981) Structure and mechanics of cork. *Proc. Roy. Soc.* **A377**, 99–117.

Hooke R (1665) *Micrographica.* London: Royal Society, p. 112.

Horace Q (27 BC) *Odes*, book III, ode 8, line 10.

Lewis FT (1928) The shape of cork cells: a simple demonstration that they are tetrakaidecahedral. *Science*, **68**, 625–6.

Pereira H (2007) *Cork: Biology, Production and Uses.* Amsterdam: Elsevier.

Pliny C (AD 77) *Natural History*, vol. 16, section 34.

Plutarch (AD 100) *Life of Camillus. Parallel Lives*, vol. II, ch. xxv, p. 154.

Rosa ME and Fortes MA (1988a) Stress relaxation and creep of cork. *J. Mat. Sci.* **23**, 35–42.

Rosa ME and Fortes MA (1988b) Rate effects on the compression and recovery of dimensions in cork. *J. Mat. Sci.* **23**, 879–85.

Rosa ME and Fortes MA (1991) Deformation and fracture of cork in tension. *J. Mat. Sci.* **26**, 341–8.

Zimmerman MH and Brown CL (1971) *Trees Structure and Function.* Berlin: Springer, p. 88.

5 Foam-like materials in nature

5.1 Introduction

Trabecular bone and subcutaneous fat are examples of tissues in the body with a foam-like structure. Trabecular bone exists at the ends of the long bones, in the vertebrae, and in the cores of shell-like bones such as the skull and pelvis. The mechanics of trabecular bone are of interest in understanding how it behaves in both healthy and diseased states. In patients with osteoporosis, for example, bone mass decreases dramatically with increasing age, leading to an increased risk of fracture, especially of the hip and vertebrae; at both sites, the loads are carried primarily by trabecular bone. The behavior of subcutaneous fat, or adipose tissue, is relevant to the development of needle-free injection systems for drug delivery.

Plant parenchyma and adipose tissue are both fluid-filled closed-cell foams. In parenchyma, the fluid is water while in adipose tissue it is lipid. The mechanics of parenchyma are of interest in understanding the behavior of plant leaves and stems, described in more detail in Chapter 6. Coral and sponge are two examples of marine organisms with a foam-like structure. In this chapter, we describe the structure and mechanics of each of these tissues.

5.2 Trabecular bone

5.2.1 Introduction

Superficially, bones look fairly solid. But looks are deceptive. Most bones are an elaborate construction, made up of an outer shell of dense compact bone, enclosing a core of porous cancellous, or trabecular bone (trabecula means 'little beam' in Latin). Examples are shown in Fig. 5.1: cross-sections of a femur, tibia and vertebra. In some instances (as at joints between vertebrae or at the ends of the long bones) this configuration minimizes the weight of bone while still providing a large bearing area, a design which reduces the bearing stresses at the joint. In others (as in the vault of the skull or the iliac crest) it forms a low weight sandwich shell, analyzed in more detail in Chapter 6. In either case the presence of the trabecular bone reduces the weight while still meeting its primary mechanical function.

An understanding of the mechanical behavior of trabecular bone has relevance for several biomedical applications. In elderly patients with osteoporosis the mass

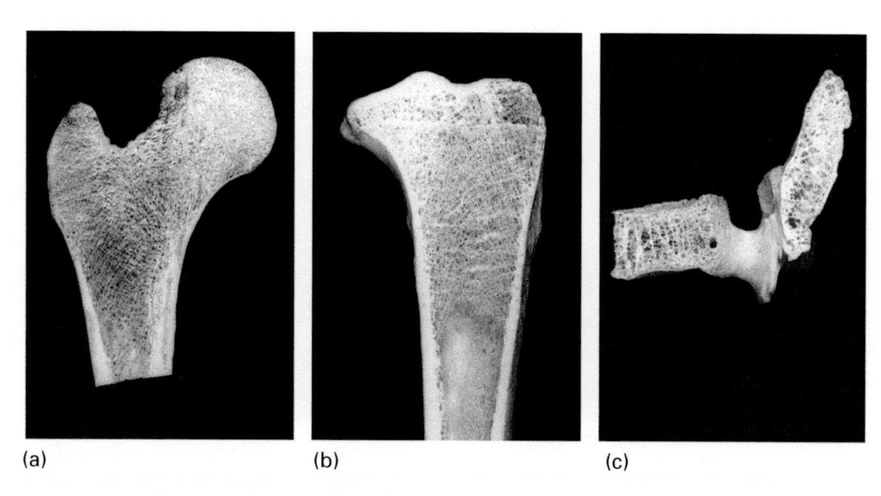

Fig. 5.1 Cross-sectional views of (a) the head of a femur, (b) the tibia and (c) a lumbar vertebra. In each case there is an outer shell of almost fully dense compact bone surrounding a core of foam-like, low-density trabecular bone.

of bone in the body decreases over time to such an extent that fractures can occur under loads that, in healthy people, would be considered normal. Such fractures are common in the vertebrae, hip and wrist, and are due in part to a reduction in the amount of trabecular bone in these areas. The degree of bone loss in a patient can be measured using non-invasive techniques, so an understanding of the relationship between bone density and strength helps in predicting when the risk of a fracture has become high. It helps, too, in the design of artificial hips. Most of the bone replaced by an artificial hip is trabecular; an improved understanding of the structure–property relationships for trabecular bone allows the design of artificial hips with properties that more closely match those of the bone they replace. The mismatch of properties between current artificial hips and the surrounding bone is thought to be one reason that they work loose, an unpleasant development from the patient's point of view because replacement is difficult. Trabecular bone may also play a role in osteoarthritis, which is thought to be related to a breakdown in the lubrication process at joints. The distribution of forces acting across a joint is directly related to the mechanical properties of the underlying trabecular bone, so changes in its structure (and hence properties) may change the distribution of forces and cause damage to the lubrication system.

5.2.2 The structure of trabecular bone and the mechanical properties of the solid tissue

The cellular structure of trabecular bone is shown in Fig. 5.2. Low-density trabecular bone has a rod-like structure, much like that of an open-cell foam. As the density increases, the solid becomes more like perforated plates. In practice, the relative density of trabecular bone varies from about 0.05 to 0.5 (technically, any bone with a relative density less than 0.7 is classified as "trabecular").

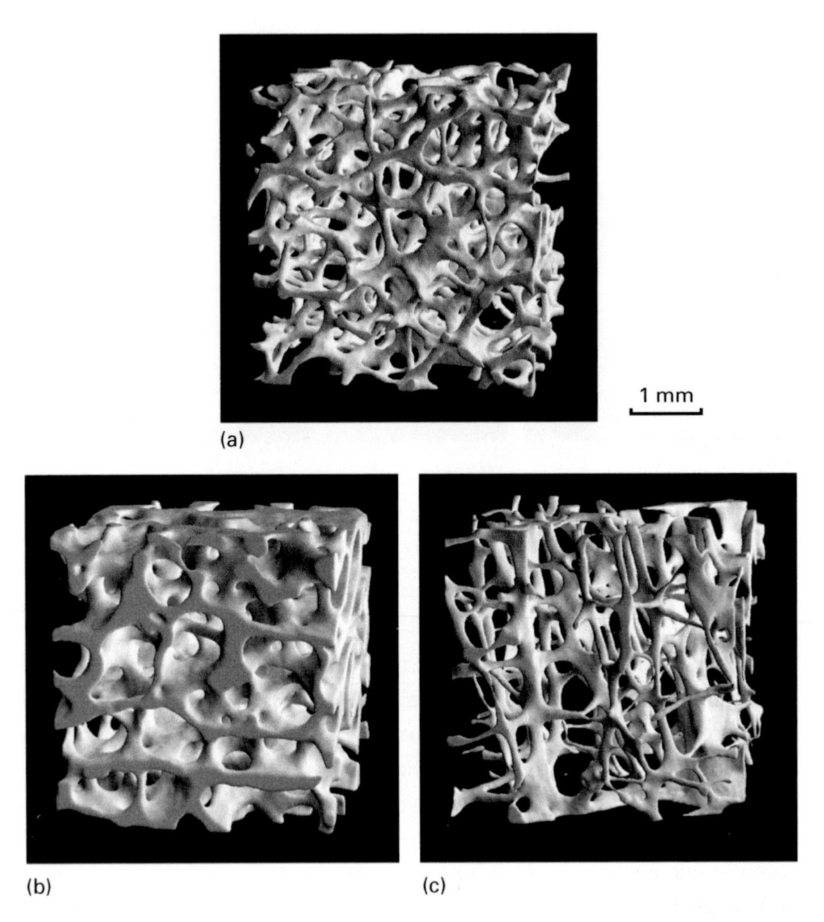

Fig. 5.2 Micro-computed tomography images showing the cellular structure of trabecular bone: (a) the lumbar spine of a 42-year-old male, with rod-like structure $\rho^*/\rho_s = 11.1\%$; (b) the femoral head of a 37-year-old male, with more plate-like structure, $\rho^*/\rho_s = 25.6\%$; (c) the lumbar spine of a 59-year-old male, $\rho^*/\rho_s = 6.1\%$. Each image represents a 4 mm cube. (All images courtesy of Professor Ralph Muller, ETH Zurich, Switzerland.)

Bone grows in response to mechanical loading (Carter and Beaupre, 2001; Currey, 2002) and there is substantial evidence that the trabeculae are oriented in the directions of the principal stresses. This is especially apparent in bones that are loaded as cantilevers, such as the ischium and the spinous process of the thoracic vertebra of the horse (Currey, 2002). Trabecular bone in human vertebrae are loaded primarily along the length of the spine, and, to a lesser extent, from the constraint of the cortical shell, radially; most vertebral trabeculae are oriented roughly in the longitudinal or transverse directions (Silva and Gibson, 1997a). Other evidence supporting this idea derives from finite element studies of the stresses in the human patella (Hayes and Snyder, 1981) and from in vivo measurements of the strains in the cortex of the calcaneous of sheep (Lanyon, 1974) and of the kangaroo-like marsupial, *Potorous tridactylus* (Biewener *et al.*, 1996). It is generally accepted that the density of trabecular bone depends on the magnitude of the stresses that it experiences while the microstructural anisotropy depends on the ratio and direction of the principal stresses.

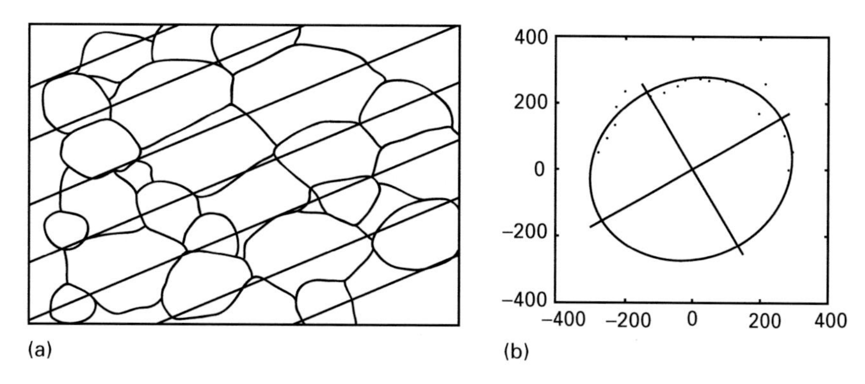

Fig. 5.3 (a) Measurement of mean intercept length of a polymer foam at a particular orientation using a set of equidistant, parallel lines; (b) an ellipse of the mean intercept length measured for different orientations on one plane. (With kind permission from Springer Science+Business Media: Huber and Gibson (1988) Fig. 9.)

One standard stereological method of characterizing microstructural anisotropy is to measure the mean intercept length (Underwood, 1970), the inverse of the number of intersections of a microstructural feature with a line or a set of parallel lines (Fig. 5.3). A polar plot of the mean intercept lengths measured for incrementally increasing angles of orientation can be fit to an ellipse with the major and minor axes of the ellipse characterizing the structural anisotropy on that plane. The first such measurements on trabecular bone were performed by Whitehouse (1974). Harrigan and Mann (1984) later observed that the ellipses on three orthogonal planes give an ellipsoid defined by a positive definite second-rank tensor. Another, related, measure of anisotropy is the fabric tensor, which is defined as the inverse square root of the mean intercept length tensor. With increasingly accurate methods of imaging trabecular bone, such as micro-computed tomography or automated serial sectioning, the fabric tensor has become the standard method of quantifying microstructural anisotropy in trabecular bone (see, for instance, Cowin, 1986; Turner *et al.*, 1990; Kabel *et al.*, 1999; Zysset, 2003).

The mechanical properties of the individual trabeculae making up the cell walls of trabecular bone were long assumed to be the same as those of cortical bone (McElhaney *et al.*, 1970a,b; Townsend *et al.*, 1975a,b; Carter and Hayes, 1977 and Gibson, 1985). There have been many attempts to determine the Young's modulus of the trabeculae using a number of different techniques: direct mechanical testing of a single trabecula in tension, bending or buckling (Townsend *et al.*, 1975a; Runkle and Pugh, 1975; Ryan and Williams, 1989; and Choi *et al.*, 1990); ultrasonic wave propagation in trabecular bone specimens (Ashman and Rho, 1988; Rho *et al.*, 1993; Turner *et al.*, 1999; Jorgensen and Kundu, 2002); finite element analysis (Pugh *et al.*, 1973b; Williams and Lewis, 1982; Mente and Lewis, 1987; van Rietbergen *et al.*, 1995; Bayraktar *et al.*, 2004; Chevalier, 2007) and, more recently, nanoindentation (Rho *et al.*, 1997; 1999; Roy *et al*, 1999; Turner *et al.*, 1999; Zysset *et al.*, 1999; Chevalier *et al.*, 2007).

Direct tension, bending and buckling tests are difficult: unmachined specimens have varying cross-sectional areas and are curved along their lengths while machined specimens may have significant surface defects introduced by the machining. In addition, the deformations are typically small so that any additional displacement introduced as an artifact of the testing technique (e.g. slippage or stress concentrations at loading points) results in a lower measured value than the true value. The results from direct tension and bending tests are lower than the moduli of relatively dense trabecular bone (up to 6 GPa (Currey, 2002)), suggesting that they are not reliable. Buckling tests on an individual trabecula give a somewhat higher value of E_s = 11.4 GPa for wet bone (Townsend *et al.*, 1975a), but still significantly lower than more recent data, described below.

In ultrasonic testing, the Young's modulus of the trabeculae can be found by measuring the speed of a high frequency ultrasonic wave on a specimen of porous trabecular bone, eliminating the need for testing of individual trabeculae. Ultrasonic tests give a value of E_s of 14.8 GPa for wet bone (Rho *et al.*, 1993) and somewhat higher values, 17.5 GPa (Turner *et al.*, 1999) and 19.9 GPa (Jorgensen and Kundu, 2002) for thin sections (500 or 700 μm thick) of embedded bone, initially dried, but rewetted by submerging in water.

Finite element analysis of the trabecular architecture of a specimen of trabecular bone for which the modulus has been experimentally measured can be used to back-calculate the modulus of the individual trabeculae. The latest attempts to do this use micro-computed tomography to determine the trabecular architecture as input to the finite element model. One such study found a value of E_s = 18.0 GPa for wet trabecula, in comparison with a value of 19.9 GPa for the modulus of cortical bone (Bayraktar *et al.*, 2004) while another found values of E_s in the range of 16.0 to 22.5 GPa for dry bone, which were in good agreement, although generally slightly lower, than the values of E_s measured in the same study using nanoindentation (Chevalier *et al.*, 2007).

Finally, nanoindentation has been used to measure the Young's modulus and hardness of trabeculae. This technique allows properties to be mapped across a surface with a resolution of the order of 1 μm. Values of E_s for human bone range from 13.4 to 22.7 GPa for dry trabeculae (Rho *et al.*, 1997; 1999; Roy *et al.*, 1999; Turner *et al.*, 1999; Chevalier *et al.*, 2007), with the higher values (19.4 to 22.7 GPa) corresponding to indents that primarily load the trabeculae in the longitudinal direction. A lower value was reported for wet human bone (11.4 GPa) (Zysset *et al.*, 1999), but this may be associated with a surface liquid film that can cause difficulty in identifying the first point of contact of the indenter with the surface (Rho *et al.*, 1997). Nanoindentation on submerged and dry bovine cortical bone specimens indicated that drying increased the modulus of the interstitial lamellae by 10% and that of osteons by 15% (Rho and Pharr, 1999).

The data are assembled in Table 5.1. The ultrasound data suggest a value for $E_s \sim$ 17 GPa for wet trabeculae, consistent with Bayraktar *et al.*'s (2004) finite element back-calculation (wet E_s = 18 GPa). The nanoindentation results suggest a value of E_s = 21 GPa for dry trabeculae for indents loading primarily in the longitudinal direction. The nanoindentation results underestimate the longitudinal modulus slightly, as the lower transverse moduli also contribute to the indentation unloading stiffness. Drying bone increases the modulus by about 10–25% (Evans and Lebow, 1951; Townsend *et al.*, 1975a; Rho and Pharr, 1999). Taking these results together, we assume that the Young's modulus for wet human trabeculae in the longitudinal

Table 5.1 Data for solid cell-wall properties for trabecular bone

Method	Type of bone; loading direction	E_s (GPa) – wet	E_s (GPa) – dry	Hardness, H (MPa)	Reference
Ultrasound	Human tibia	14.8			Rho *et al.*, 1993
Ultrasound	Human femur	17.5			Turner *et al.*, 1999
Ultrasound	Canine femur; longitudinal	19.9			Jorgensen and Kundu, 2002
Finite element back-calculation	Human femur	18			Bayraktar *et al.*, 2004
Finite element back-calculation	Human femur (various sites)		16.0–22.5		Chevalier *et al.*, 2007
Nanoindentation	Human vertebra; transverse		13.4	468	Rho *et al.*, 1997
Nanoindentation	Human vertebra;				Rho *et al.*, 1999
	longitudinal,		19.4	618	
	transverse		15.8	540	
Nanoindentation	Human vertebra;				Roy *et al.*, 1999
	longitudinal,		22.7	664	
	transverse		17.0	559	
Nanoindentation	Human femur		18.4		Turner *et al.*, 1999
Nanoindentation	Human femur; transverse	11.4		150–300	Zysset *et al.*, 1999
Nanoindentation	Human femur; longitudinal		19.6–22.3 mean = 20.9		Chevalier *et al.*, 2007

direction is the same as the generally accepted value of 18 GPa for wet human cortical bone loaded in the longitudinal direction (Martin *et al.*, 1998; Currey, 2002).

Most of the hardness data from nanoindentation give values around 550 MPa; one study gives substantially lower values (150–300 MPa). Assuming that the latter is an outlier and averaging the results of the other studies gives the hardness of the trabeculae of 564 MPa, suggesting a strength of 188 MPa. We assume that the compressive and tensile strengths of the trabeculae are also the same as those of cortical bone: 182 MPa and 115 MPa, respectively, for loading in the longitudinal direction (Martin *et al.*, 1998). We take the density of the trabeculae, to be 1800 kg/m^3, similar to that for human cortical bone (Currey, 2002). The solid trabecular properties are summarized in Table 5.2.

5.2.3 Young's modulus and strength of trabecular bone

The compressive stress–strain curve of trabecular bone is typical of that of a cellular solid. Figure 5.4 shows the three familiar regimes of behavior described in Chapter 3. In trabecular bone, the small strain, linear-elastic response usually derives

Table 5.2 Solid cell-wall properties for human trabecular bone

Property	Value
Density, ρ_s	1800 kg/m^3
Young's modulus, E_s	18 GPa
Compressive strength, $\sigma_{ys\ comp}$	182 MPa
Tensile strength, $\sigma_{ys\ ten}$	115 MPa

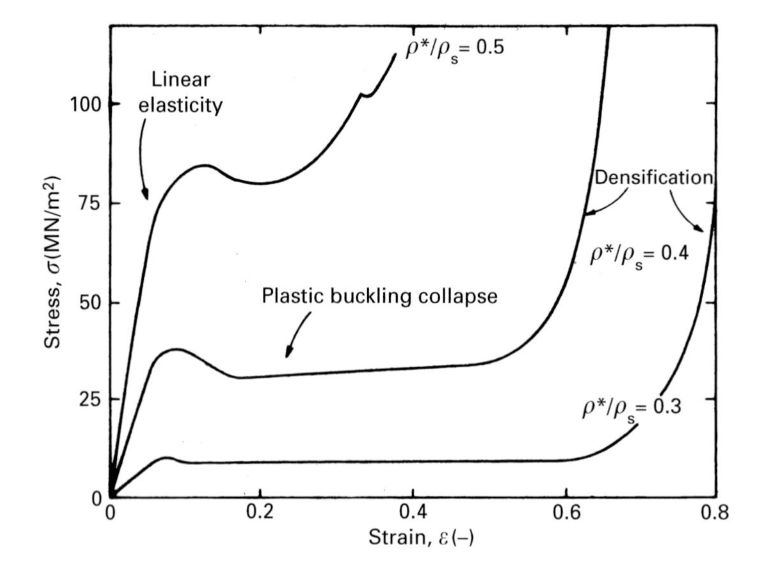

Fig. 5.4 Compressive stress–strain curves for several relative densities of wet trabecular bone. (Reprinted with kind permission of John Wiley and Sons, from Hayes and Carter, 1976, Fig. 1a.)

from elastic bending of the trabeculae. Axial or membrane stresses may be present, too, but the deformations resulting from these are small compared with those from bending, unless the trabeculaer are nearly perfectly aligned with the applied stress, or the relative density is particularly high. The linear-elastic regime typically ends when a localized band of cells begins to collapse by inelastic buckling, which then propagates to neighboring bands, giving the long, horizontal plateau of the stress–strain curve. As with foams, densification occurs when opposing cell walls meet and touch, causing the stress to rise steeply. It is instructive to compare Fig. 5.4, which shows the stress–strain curve for trabecular bone, with Fig. 3.2b for a rigid polymer foam.

Bending and buckling were long assumed to be the primary mechanisms of deformation and failure in trabecular bone under compression (Pugh *et al.*, 1973a,b; Behrens *et al.*, 1974; Townsend *et al.*, 1975a). The first scanning electron micrograph of buckled trabeculae was reported by Hayes and Carter (1976). With the development of a micro-compression device capable of loading specimens within a micro-computed tomography machine, direct observations of bending and inelastic buckling of trabeculae within specimens of trabecular bone are now available (Muller *et al.*, 1998; Nazarian and Muller, 2004) (Fig. 5.5).[1] Recent high-speed photography of uniaxially compressed specimens of human vertebral bone also shows bending and buckling of trabeculae (Thurner *et al.*, 2006a). Although Muller's group has reported bending and buckling in both rod-like and plate-like trabecular bone taken from the whale vertebrae (Muller *et al.*, 1998; Nazarian and Muller, 2004), they have also reported that they were unable to observe the deformation in some plate-like trabecular specimens from bovine tibia. It is possible that these were highly oriented specimens that deformed primarily axially, rather than in bending, and that the axial deformations were too small to detect (Muller *et al.*, 1998). We note that Williams and Lewis' (1982) modulus and strength data for oriented bone from the human tibia also suggest axial deformation for loading in the longitudinal direction. As the strains within trabeculae approach the yield point of the solid tissue, microcracking occurs, producing permanent deformation (Arthur Moore and Gibson, 2002; Thurner *et al.*, 2006b).

Finite element models for the linear elastic compressive response of trabecular bone from the proximal human tibia have also found that deformation is by bending of trabeculae (van Reitbergen *et al.*, 1995). Finite element modeling of failure of specimens of trabecular bone from bovine tibia indicated that failure was often found to involve one edge of a trabecula, consistent with bending or buckling (Niebur *et al.*, 2000). Failure involving the entire cross-section of a trabecular, suggestive of axial deformation, was also found, possibly due to the high relative density and oriented structure of trabecular bone from the bovine tibia.

A tensile stress–strain curve for wet trabecular bone is shown in Fig. 5.6. The initial linear-elastic portion of the curve results from the elastic bending or, in limited cases, extension of trabeculae, as in compression. At strains of about 1%, the stress–strain curve becomes non-linear as the trabeculae start to deform irreversibly and crack. Beyond the peak, the curve falls gradually as the trabeculae progressively fail by tearing and fracturing (Carter *et al.*, 1980). The yield strength of trabecular bone is lower in tension than in compression and the difference in strengths increases with relative density (Stone *et al.*, 1983; Kaplan *et al.*, 1985; Keaveny *et al.*, 1994; Kopperdahl and Keaveny, 1998; Keaveny *et al.*, 2001; Morgan and Keaveny, 2001).

Data for the Young's modulus and the compressive and tensile strengths of trabecular bone are plotted against density in Figs. 5.7 to 5.9. Data for the compressive strength (Fig. 5.8) is separated into two plots, from early and later studies, to allow individual data points to be distinguished. The data are normalized by the values for the solid trabeculae for human bone given in Table 5.2. The values for bovine bone are slightly higher (Martin *et al.*, 1998; Currey, 2002); on the log–log plots the error in using the human values is small. As one would expect for a cellular solid, the density has a profound influence in determining the stiffness and strength of trabecular bone.

[1] Animations of the deformations in trabecular bone at increasing strains, from Muller's group, available on the Cambridge University Press website, are particularly impressive, see www.cambridge.org/cellular materials.

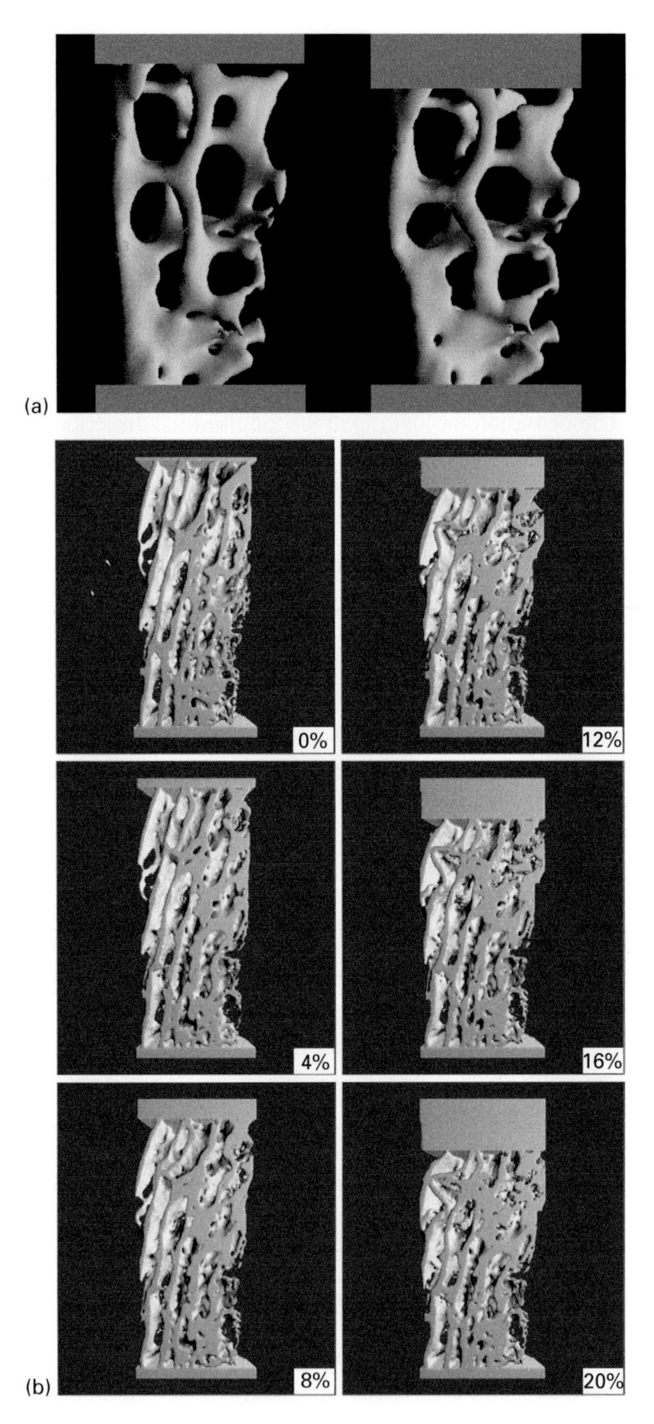

Fig. 5.5 Micro-computed tomography images of bending and buckling deformation in whale vertebral trabecular bone under uniaxial compression. (a) Comparison of a specimen unloaded (left) and loaded (right). (b) A series of images of a specimen compressed up to 20% strain. (a, Reprinted from Muller *et al.* (1998) with permission from IOS Press; b, reprinted from Nazarian and Muller (2004), with permission of Elsevier.)

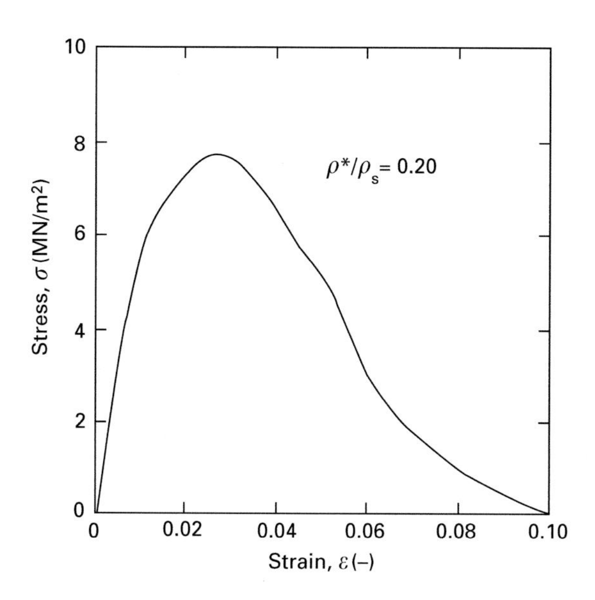

Fig. 5.6 Tensile stress–strain curve for wet trabecular bone. (Reprinted from Carter *et al.* (1980), with permission from Taylor and Francis (http://www.informaworld.com).)

The spread in the data is large, for several reasons. Trabecular bone is anisotropic; a great part of the scatter arises from anisotropy in the trabecular architecture and in the directions of loading of the samples. Aligning the direction of loading during mechanical testing with the trabecular orientation reduces the scatter in the measured properties (Turner and Cowin, 1988); early studies made no attempt to do this. For instance, Keaveny *et al.*'s (2001) data, using specimens with the trabecular orientation aligned with the direction of loading, show less scatter than some of the earlier studies plotted on Fig. 5.7. Cowin (1985, 1986) has derived a relationship between the fabric tensor and the elasticity tensor which is widely used to describe the elastic constants of trabecular bone in terms of both the density and the microstructural anisotropy. A smaller part of the scatter in the data may arise from variations in the properties of the bone making up the cell wall as a result of small differences in its porosity and inorganic content. And the data shown derive from tests carried out at strain-rates varying over five orders of magnitude; it is known that both Young's modulus and strength depend on strain-rate (Carter and Hayes, 1977). Finally, the moisture content can be important. Some of the scatter in the early strength data, particularly, can be traced to different levels of dryness of the bone; more recent studies usually maintain specimens wet.

Most trabecular architectures are foam-like structures. As we saw in Chapter 3, foam-like structures deform in the linear-elastic regime by bending and, in compression, collapse by buckling, plastic yielding or brittle fracture, depending on the nature of the cell wall material. There is substantial evidence that, in general, trabecular bone deforms linearly elastically by bending and fails in compression by inelastic buckling: scanning electron micrographs of failed sections, micro-computed tomography images of deforming specimens (Fig. 5.5) and optical high-speed video of compressed specimens all show trabeculae bending and buckling. The models for open-cell foam

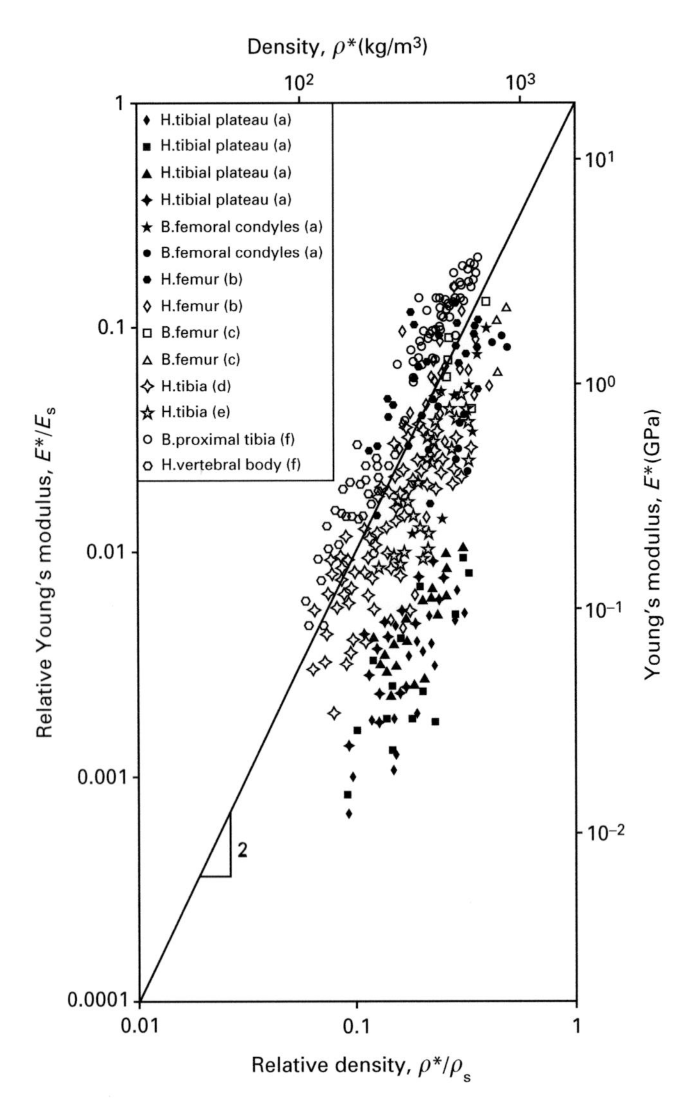

Fig. 5.7 Young's modulus of trabecular bone plotted against density. The relative moduli are normalized by E_s = 18 GPa and the relative densities are normalized by ρ_s = 1800 kg/m³. The data lie very roughly around a line of slope 2. Data from (a) Carter and Hayes, 1977; (b) Carter *et al.*, 1980; (c) Bensusan *et al.*, 1983; (d) Hvid *et al.*, 1989; (e) Linde *et al.*, 1991; (f) Keaveny *et al.*, 2001. H = human, B = bovine.

behavior developed in Chapter 3 indicate that the modulus and compressive strength then vary with the square of the relative density (see (3.16) and (3.19)), although inelastic buckling in trabecular bone is expected to depend on the tangent modulus rather than the Young's modulus of the solid trabeculae. (Since data for the solid tangent modulus are unavailable, we have normalized the compressive strengths of trabecular bone by the compressive strength of cortical bone, which has a value of roughly $0.01E_s$.) More detailed finite element models of 141 trabecular bone specimens found that the

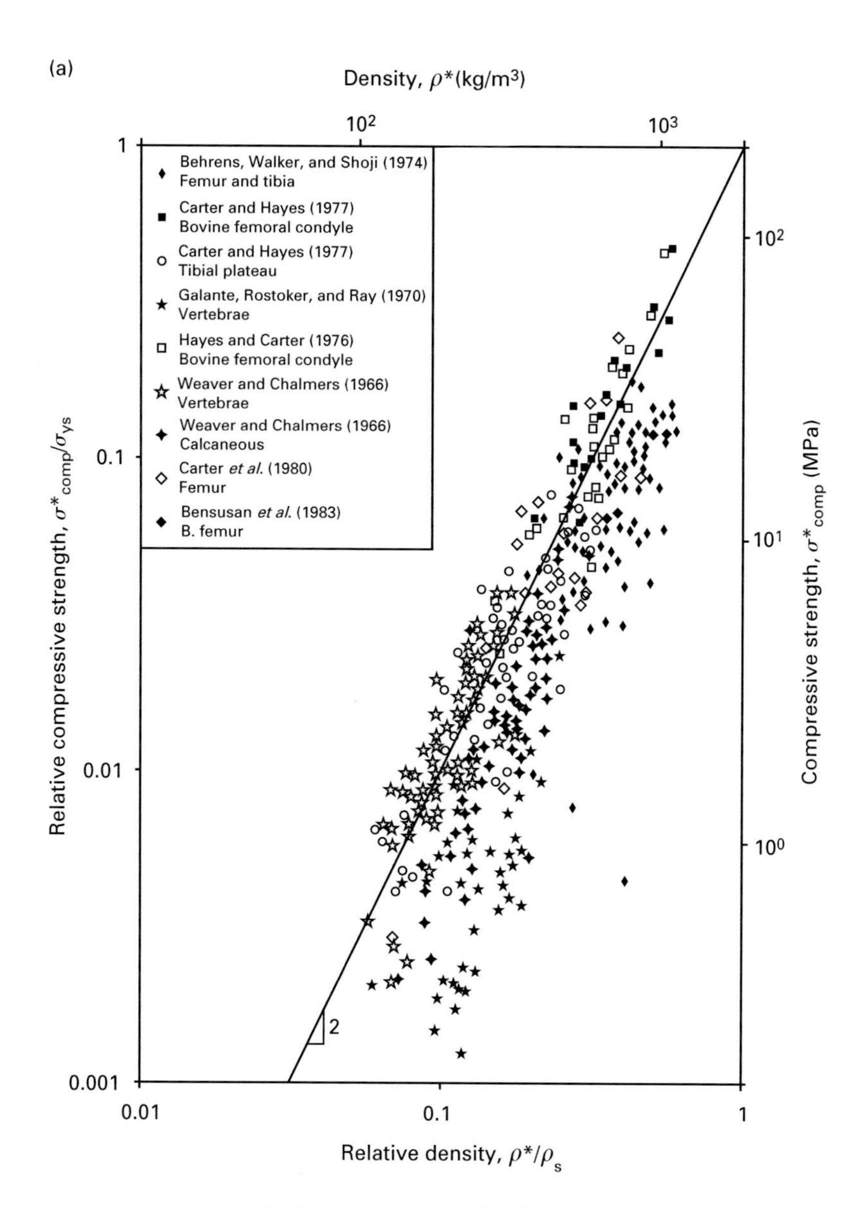

Fig. 5.8 Compressive strength of trabecular bone plotted against density. The data are separated into two plots from (a) early and (b) later studies, to allow individual data points to be distinguished. The relative compressive strengths are normalized by $\sigma_{ys\,comp} = 182$ MPa and the relative densities are normalized by $\rho_s = 1800$ kg/m^3. The data lie very roughly around a line of slope 2. All data for human bone unless otherwise indicated.

Young's moduli in the three principal directions and the shear moduli on the three principal planes all varied with approximately $(\rho^*/\rho_s)^2$ (the values of the powers ranged from 1.80 to 2.04) (Yang *et al.*, 1999).

The data for the Young's modulus and compressive strength from a number of studies are generally consistent with this, lying along a line of slope 2 on the log–log plots (Figs. 5.7 and 5.8). Other data sets lie below the line of slope 2 and have considerable

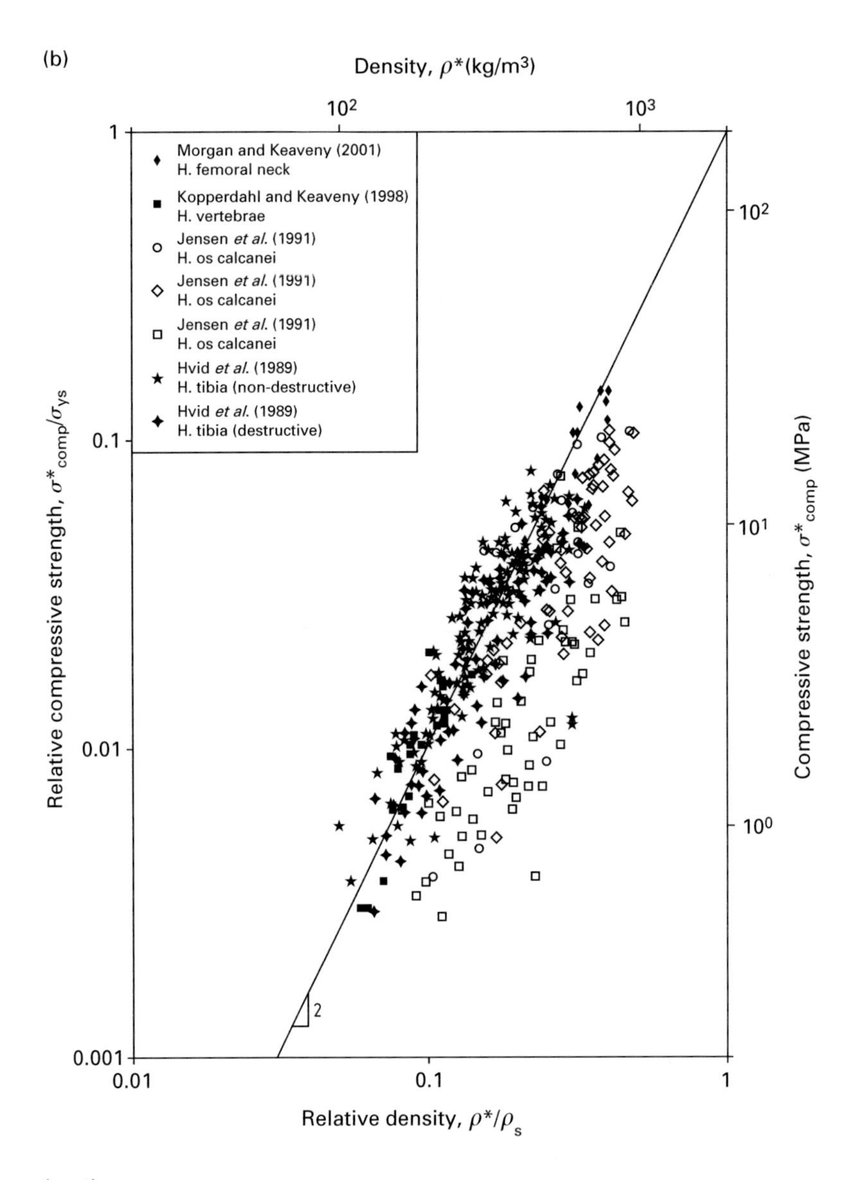

Fig. 5.8 *(cont.)*

scatter. For instance, the compressive strength data of Galante *et al.* (1970) (Fig. 5.8a) vary by a factor of about ten, with little variation in density, possibly due to off-axis loading of highly anisotropic specimens. A statistical analysis of data available in the literature concluded that both modulus and compressive strength are proportional to density squared (Rice *et al.*, 1988).

In tension, trabecular bone fails by the formation of plastic hinges in bent trabeculae. Models for open-cell foams suggest that the tensile strength should then vary with relative density raised to the power 3/2 (3.22). The data for the tensile strength lie between lines of slope 3/2 and 2 on the log–log plot of Fig. 5.9.

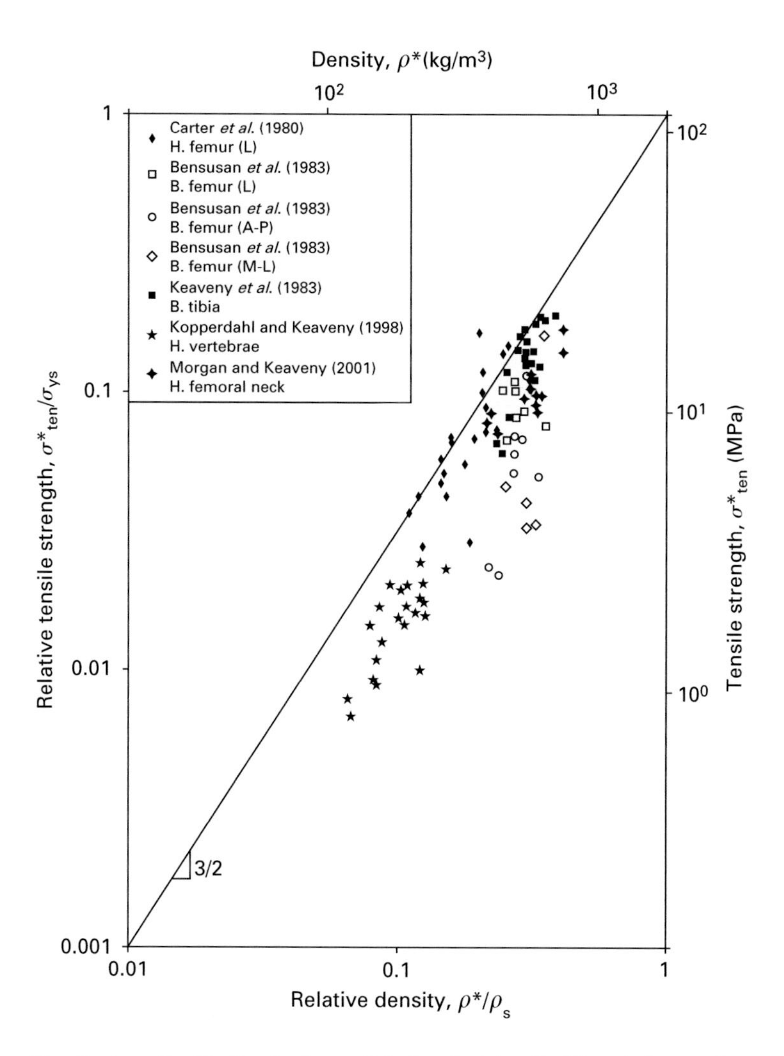

Fig. 5.9 Tensile strength of trabecular bone plotted against density. The relative tensile strengths are normalized by $\sigma_{ys\ tension}$ = 115 MPa and the relative densities are normalized by ρ_s = 1800 kg/m^3. The long axis of the specimens coincided with the bone material axis. Some specimens were loaded in the longitudinal (L) direction, the anterior–posterior direction (A–P), or the medial–laterial direction (M–L). The data for loading in the longitudinal direction lie between lines of slopes 3/2 and 2. H = human, B = bovine.

One consequence of both modulus and compressive strength having the same dependence on density is that the compressive yield strain for trabecular bone is roughly constant. Keaveny and co-workers found that the yield strain corresponding to the 0.2% offset yield stress is roughly 0.7%, with only a slight, if any, dependence on density (Keaveny *et al.*, 1994; Kopperdahl and Keaveny, 1998). They also found that the yield strains are higher in compression than in tension (Keaveny *et al.*, 1994; Kopperdahl and Keaveny, 1998), that they are isotropic (Turner, 1989; Chang *et al.*, 1999) and that they are relatively uniform within a particular anatomical site but that they vary across

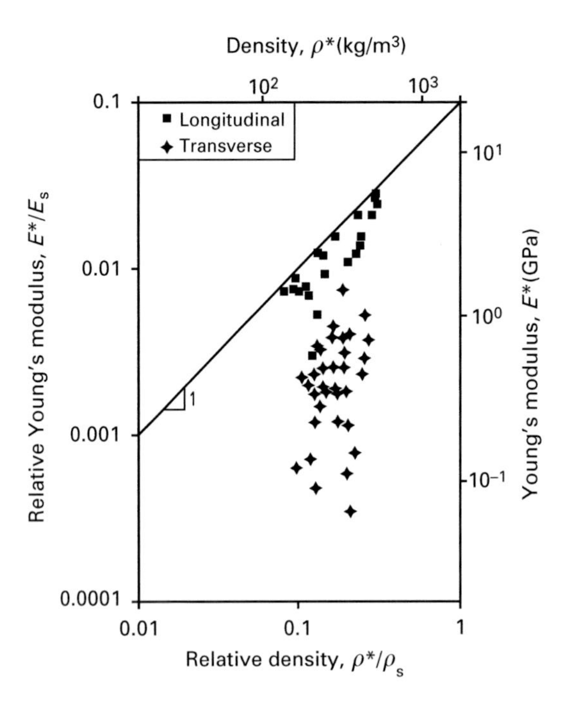

Fig. 5.10 Young's modulus of highly oriented trabecular bone from the human proximal tibial epiphysis plotted against density. The relative moduli are normalized by E_s = 18 GPa and the relative densities are normalized by ρ_s = 1800 kg/m^3. The data for loading along the principal direction of trabecular orientation fall on a line of slope 1. All data from Williams and Lewis (1982).

sites (Kopperdahl and Keaveny, 1998; Morgan and Keaveny, 2001). Collectively, these results suggest that failure of trabecular bone is based on strain.

Some trabecular architectures are highly aligned, presumably as a result of loading predominantly in one direction in vivo (Fig. 5.2c). If the degree of alignment is sufficiently high, these structures deform in the linear elastic regime by axial shortening, rather than bending; their moduli in the axial direction then depend linearly on density. They may fail by either buckling, or, if sufficiently dense, by yielding, which again produces a linear dependence of strength on density. Limited data for highly oriented bone, with trabeculae aligned along the direction of principal stress, indicate that the modulus and strength are linearly related to relative density for loading in the longitudinal direction (Figs. 5.10 and 5.11), suggesting that axial deformations and failure are important in this case (Williams and Lewis, 1982).

The data for the modulus and density of trabecular bone and its constituents are summarized in Fig. 5.12; the slope of 2 represents the dependence of the modulus of trabecular bone on density squared. The data for collagen and hydroxyapatite are given in Table 2.2. The envelope of composites that could be made from collagen and hydroxyapatite is obtained from simple upper and lower bound estimates. Human compact (or cortical) bone has a density of 1800–2000 kg/m^3 and a Young's modulus of 17–20 GPa (Martin *et al.*, 1998; Currey, 2002). The envelope for trabecular bone is obtained from the data on

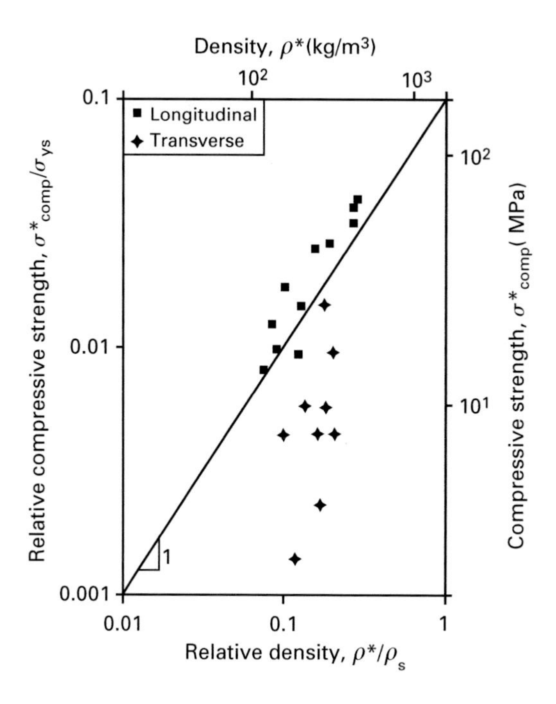

Fig. 5.11 Compressive strength of highly oriented trabecular bone from the human proximal tibial epiphysis plotted against relative density. The relative compressive strengths are normalized by $\sigma_{ys\,comp} = 182$ MPa and the relative densities are normalized by $\rho_s = 1800$ kg/m^3. The data for loading in the longitudinal direction fall roughly along a line of slope 1. All data from Williams and Lewis (1982).

Fig. 5.7. The plot shows the contributions of both the cellular structure and the composite nature of the trabeculae to the modulus of trabecular bone.

With aging, the structure of trabecular bone changes due to reductions in bone mass. For instance, there is roughly a 50% reduction in the relative density of vertebral trabecular bone between 20 and 80 years of age (Moskilde, 1989). The trabeculae at first thin, and then, when the thickness reaches a critical dimension (presumably related to the size of the bone cells, of the order of tens of microns), they resorb entirely, reducing the connectivity of the structure (Parfitt, 1992) (Fig. 5.13a,b). Resorption of trabeculae has a dramatic effect on the modulus and strength of trabecular bone (Silva and Gibson, 1997a,b; Guo and Kim, 1999, 2002; Vajjhala *et al.*, 2000). For instance, random removal of 10% of the struts in a model tetrakaidecahedral structure reduces the modulus and strength by about 40% (compared to a 20% reduction in modulus for uniformly thinning the struts) (Fig. 5.13c). Similar studies on random polyhedral cells indicate that the results are insensitive to the exact three-dimensional cell geometry.

5.2.4 Time-dependent behavior

The creep of cortical bone has been well studied (Smith and Walmsley, 1959; Sedlin, 1965; Knets and Vils, 1975; Fondrk *et al.*, 1988; Caler and Carter, 1989; Rimnac *et al.*,

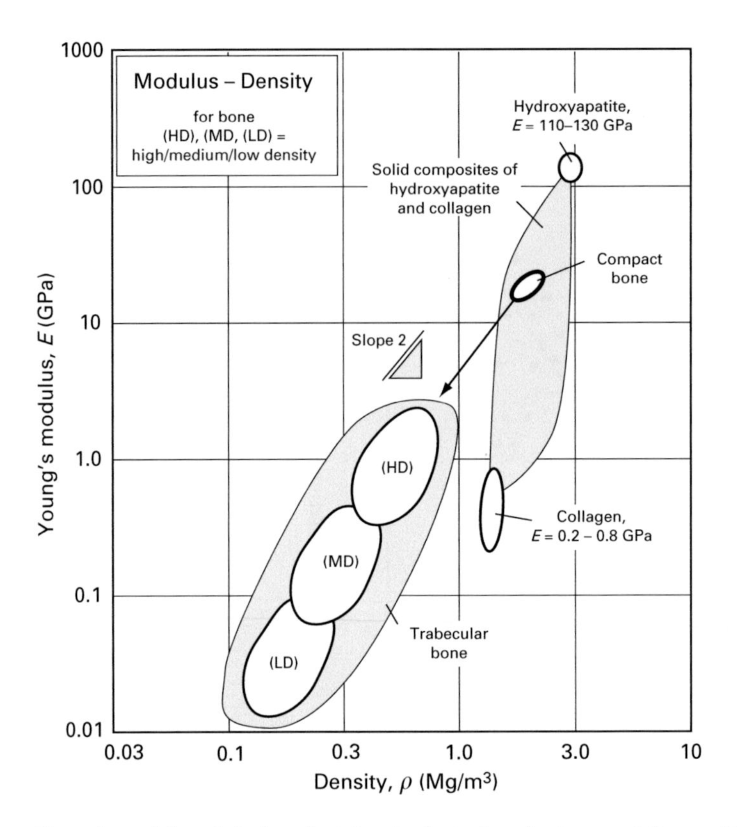

Fig. 5.12 Young's modulus plotted against density for trabecular bone and its constituents.

1993). Plots of strain against time show three distinct regimes of primary, secondary and tertiary creep, with a constant, steady-state strain-rate during secondary creep. Experimental results, summarized in Bowman *et al.* (1999), reveal that power-law relationships exist between steady-state, secondary creep rate, applied stress and time to failure.

Trabecular bone, too, has been shown to follow power-law creep (Bowman *et al.*, 1999). A typical plot of strain against time for a creep test on trabecular bone is shown in Fig. 5.14a. The steady–state, secondary creep rate is plotted against stress (normalized by the Young's modulus of each specimen) in Fig. 5.14b, along with a line representing data for cortical bone from Fondrk *et al.* (1988) and data for demineralized cortical bone. The best-fit linear regressions are as follows:

Cortical bone: $$d\varepsilon/dt = 3.38 \times 10^{38} (\sigma/E)^{18.9}$$ (Fondrk *et al.*, 1988) (5.1a)

Trabecular bone: $$d\varepsilon/dt = 1.18 \times 10^{28} (\sigma/E)^{15.6}$$ (Bowman *et al.*, 1998) (5.1b)

Demineralized bone: $$d\varepsilon/dt = 6.74 \times 10^{12} (\sigma/E)^{15.5}$$ (Bowman *et al.*, 1999) (5.1c)

where the unit of the constant in each equation is s^{-1}. All three tissues have similar creep exponents. If the creep properties of individual trabeculae are the same as those

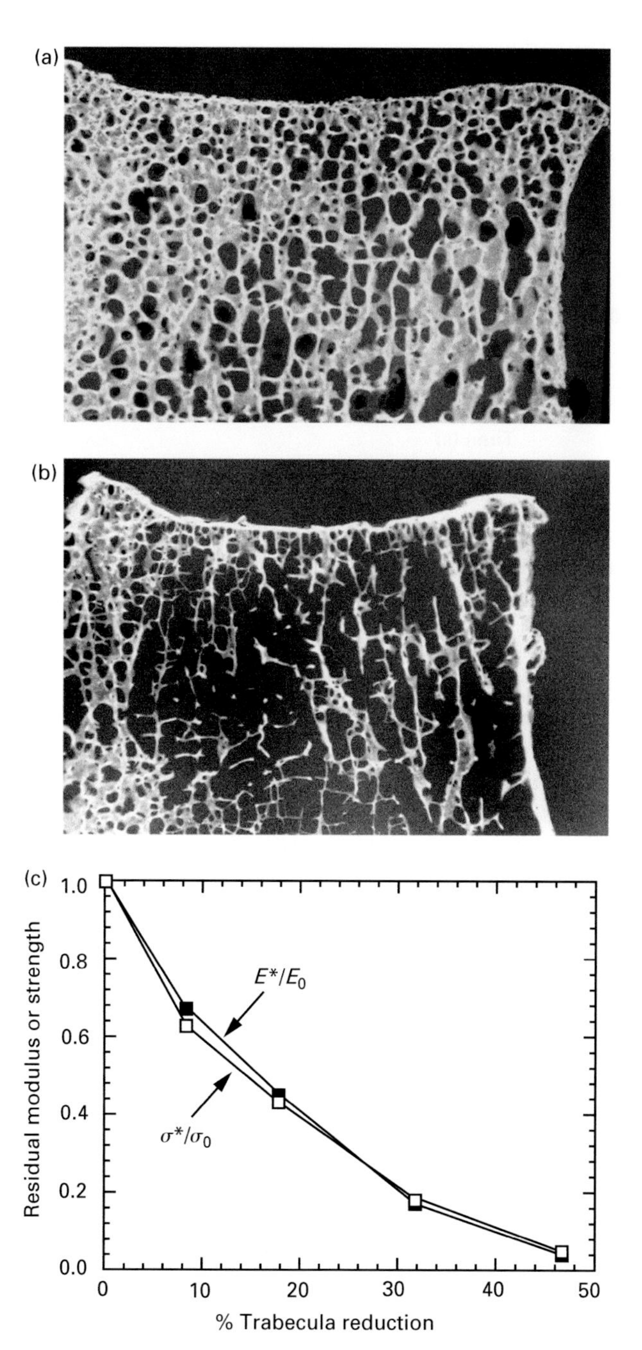

Fig. 5.13 Micrographs of cross-sections of lumbar vertebrae from (a) a 55-year-old woman ($\rho^*/\rho_s = 0.17$) and (b) an 86-year-old woman ($\rho^*/\rho_s = 0.07$). The trabeculae thin and resorb with aging. (c) The reduction in the Young's modulus and compressive strength of a random Voronoi model of trabecular bone, with resorption of trabeculae. (a,b, Reprinted from Vajjhala *et al.*, 2000, with permission of American Society of Mechanical Engineers; c, reprinted from Guo and Kim 1999, with permission of the American Society of Mechanical Engineers.)

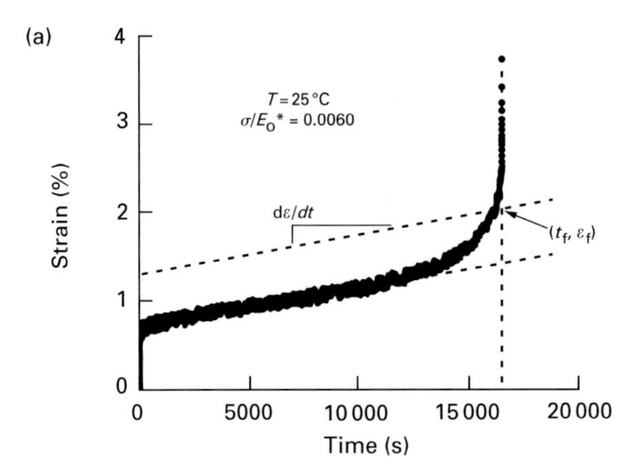

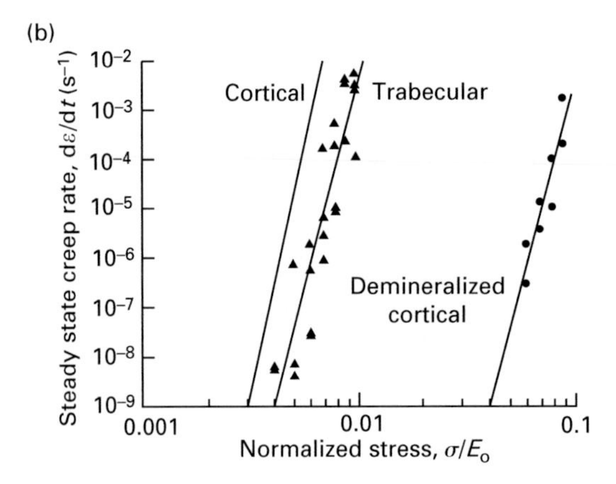

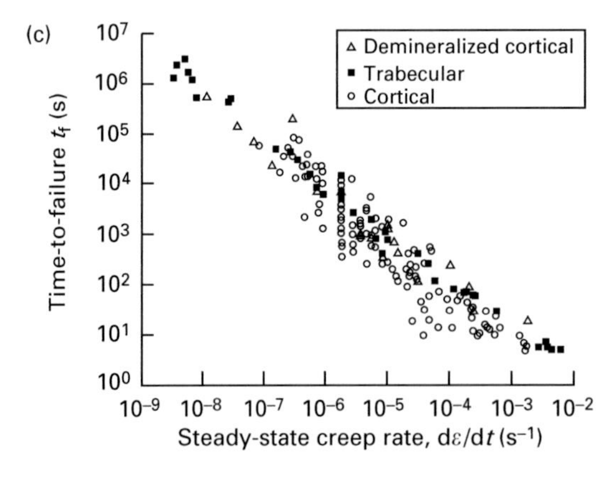

Fig. 5.14 (a) Strain plotted against time for a creep test on trabecular bone. (b) Steady-state, secondary creep rate plotted against applied stress normalized by the Young's modulus of each specimen. (c) Time to failure plotted against the steady-state, secondary creep rate. Data for cortical bone, trabecular bone and demineralized bone. (a, Reprinted from Bowman *et al.*, 1998, with permission of American Society of Mechanical Engineers; b, c Reprinted from Bowman *et al.*, 1999 with permission of American Society of Mechanical Engineers.)

of cortical bone, we expect the exponents to be identical (3.30); the data indicate that they are close. The similarity of the exponents for cortical and trabecular bone and of demineralized bone (i.e. collagen) suggests that demineralized bone is the phase responsible for the creep of both cortical and trabecular bone. In both cortical and trabecular bone, the mineral acts as a stiff, elastic inclusion.

The above power laws, with their large exponents, suggest that the strain rate increases dramatically with normalized stress. An exponential fit of the data is more realistic. The results, for Fig. 5.14b, give

$$\text{Cortical bone:} \quad \frac{d\varepsilon}{dt} = 5.7 \times 10^{-15} \exp\left(4030\frac{\sigma}{E_o}\right) \tag{5.2a}$$

$$\text{Trabecular bone:} \quad \frac{d\varepsilon}{dt} = 1.7 \times 10^{-14} \exp\left(2680\frac{\sigma}{E_o}\right) \tag{5.2b}$$

$$\text{Demineralized bone:} \quad \frac{d\varepsilon}{dt} = 2.3 \times 10^{-14} \exp\left(260\frac{\sigma}{E_o}\right) \tag{5.2c}$$

The term multiplying stress, which can be thought of as a stress concentration factor, changes in a systematic way while pre-exponentials do not vary much.

Data for the time to failure plotted against the steady-state, secondary creep strain rate, are plotted in Fig. 5.14c: the time to failure for all three tissues is well described by the Monkman–Grant relationship (3.31):

$$t_f = 0.03\dot{\varepsilon}^{-0.91} \tag{5.3}$$

Limited creep tests on human vertebral trabecular bone, at two normalized stress levels of $\sigma/E = 0.000\,75$ and 0.0015, also confirm its non-linear creep behavior, although there is insufficient data to indicate power-law creep (Yamamoto *et al.*, 2006). Interestingly, the measured secondary creep strain rates are much higher than expected from (5.1b). We note that (5.1b) is based on data for trabecular bone from young bovine tibia (average apparent density 0.61 g/cm^3) while Yamamoto *et al.* (2006) used human vertebral trabecular bone from cadavers of mean age 76 years (range 58–92 years) (apparent density not reported). Similarly aged (mean 74 years) human vertebral bone in another study had a mean apparent density of 0.23 g/cm^3 (Haddock *et al.*, 2004). The increased creep strain rates reported by Yamamoto *et al.* (2006) may be in part due to the lower apparent density and in part due to resorption of trabeculae in osteoporotic specimens. As described above, finite element models have demonstrated that resorption of trabeculae can produce a dramatic reduction in the modulus and uniaxial compressive strength of trabecular bone (Fig. 5.13c); similarly, we would expect a significant increase in the creep rate.

Several other studies on the time-dependent behavior of trabecular bone have focused on stress relaxation, rather than creep. Like creep, stress relaxation is found to be non-linear (Deligiannni *et al.*, 1994; Quaglini *et al.*, 2009), although some studies have modeled it as linear viscoelastic at low stresses (Bredbenner and Davy, 2006). Quaglini *et al.* (2009) suggest a relaxation function describing two processes: one short-term, which depends non-linearly on stress, and the other, long-term, which depends linearly on stress. Stress relaxation tests at a single, constant strain above the yield strain of trabecular bone induced microcracks to propagate with time (Nagaraja *et al.*, 2007).

5.2.5 Fatigue behavior

It is estimated that half of all vertebral fractures and 10% of hip fractures are the result of repeated loadings due to daily living, rather than trauma (Myers and Wilson, 1997; Greenspan *et al.*, 1998). In both cases, most of the load in the vertebrae, or in the femoral head, is carried by trabecular bone in compression. For this reason, measurements of the fatigue behavior of trabecular bone have focused on compressive loading (Michel *et al.*, 1993; Bowman *et al.*, 1998; Moore and Gibson, 2003a, b; Haddock *et al.*, 2004; Rapillard *et al.*, 2006; Yamamoto *et al.*, 2006; Dendorfer *et al.*, 2008).

Under cyclic compressive loading, trabecular bone shows a decrease in the Young's modulus and an increase in the residual plastic strain, similar to the aluminum foams discussed in Section 3.3.6. (Fig. 5.15a,b). Failure is typically defined as the number of cycles to reduce the modulus to a given fraction of its initial value, for instance, $E/E_o = 0.9$. Fatigue life data are usually shown by plotting the applied stress range (typically normalized by the initial modulus of the specimen) against the number of cycles to failure on a log–log plot to give an *S–N* curve. Data from a number of studies, assembled in Fig. 5.15c, can be described by

$$N_f = C\left(\frac{\Delta\sigma}{E_o}\right)^{-n} \tag{5.4}$$

with n typically about 12. The data have different intercepts, possibly as a result of using different species (human vs. bovine), sites (vertebral vs. proximal tibia) and testing geometries (cylinders vs. cubes). Haddock *et al.* (2004) have shown that if σ/E is normalized by the mean monotonic yield strain for trabecular bone from different species and sites, then a single *S–N* curve can be used to describe the combined data of their study and that of Bowman *et al.* (1998). Rapillard *et al.* (2006) have suggested a relationship between normalized stress and number of cycles to failure that includes both the volume fraction and fabric of the trabecular bone; experimental results show a high correlation with this relationship. Dendorfer *et al.* (2008) have shown that the *S–N* curve is sensitive to the direction of loading.

Some studies have suggested that the increasing residual strain with increasing numbers of cycles of loading in fatigue tests is due to creep (Michel *et al.*, 1993; Bowman *et al.*, 1998; Yamamoto *et al.*, 2006). Upper bound estimates of creep strains in fatigue tests on bovine trabecular bone, based on the maximum stress in the fatigue test and the best-fit regression of the creep data for bovine trabecular bone (5.1b), indicate that the creep strains are much lower than the residual strains, so that, in this case, creep does not contribute to fatigue (Moore *et al.*, 2004). As indicated above, however, creep strain rates in human vertebral bone from elderly cadavers are much higher than those in young bovine trabecular bone, so that creep may contribute to fatigue in this case.

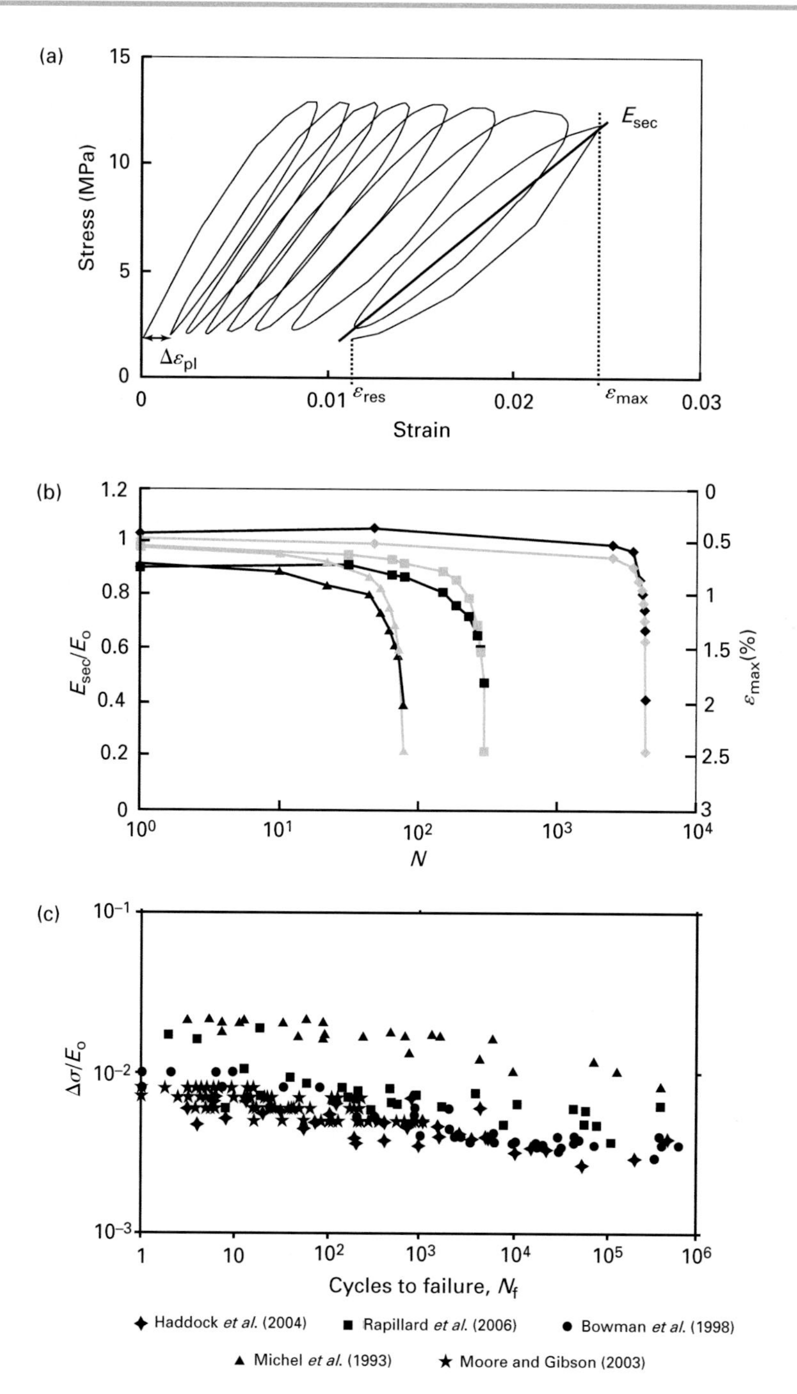

Fig. 5.15 (a) Stress–strain loops for a typical compression fatigue test on trabecular bone ($\Delta\sigma/E$ = 0.007). The secant modulus, residual strain and maximum strain in the test are shown. (b) Normalized secant modulus E_{sec}/E_o (black) and ε_{max} (gray) plotted against number of cycles of loading, N, for $\Delta\sigma/E = 0.005$. Specimens stopped at a given maximum strain will have similar normalized modulus, but those stopped at a preset number of cycles may have different modulus reductions. (c) Compressive fatigue ($\Delta\sigma/E$)–N curve for trabecular bone. The Moore and Gibson (2003a) study took failure as ε_{max} = 1.3% while the other studies took failure as a specified reduction in secant modulus. (a,b, Reprinted from Moore and Gibson, 2003a, with permission of the American Society of Mechanical Engineers.)

The progression of microdamage in trabecular bone subjected to compressive fatigue has been observed using chelating flurochrome stains that attach to exposed calcium on the surface of the specimen (Moore *et al.*, 2003b). A specimen is first stained (e.g. with alizarin complexone) after machining and before testing, and then stained again after testing with a stain of a different color (e.g. calcein green) to identify any damage in the form of microcracks that occurred during testing. Damage was observed in the form of diffusely stained areas without visible cracks, single cracks, parallel cracks, cross-hatched cracks and complete trabecular fracture. Above a threshold of about 0.5% maximum strain, microdamage (the number of damaged trabeculae per unit area and the damaged area fraction) increases with increasing maximum strain during a test. Localized regions of damage were observed at strain levels as low as 0.8% and damage bands extending across the specimens were visible at strains of 1.3%. The width of the damage band increased up to strains of 1.65%, beyond which the density of damage within the band increased.

Initial models for compressive fatigue of trabecular bone analyzed crack growth within individual trabeculae in a regular hexagonal honeycomb (Guo *et al.*, 1994) or in a random foam model (Makiyama *et al.*, 2002). In both studies, the initial crack lengths were randomly distributed and crack growth was assumed to follow a Paris law, where the rate of change of crack length, a, with the number of cycles of loading, N, is proportional to the change in the stress intensity factor, ΔK, at the crack tip during a fatigue cycle raised to a power, n_p:

$$\frac{da}{dN} = C'(\Delta K)^{n_p} \tag{5.5}$$

where C' is a constant. Finite element analysis was used to calculate the stress intensity at each crack tip and to model the crack growth. Once a crack extended completely across a strut, that member was removed from the structure, the reduced modulus of the honeycomb or foam was calculated, and the stress intensity factors at each crack tip recalculated for the next iteration. One interesting result is that the first few trabeculae to fail were those with the largest cracks, but the fracture sequence of the remaining trabeculae was determined both by the initial crack length and the stress concentration arising from the previously fractured trabeculae. Fracture of only a few trabeculae led to substantial loss of modulus of the honeycomb or foam; for instance, fracture of 1–2% of the members in both models resulted in a 10% drop in modulus. Both models gave the following result:

$$N_f = C'\left(\frac{\Delta\sigma}{E_o}\right)^{-n_p} \tag{5.6}$$

In the models, the value of n_p for the solid trabeculae was taken as that for cortical bone, $n_p = 4.5$, which differs substantially from the slopes of the data shown in Fig. 5.15c. One limitation of both models was that they neglected the reduction in modulus of an individual strut with increasing crack length.

More recent models have analyzed the effect of local modulus reduction within trabeculae using continuum damage mechanics and have also modeled more realistic trabecular architectures using two-dimensional slices of trabecular bone from micro-computed tomography (Taylor *et al.*, 2002; Kosmopoulos *et al.*, 2008). Taylor and co-workers find that including the local modulus reduction increases the predicted fatigue life, as the modulus reduction decreased the local stress within the trabeculae (redistributing it to the adjacent material), decreasing the stress intensity factor at the crack tip and the rate of crack growth. Interestingly, the local modulus reduction had little effect on the overall modulus of the model, as only a small fraction of the solid is near crack tips. Kosmopoulos and co-workers do not assume an initial crack distribution but instead analyze both the initiation and accumulation of microdamage. The local modulus of each element is assumed to depend on the strain intensity. They find that

$$N_{\mathrm{f}} = C''\left(\frac{\Delta\sigma}{E_{\mathrm{o}}}\right)^{-12.2} \tag{5.7}$$

with the slope of the model *S–N* curve being in excellent agreement with the data of Fig. 5.15c. They further conclude that microdamage, without complete trabecular fracture, can lead to the failure of high-density trabecular bone.

The mechanics of trabecular bone are important in understanding fracture risk in osteoporosis, stress distributions around implants, and in designing replacement materials for bone. Micromechanical models of the cellular structure of trabecular bone give insight into its stiffness and strength, and the way they depend on bone density, architecture and tissue properties.

5.3 Plant parenchyma

Plants have three main types of tissues: the dermal tissue system, consisting of the epidermis, which provides an outer layer of protection and regulates air and water loss through the stomata; the primary vascular system, consisting of xylem and phloem, which conduct water and sap, respectively; and the ground tissue system, which consists of parenchyma cells which store sugars and typically make up the bulk of non-woody plants (Niklas, 1992).

The cells of storage parenchyma are roughly tetrakaidecahedra, with relatively thin cell walls (Fig. 5.16). The cell walls typically consist only of a primary layer; unlike woody plants, they lack lignified secondary wall layers. Within the cell wall, cellulose fibers reinforce a matrix of hemicellulose, pectins and glycoproteins (Niklas, 1992). The living protoplasm in the cell consists of the water along with the cell organelles and molecules within it. In most parenchyma tissues, especially those likely to experience high compressive or bending stresses, such as stems and leaf petioles, the cells are densely packed together: they can be thought of as a pressurized, liquid-filled closed-cell foam. There are some specialized parenchyma tissues, such as aerenchyma, with large volumes of interstitial air between the cells; such tissues provide little mechanical support (Niklas, 1992). Here, we consider only densely packed parenchyma cells.

The mechanical response of plant parenchyma cells depends on the turgor pressure within the cells. The effect of turgor pressure has been studied by placing specimens

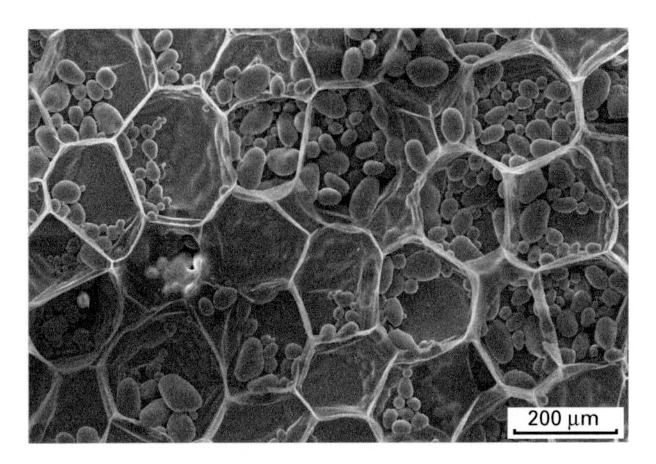

Fig. 5.16 Parenchyma cells in potato. The oblong particles within the cells are starch.

(often apple or potato tissue, which are nearly entirely parenchyma cells) into a bath of mannitol solution of known molarity. Water moves into or out of the plant tissue, depending on the difference in concentration of solutes between the protoplasm within the cell and the mannitol bath, correspondingly increasing or decreasing the turgor pressure. Turgor pressures for fresh plant tissue are typically in the range of 0.1 to 1.0 MPa.

Typical uniaxial compressive and tensile stress–strain curves for apple parenchyma at different turgor pressures are shown in Fig. 5.17. As the turgor pressure increases, the length of the initial "toe" region decreases, but the slope of the curve beyond the "toe" region is roughly constant. The peak load decreases in compression and increases in tension, although by small amounts in both cases. At low turgor pressures, for fresh tissue, the slope of the compressive stress–strain curve at 80% of the ultimate strength (roughly the maximum slope) is about 50 times that corresponding to the initial modulus. At normal turgor pressures the ratio is about 10, and at high turgor pressures it is about 4. The initial Young's modulus (the slope of the stress–strain curve at 0% strain) of potato parenchyma loaded in uniaxial compression, reflecting the slope of the "toe" region, is plotted against turgor pressure for potato parenchyma in Fig. 5.18. The initial modulus increases from a value of about 300 kPa at zero turgor pressure to a roughly constant value of 5500 kPa at turgor pressures greater than 0.3 MPa. The compressive yield strength of potato parenchyma is plotted against turgor pressure in Fig. 5.19: the strength decreases with increasing turgor pressure, as would be expected if the cell membrane has a constant strength.

The effect of turgor pressure on the stress–strain response of parenchyma cells can be understood by considering the response of a fluid-filled closed-cell foam (Section 3.3.8) (Niklas, 1989; Warner *et al.*, 2000). At low turgor pressures, some of the thin cell walls are curved (Figs. 5.16 and 5.20a). Under a uniaxial stress, the cell walls respond by bending: the initial modulus depends on the relative density, ρ^*/ρ_s, raised to a power of about 3. (Here, we exclude the contribution of the fluid to the parenchyma density.) For thin-walled parenchyma cells with low relative densities (typically less than 0.10), this gives the low value of the

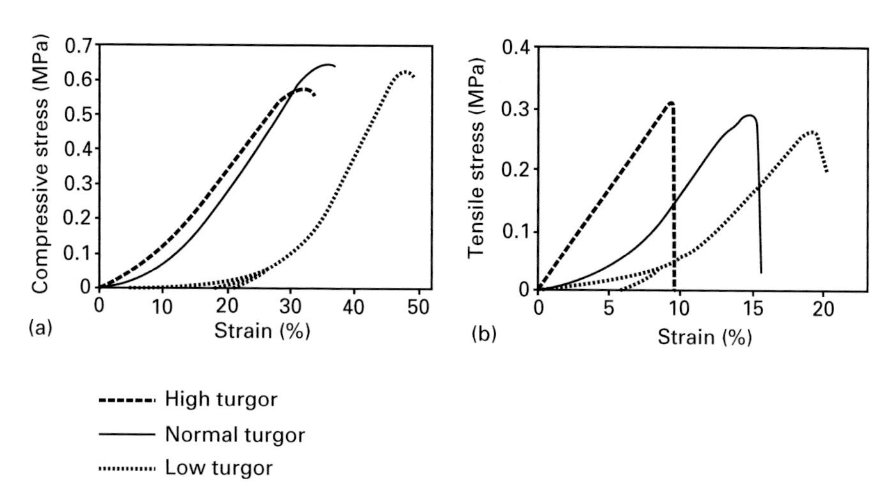

Fig. 5.17 Uniaxial (a) compressive and (b) tensile stress–strain curves for fresh apple parenchyma at different turgor pressures (Reprinted from Oye *et al.*, 2007, with permission of Elsevier.)

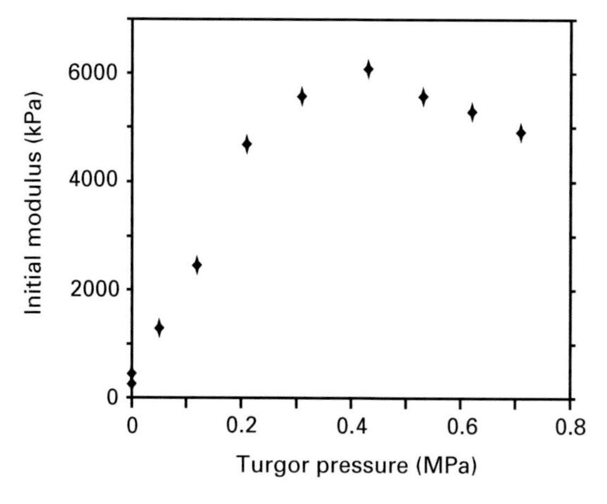

Fig. 5.18 The initial Young's modulus of potato parenchyma loaded in uniaxial compression plotted against turgor pressure. (After Lin and Pitt, 1986, with permission of Wiley-Blackwell.)

initial modulus. As the uniaxial deformation increases, the cell walls straighten with increasing stiffness (corresponding to increased stretching and decreased bending), giving the long "toe" region of the stress–strain curve. Once the cell walls become taut, further deformation is at constant volume, so that any reduction in cell volume from bending must be compensated for by an increase in cell volume from cell-wall stretching; since stretching requires a higher stress within the cell wall compared to bending, the Young's modulus is then dominated by the stretching response. At this point, if the cell-wall material is linear elastic, the stress–strain curve becomes linear; further uniaxial strain simply induces further stretching of the cell walls. The Young's modulus is then constant with increasing strain: it depends nearly linearly on relative density (3.36) and is much higher than the initial modulus.

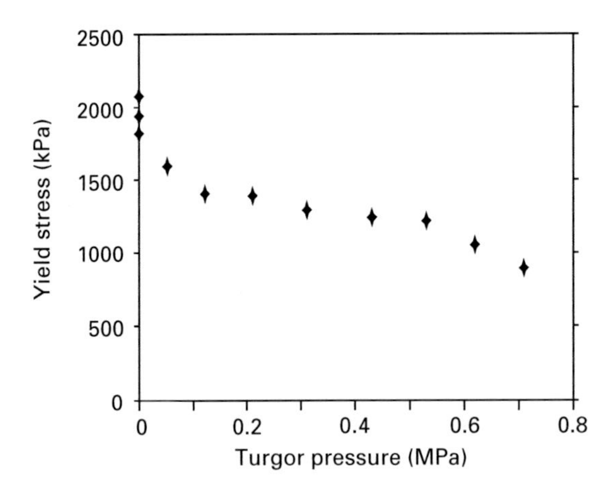

Fig. 5.19 The yield strength of potato parenchyma plotted against turgor pressure. (After Lin and Pitt, 1986, with permission of Wiley-Blackwell.)

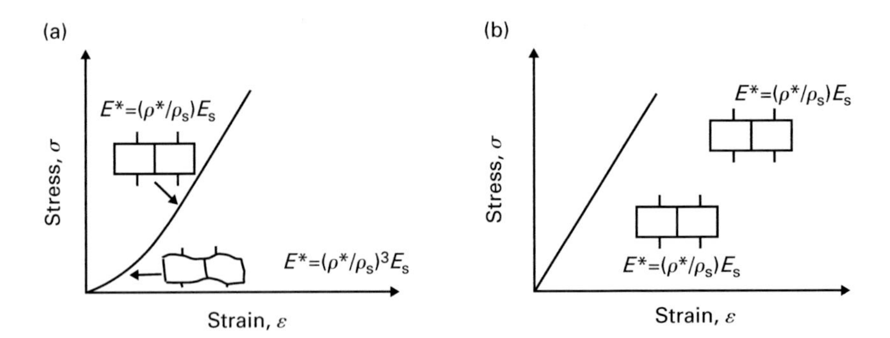

Fig. 5.20 (a) Schematic stress–strain curve for parenchyma at low initial turgor pressure. At low uniaxial strains, the curved cell walls bend. As the strain increases, the bent walls straighten out. Once straight, the walls resist further uniaxial strain primarily by stretching. (b) Schematic stress–strain curve for parenchyma at high initial turgor pressure. The deformation is stretching-dominated throughout.

As the initial turgor pressure in a specimen is increased, the initial curvature in the cell walls decreases, so that the transition from bending-dominated deformation to stretching-dominated deformation occurs at a lower uniaxial strain, decreasing the length of the initial "toe" region of the stress–strain curve (Fig. 5.17). Once the cell walls become taut and the modulus is dominated by cell-wall stretching, it is again roughly independent of the turgor pressure (assuming that the cell wall itself is linear-elastic): Kamiya *et al.* (1963) and Lin and Pitt (1986) find that this occurs at a turgor pressure of about 0.3–0.4 MPa (Fig. 5.18). Increasing the turgor pressure increases the tensile prestress in the cell walls of the tissue, decreasing the yield strength measured in a uniaxial test (Fig. 5.19).

The limited data on the dependence of the elastic moduli of parenchyma on density are plotted in Fig. 5.21. The data are from Vincent (1989), who tested fresh apple parenchyma in torsion. The densities lie between about 600 and 900 kg/m^3, well below

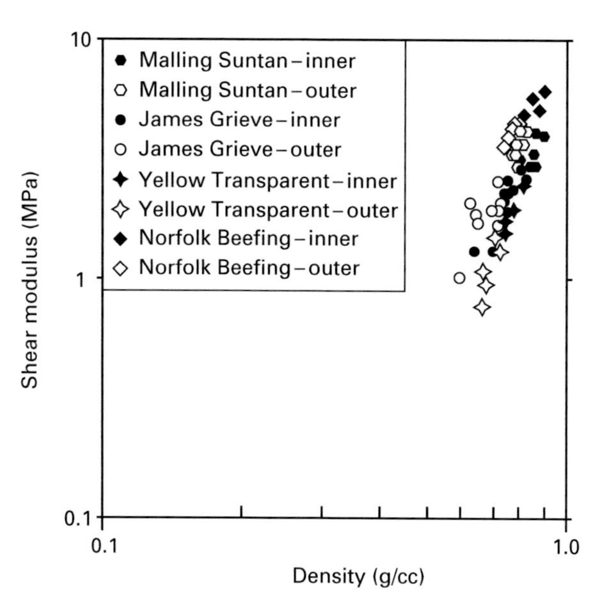

Fig. 5.21 The shear modulus of several cultivars of apple parenchyma plotted against density. Inner and outer refer to the radial location of the specimen in the apple. Data from Vincent (1989).

the values for the fluid (ρ_{water} = 1000 kg/m^3) and the solid cell wall ($\rho_s \sim$ 1300 kg/m^3), indicating that there is a significant volume fraction of air between the cells, making a direct comparison with models for cell-wall stretching difficult. Vincent suggests that the moduli depend both on the cell-wall deformation and the relative contact area between the cells. Additional data for the elastic moduli of parenchyma from apple, potato and carrot are listed in Table 5.3: they range from 0.31 to 14 MPa, with most falling within the range of about 0.5 to 6 MPa. The compressive strengths are in the range of 0.25 to 1.3 MPa.

Many types of long, thin monocotyledon leaves (e.g. iris, cattail) have a sandwich-beam like structure, with two faces reinforced by stiff, rib-like sclerenchyma separated by parenchyma in the core. And many plant stems have what botanists refer to as a "core-rind" structure, with the stiff, dense outer layer of epidermis and sclerenchyma (the "rind") supported by a layer of parenchyma cells (the "core"), which act as an elastic foundation, increasing the local buckling resistance of the stem. Both leaf sandwich structures and stem core-rind structures are analyzed in more detail in Chapter 6. The mechanical behavior of plant parenchyma is relevant in both cases.

5.4 Adipose tissue

Healthy stem and leaf cells of plants derive their stiffness in part from the intrinsic stiffness of the cellulose-based materials of the cell walls and in part from the osmotic turgor pressure of the liquid they contain. Adipose tissue (subcutaneous fat) of mammals has a somewhat similar structure: collagen-rich closed cells containing

Table 5.3 Elastic moduli and compressive strengths for parenchyma

Plant material	Young's modulus, E* or shear modulus, G* (MPa)	Compressive strength, σ_{comp}^* (MPa)	Reference
Apple	$E^* = 0.31–3.46$	0.66	Oye *et al.*, 2007
Apple	$E^* = 2.8–5.8$	0.25–0.37	Lin and Pitt, 1986
Apple	$G^* = 1–6$		Vincent, 1989
Potato	$E^* = 3.6$	1.3	Lin and Pitt, 1986
Potato	$E^* = 3.5$		Scanlon *et al.*, 1996
Potato	$E^* = 5.5$	0.27	Hiller and Jeronimides, 1996
Potato	$G^* = 0.5–1$		Scanlon *et al.*, 1996; 1998
Carrot	$E^* = 2–14$		Georget *et al.*, 2003

Data for fresh, wet tissue, at normal turgor.

a fluid-like lipid, which, in healthy tissue, is at slightly higher than atmospheric pressure.

Adipose tissue is a connective tissue made up of closed lipid-filled cells called adipocytes. The lipid is a mixture of triacylglycerols with a molecular weight of order 900 g/mol. Triacylglycerols are fatty acids esterified to glycerol. Each fatty acid is a carboxylic acid with a hydrocarbon tail; the number of carbons in the tail is reported as C_x. The most common of these are stearic (C_{18}-saturated), oleic (C_{18}-unsaturated) and palmitic (C_{16}-saturated); the proportions are species-dependent. The adipocyte cells are about 80 microns in diameter and occupy a volume fraction of about 0.1. The cell faces are a phospholipid bi-layer reinforced with a mesh of type IV collagen fibers called the basement membrane interwoven by a type I collagen fiber network called the interlobular septa containing blood vessels (Fig. 5.22). Between 60 and 80% of adipose tissue is lipid, some 20% is water and 2–3% is protein (Greenwood and Johnson, 1983; Samani, 2003). The histology of adipose tissue suggests that it is approximately equiaxed in structure. The large lipid content enforces incompressibility, meaning that the cell structure deforms, when compressed, at constant volume.

The response of adipose tissue to external loads is of considerable medical interest. The response to near-static loads is a factor in treating patients with impaired mobility. The behavior under high rates of strain is relevant for the design of needle-free injection systems and in understanding the consequences of exposure to impact or blast (Comley and Fleck 2009; and unpublished data). In the unloaded state the cells are almost equiaxed and the properties are isotropic. Compression distorts them into an oblate or a prolate shape. Comley and Fleck demonstrate that the volume change caused by compression up to 50% is very small – less than 1% – and that the permeability of the cells is low, meaning that they are closed. The lipid that fills the cells has a measured viscosity of 40 mPa sec at 37 °C; the authors demonstrate that this is sufficiently low that it simply behaves as an incompressible fluid, suggesting that the tissue behaves mechanically like an oil-filled closed-cell foam. A

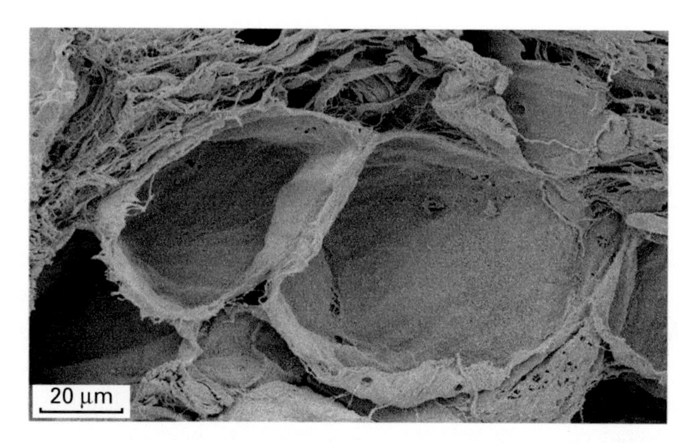

Fig. 5.22 Scanning electron micrograph of porcine type IV extracellular matrix of adipose tissue. The adipocytes fill the pouch-like areas of the extracellular matrix. (Reprinted from Comley and Fleck 2009, Fig. 2, with kind permission of Springer Science and Business Media.)

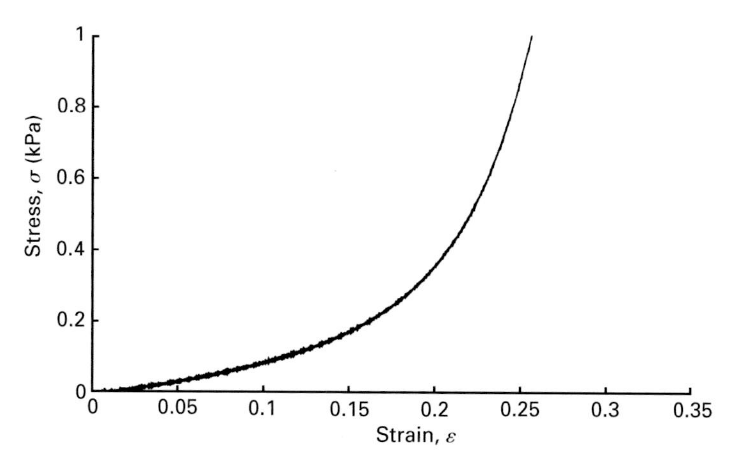

Fig. 5.23 Compressive stress–strain curve for adipose tissue tested at a strain rate of 0.002 s^{-1}. The shape of the curve is similar to that for parenchyma at low turgor pressures, for the same reason. (Reprinted from Comley and Fleck 2009, Fig. 3, with kind permission of Springer Science and Business Media.)

typical stress–strain curve for porcine adipose tissue loaded in uniaxial compression is shown in Fig. 5.23: the curve has the same shape as that for parenchyma at low turgor pressures, for the same reason.

5.5 Coral

Corals are colonial organisms made up of individual living polyps. In many corals, the polyps lift and reattach their bases periodically, creating a series of small chambers with a cellular structure (Fig. 5.24). The familiar tropical coral reefs are made up of hundreds of thousands of individual polyps and the mineral deposited beneath them.

The following description of the three main types – stony, horny and soft – is based on the books of Barnes (1987) and Levinton (2001).

The most common corals are the stony corals (Order Scleractinia) that produce an external calcium carbonate skeleton. Horny corals (Order Gorgonacea), on the other hand, have an internal skeleton, consisting of a central axial rod that is a composite of gorgonin (which is itself made up of protein and mucopolysaccharides) impregnated with calcium carbonate. The living polyps grow perpendicular to the axial rod. As a result of the axial skeleton, branched morphologies are common in horny corals. Whip corals, with their long cylindrical filaments, are only slightly branched while sea fan corals are branched in a plane, with interconnecting struts that form a cellular structure (Fig. 5.25). The branched morphology minimizes the surface area required to attach the coral to a substrate while maximizing the surface area available for feeding. In some species, there can be unmineralized sections along the central axial rod, allowing increased flexibility, especially in currents. The third main category of coral is the soft or leather corals (order Alcyonacea) that resemble the stony corals, except that their substrate is a fleshy matrix embedded with calcareous spicules.

Within the three main types of corals, there are many species, each with its own morphology, ranging from massive to branched to table to mushroom and elkhorn. The mechanical properties of the skeletons of only a few of these have been studied. Chamberlain (1978) measured the porosity and compressive response of three species of coral (*Acropora palmata*, *Siderastrea radians* and *Montastrea annularis*) under axisymmetric loading, with a radial stress equal to 0.1 MPa, equivalent to the hydrostatic

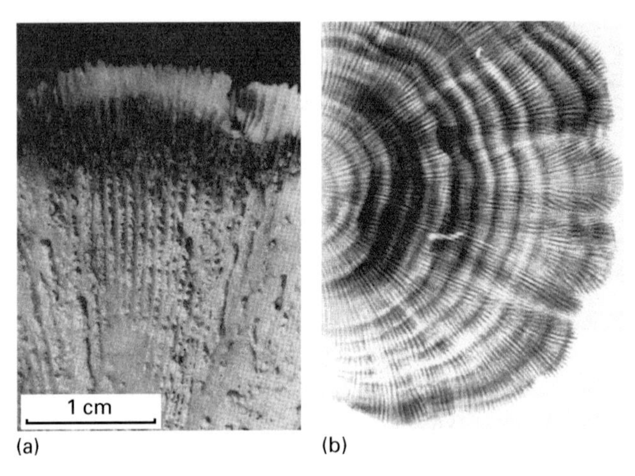

Fig. 5.24 (a) The porous structure of coral. The polyp lifts off the base, secrets calcium carbonate mineral and then reattaches, producing a porous structure. (b) Radiograph of a radial section of *Porites lobata*, showing annual growth rings, similar to those in trees. The dark bands correspond to low-density regions that form in winter while the light bands correspond to high-density regions that form in summer. (a, From Barnes/J&R Technical Services *Invertebrate Zoology*, 5E © 1987 Brooks/Cole, a part of Cengage Learning Inc. Reproduced by permission. www.cengage.com/permissions; b, reprinted from Isdale, 1977, reprinted with kind permission of the International Society for Reef Studies.)

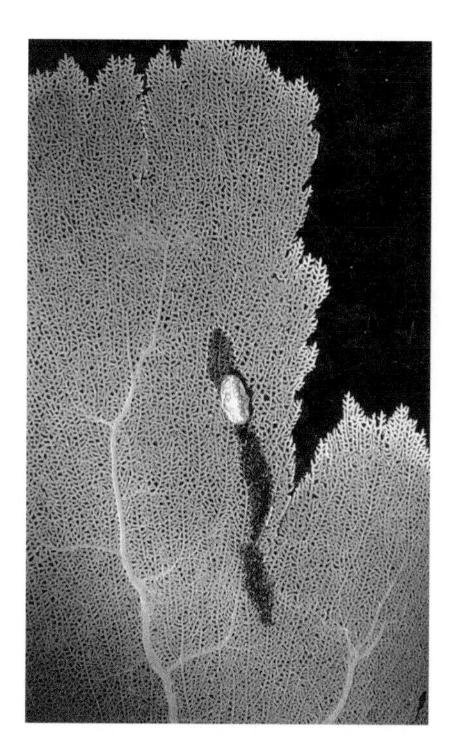

Fig. 5.25 A Gorgian sea fan coral, showing the interconnected network. (Reprinted from Plate XXVIII.3 from *Marine Biology: Function, Biodiversity and Ecology*, 2nd edn. by Levinton JS, 2001, by permission of Oxford University Press, Inc.)

pressure at a depth of about 10 m. His data are shown in Fig. 5.26. The Young's modulus varies roughly as the square of the relative density, suggesting that the solid struts in the coral deform by bending, like an open-cell foam, although the relative densities of the corals are high compared with most foams. The data for the compressive strength lie around a line of slope 2, although there is substantial scatter in the data. Both Chamberlain (1978) and Alvarez *et al.* (2002) find that the compressive strength of *Acropora palmata* is greater in the direction parallel to the polyp orientation than that perpendicular.

5.6 Sponge

Sponges belong to the phylum *Porifera*, from the Latin *porus* for pore and *ferre*, to bear; hence, pore-bearer. They have numerous pores and channels that connect to a larger, internal, central cavity. The bulk of the sponge is made up of the mesohyl, containing collagen fibrils and, in some cases, spongin, a fibrous protein similar to collagen. Some sponges also contain calcareous or siliceous spicules, which increase the stiffness of the sponge. The following description is based on the books of Barnes (1987) and Levinton (2001).

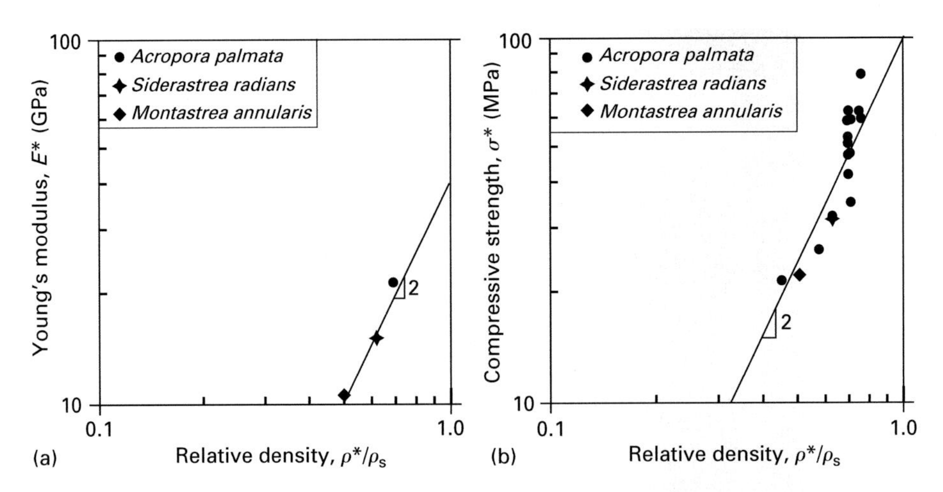

Fig. 5.26 (a) Young's modulus of several corals, plotted against relative density. (b) The compressive strength of coral plotted against relative density. (Data from Chamberlain, 1978.)

Sponges are among the simplest animals, lacking specialized tissues. Instead, they have modified cells in different areas that perform specific functions. The pores and channels are covered with flagellated collar cells, or choanocytes, that move water from the exterior of the sponge, through the channels and into the central cavity by beating their flagella. As the water moves past the collar cells, they consume extremely fine particulate organic matter in the water. The water then exits through a large opening called the osculum. Cells at the base of the sponge secrete an adhesive that attaches the sponge to a substrate such as a rock or coral.

There are over 5000 species of sponges; the vast majority (over 90%) are in the class Demospongiae. The skeletons of sponges in this class may have siliceous spicules or spongin or both. Sponges in the class Calcarea have, as the name suggests, calcium carbonate spicules. A third class, Hexactinellida, have six-pointed spicules that in some species fuse to form a framework; one of the most beautiful examples of this is the Venus' flower basket sponge (*Euplectella aspergillum*) (Fig. 5.27). Because the long spicules are siliceous, they are also known as glass sponges. Hexactinellida sponges typically live in deep waters, between 200 and 1000 m.

The structure of the spicule framework of *Euplectella aspergillum* has been described at various length scales (Aizenberg *et al.*, 2005; Weaver *et al.*, 2007). At the macroscopic scale, it is made up of a square network of struts, with diagonals bracing every other square (Fig. 5.27b). The network satisfies the Maxwell condition (3.37), so that the struts are loaded in either tension or compression, rather than bending, giving a stable, highly efficient structure. In addition, external ridges wind helically around the outside of the cylindrical tube at an angle of 45°, possibly reducing ovalization of the tube in bending. (Ovalization reduces the moment of inertia of a tube, reducing its flexural rigidity and resistance to local buckling; a more detailed discussion of ovalization is given in Sections 6.3 and 6.4.) The spicules themselves are made up of bundles of silica fibers, each of which has thick, concentric layers of mineral, separated by thin interlayers of organic material; the organic interlayers are thought to increase the resistance to crack propagation. Finally,

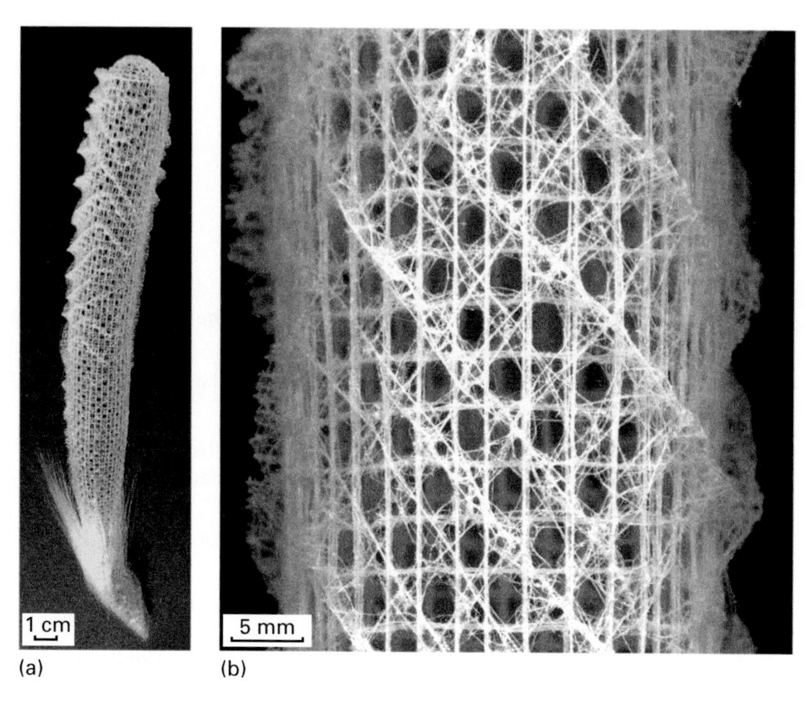

Fig. 5.27 (a) Photograph of Venus' flower basket (*Euplectella aspergillum*). (b) Higher magnification image, showing diagonal bracing on every other square frame, as well as the outer helical ridges. (From Aizenberg *et al.*, 2005, Skeleton of *Euplectella* sp: Structural hierarchy from the nanoscale to the macroscale, *Science* **309**, 275–8, reprinted with permission from AAAS.)

at the smallest scale, the mineral layers are made up of silica nanoparticles, 50–200 nm in diameter, arranged around an axially oriented organic filament. The anchor spicules of *Euplectella aspergillum*, which tie the sponge to the seabed, also have exceptional properties: excellent resistance to crack growth and fracture (Aizenberg *et al.*, 2004; Woesz *et al.*, 2006; Miserez *et al.*, 2008) and, perhaps most remarkably, fiber-optic characteristics similar to those of commercial optical fibers (Aizenberg *et al.*, 2004).

5.7 Summary

The mechanical behavior of foam-like tissues in plants and animals can be understood in terms of the models for cellular solids developed in Chapter 3. The properties depend on the relative density of the foam-like tissue, the properties of the cell-wall material (which may itself have a hierarchical structure), the cell geometry and any internal pressure within closed cells. Trabecular bone is best described as an open-cell foam that deforms and fails primarily by bending, buckling (in compression) and yield (in tension). The stress–strain response of plant parenchyma and adipose tissue can be understood in terms of fluid-filled closed-cell foams. The little data available on the mechanical properties of coral indicate that it, too, behaves like open-cell foam. The diagonal spicules of the sponge, *Euplectella aspergillum*, give a two-dimensional network that satisfies Maxwell's condition, so that it deforms by axial stretching or compression of the members, rather than bending, giving a highly efficient structure.

References

Trabecular bone

Arthur Moore TL and Gibson LJ (2002) Microdamage accumulation in bovine trabecular bone in uniaxial compression. *J. Biomech. Eng.* **124**, 63–71.

Ashman RB and Rho JY (1988) Elastic modulus of trabecular bone material. *J. Biomech.* **21**, 177–81.

Bayraktar HH, Morgan EF, Niebur GL, Morris GE, Wong EK and Keaveny TM (2004) Comparison of the elastic and yield properties of human femoral trabecular and cortical bone tissue. *J. Biomech.* **37**, 27–35.

Behrens JC, Walker PS and Shoji H (1974) Variations in strength and structure of cancellous bone at the knee. *J. Biomech.* **7**, 201–7.

Bensusuan JS, Davy DT, Heiple KG and Verdin PJ (1983) Tensile, compressive and torsional testing of cancellous bone. *19th Annual Orthopaedic Research Society Meeting* **8**, 132.

Biewener AA, Fazzalari NL, Konieczynski DD and Baudinette RV (1996) Adaptive changes in trabecular architecture in relation to functional strain patterns and disuse. *Bone* **19**, 1–8.

Bowman SM, Guo XE, Cheng DW, Keaveny TM, Gibson LJ, Hayes WC and McMahon TA (1998) Creep contributes to the fatigue behavior of bovine trabecular bone. *J. Biomech. Eng.* **120**, 647–54.

Bowman SM, Gibson LJ, Hayes WC and McMahon TA (1999) Results from demineralized bone creep tests suggest that collagen is responsible for the creep behavior of bone. *J. Biomech. Eng.* **121**, 253–8.

Bredbenner TL and Davy DT (2006) The effect of damage on the viscoelastic behavior of human vertebral trabecular bone. *J. Biomech. Eng.* **128**, 473–80.

Caler WE and Carter DR (1989) Bone creep-fatigue damage accumulation. *J. Biomech.* **22**, 625–35.

Carter DR and Beaupre GS (2001) *Skeletal Function and Form: Mechanobiology of Skeletal Development, Aging and Regeneration*. Cambridge: Cambridge University Press.

Carter DR and Hayes WC (1977) Compressive behavior of bone as a two-phase porous structure. *Bone Joint Surg.* **59A**, 954–62.

Carter DR, Schwab GH and Spengler DM (1980) Tensile fracture of cancellous bone. *Acta Orthopaed. Scan.*, **51**, 733–41.

Chang WCW, Christensen TM, Pinilla TP and Keaveny TM (1999) Dependence of yield strain on anatomic site for bovine trabecular bone. *J. Orthop. Res.* **17**, 582–5.

Chevalier Y, Pahr D, Allmer H, Charlebois M and Zysset P (2007) Validation of a voxel-based FE method for prediction of the uniaxial apparent modulus of human trabecular bone using macroscopic mechanical tests and nanoindentation. *J. Biomech.* **40**, 3333–40.

Choi K, Kuhn JL, Ciarelli MJ and Goldstein SA (1990) The elastic moduli of human subchondral trabecular and cortical bone tissue and the size-dependency of cortical bone modulus. *J. Biomech.* **23**, 1103–13.

Cowin SC (1985) The relationship between the elasticity tensor and the fabric tensor. *Mech. Mat.* **4**, 137–47.

Cowin SC (1986) Wolff's law of trabecular architecture at remodeling equilibrium. *J. Biomech. Eng.* **108**, 83–8.

Currey JD (2002) *Bones: Structure and Mechanics*. Princeton, NJ: Princeton University Press.

Deligianni DD, Maris A and Missirlis YF (1994) Stress relaxation behaviour of trabecular bone specimens. *J. Biomech.* **27**, 1469–76.

Dendorfer S, Maier HJ, Taylor D and Hammer J (2008) Anisotropy of the fatigue behaviour of cancellous bone. *J. Biomech.* **41**, 636–41.

Evans FG and Lebow M (1951) Regional differences in some of the physical properties of the human femur. *J. Appl Physiol.* **3**, 563–72.

Fondrk M, Bahniuk E, Davy DT and Michaels C (1988) Some viscoplastic characteristics of bovine and human cortical tissue. *J. Biomech.* **21**, 623–30.

Galante J, Rostoker W and Ray RD (1970) Physical properties of trabecular bone. *Calc. Tiss. Res.* **5**, 236–46.

Gibson LJ (1985) Mechanical behavior of cancellous bone. *J. Biomech.* **18**, 317–28.

Greenspan SL, Myers ER, Kiel DP, Parker RA, Hayes WC and Resnick NM (1998) Fall direction, bone mineral density and function: risk factors for hip fracture in frail nursing home elderly. *Am. J. Medicine* **104**, 539–45.

Guo XE, McMahon TA, Keaveny TM, Hayes WC and Gibson LJ (1994) Finite element modeling of damage accumulation in trabecular bone under cyclic loading. *J. Biomech.* **27**, 145–55.

Guo XE and Kim CH (1999) Effects of age-related bone loss: a 3D microstructural simulation. *Proc. 1999 Bioengineering Conference*, ASME BED-Vol. **42**, 327–8.

Guo XE and Kim CH (2002) Mechanical consequences of trabecular bone loss and its treatment: a three dimensional model simulation. *Bone* **30**, 404–11.

Haddock SM, Yeh OC, Mummaneni PV, Rosenberg WS and Keaveny TM (2004) Similarity in the fatigue behavior of trabecular bone across sites and species. *J. Biomech.* **37**, 181–7.

Harrigan T and Mann R (1984) Characterization of microstructural anisotropy in orthotropic materials using a second rank tensor. *J. Mat. Sci.* **19**, 761–7.

Hayes WC and Carter DR (1976) Postyield behavior of subchondral trabecular bone. *J. Biomed. Mat. Res. Symp.* **7**, 537–44.

Hayes WC and Snyder B (1981) Toward a quantitative formulation of Wolff's Law in trabecular bone. In *Mechanical Properties of Bone*, AMD, vol. 45, ed. Cowin SC. American Society of Mechanical Engineers, pp. 43–80.

Huber AT and Gibson LJ (1988) Anisotropy of foams. *J. Mat. Sci.* **23**, 3031–40.

Hvid I, Bentzen SM, Linde F, Mosekilde L and Pongsoipetch B (1989) X-ray quantitative computed tomography: the relations to physical properties of proximal tibial trabecular bone specimens. *J. Biomech.* **22**, 837–44.

Jensen NC, Madsen LP and Linde F (1991) Topographical distribution of trabecular bone strength in the os calcanei. *J. Biomech.* **24**, 49–55.

Jorgensen CS and Kundu T (2002) Measurement of material elastic constants of trabecular bone: a micromechanical analytic study using a 1 GHz acoustic microscope. *J. Orthopaedic Res.* **20**, 151–8.

Kabel J, van Rietbergen B, Odgaard A and Huiskes R (1999) Constitutive relationships of fabric, density and elastic properties in cancellous bone architecture. *Bone* **25**, 481–6.

Kaplan SJ, Hayes WC and Stone JL (1985) Tensile strengthofbovine trabecular bone. *J. Biomech.* **18**, 723–7.

Keaveny TM, Wachtel EF, Ford CM and Hayes WC (1994) Differences between the tensile and compressive strengths of bovine tibial trabecular bone depend on modulus. *J. Biomech.* **27**, 1137–46.

Keaveny TM, Morgan EF, Niebur GL and Yeh OC (2001) Biomechanics of trabecular bone. *Ann. Rev. Biomed. Eng.* **3**, 307–33.

Knets IV and Vilks YK (1975) Creep of compact human bony tissue under tension (translation). *Mekhanika Polimerov* **4**, 634–8.

Kopperdahl DL and Keaveny TM (1998) Yield strain behavior in trabecular bone. *J. Biomech* **31**, 601–8.

Kosmopoulos V, Schizas C and Keller TS (2008) Modeling the onset and propagation of trabecular bone microdamage during low-cycle fatigue. *J. Biomech.* **41**, 515–22.

Lanyon LE (1974) Experimental support for the trajectorial theory of bone structure. *J. Bone Joint Surg.* **56B**, 160–6.

Linde F, Norgaard P, Hvid I, Odgaard A and Soballe K (1991) Mechanical properties of trabecular bone: dependency on strain rate. *J. Biomech.* **24**, 803–9.

Makiyama AM, Vajjhala S and Gibson LJ (2002) Analysis of crack growth in a 3D Voronoi structure: a model for fatigue in low density trabecular bone. *J. Biomech. Eng.* **124**, 512–20.

Martin RB, Burr DB and Sharkey NA (1998) *Skeletal Tissue Mechanics*. New York: Springer.

McElhaney JH, Fogle JL, Melvin JW, Haynes RR, Roberts VL and Alem NM (1970a) Mechanical properties of cranial bone. *J. Biomech.* **3**, 495–6.

McElhaney JH, Alem N and Roberts V (1970b) *A Porous Block Model for Cancellous Bone*. ASME publication 70-WA/BHF-2. Presented at the ASME Winter Annual Meeting, Biomechanical and Human Factors Division, Nov 29–Dec 3, 1970, New York, NY.

Mente P and Lewis J (1987) Young's modulus of trabecular bone tissue. *Trans Orthop. Res. Soc.* **12**, 49.

Michel MC, Guo XE, Gibson LJ, McMahon TA and Hayes WC (1993) Compressive fatigue behavior of bovine trabecular bone. *J. Biomech.* **26**, 453–63.

Moore TLA and Gibson LJ (2003a) Fatigue of bovine trabecular bone. *J. Biomech. Eng.* **125**, 761–8.

Moore TLA and Gibson LJ (2003b) Fatigue microdamage in bovine trabecular bone. *J. Biomech. Eng.* **125**, 769–76.

Moore TLA, O'Brien FJ and Gibson LJ (2004) Creep does not contribute to fatigue of bovine trabecular bone. *J. Biomech. Eng.* **126**, 321–9.

Morgan EF and Keaveny TM (2001) Dependence of yield strain of human trabecular bone on anatomic site. *J. Biomech.* **34**, 569–77.

Moskilde L (1989) Sex differences in age-related loss of vertebral trabecular bone mass and structure – biomechanical consequences. *Bone* **10**, 425–32.

Muller R, Gerber SC and Hayes WC (1998) Micro-compression: a novel technique for the nondestructive assessment of bone failure. *Technology and Health Care* **6**, 433–44.

Myers ER and Wilson SR (1997) Biomechanics of osteoporosis and vertebral fracture. *Spine* **22**, 25S–31S.

Nagaraja S, Ball MD and Guldberg RE (2007) Time-dependent damage accumulation under stress relaxation testing of bovine trabecular bone. *Int. J. Fatigue* **29**, 1034–8.

Narzarian A and Muller R (2004) Time-lapsed microstructural imaging of bone failure behavior. *J. Biomech.* **37**, 55–65.

Niebur GL, Feldstein MJ, Yuen JC, Chen TJ and Keaveny TM (2000) High-resolution finite element models with tissue strength asymmetry accurately predict failure of trabecular bone. *J. Biomech.* **33**, 1575–83.

Parfitt (1992) Implications of architecture for the pathogenesis and prevention of vertebral fracture. *Bone* **13**, S41–7.

Pugh JW, Rose RM and Radin EL (1973a) Elastic and viscoelastic properties of trabecular bone: dependence on structure. *J. Biomech.*, **6**, 475–85.

Pugh JW, Rose RM and Radin EL (1973b) Structural model for the mechanical behavior of trabecular bone. *J. Biomech.* **6**, 657–70.

Quaglini V, La Russa V and Corneo S (2009) Nonlinear stress relaxation of trabecular bone. *Mech. Res. Comm.* **36**, 275–83.

Rapillard L, Charlebois M and Zysset PK (2006) Compressive fatigue behavior of human vertebral trabecular bone. *J. Biomech.* **39**, 2133–9.

Rho JY, Ashman RB and Turner CH (1993) Young's moduli of trabecular and cortical bone material – ultrasonic and microtensile measurements. *J. Biomech.* **26**, 111–19.

Rho JY, Tsui TY and Pharr GM (1997) Elastic properties of human cortical and trabecular lamellar bone measured by nanoindentation. *Biomaterials* **18**, 1325–30.

Rho JY and Pharr GM (1999) Effects of drying on the mechanical properties of bovine femur measured by nanoindentation. *J. Mat. Sci. Mat. Med.* **10**, 485–8.

Rho JY, Roy ME, Tsui Ty and Pharr GM (1999) Elastic properties of microstructural components of human bone tissue as measured by nanoindentation. *J. Biomed. Mat. Res.* **45**, 48–54.

Rice JC, Cowin SC and Bowman JA (1988) On the dependence of the elasticity and strength of cancellous bone on apparent density. *J. Biomech.* **21**, 155–68.

Rimnac CM, Petko AA, Santner TJ and Wright TM (1993) The effect of temperature, stress and microstructure on the creep of compact bovine bone. *J. Biomech.* **26**, 219–28.

Roy ME, Rho JY, Tsui TY, Evans ND and Pharr GM (1999) Mechanical and morphological variation of the human lumbar vertebral cortical and trabecular bone. *J. Biomed. Mat. Res.* **44**, 191–7.

Runkle J and Pugh J (1975) The micromechanics of cancellous bone III: determination of the elastic modulus of individual trabeculae by a buckling analysis. *Bull. Hosp. Jt. Dis.* **36**, 2–10.

Ryan SD and Williams JL (1989) Tensile testing of rodlike trabeculae excised from bovine femoral condyle. *J. Biomech.* **22**, 351–5.

Sedlin ED (1965) A rheological model for cortical bone. *Acta Orthop. Scand.* Suppl. **83**, 1–77.

Silva MJ and Gibson LJ (1997a) Modelling the mechanical behavior of vertebral trabecular bone: effects of age-related changes in microstructure. *Bone* **21**, 191–9.

Silva MJ and Gibson LJ (1997b) The effects of non-periodic microstructure and defects on the compressive strength of two-dimensional cellular solids. *Int. J. Mech. Sci.* **39**, 549–63.

Smith JW and Walmsley R (1959) Factors affecting the elasticity of bone. *J. Anat.* **93**, 503–23.

Stone JL, Beaupre GS and Hayes WC (1983) Multiaxial strength characteristics of trabecular bone. *J. Biomech.*, **16**, 743–52.

Taylor M, Cotton J and Zioupos P (2002) Finite element simulation of the fatigue of cancellous bone. *Meccanica* **37**, 419–29.

Thurner PJ, Erickson B, Schriock Z *et al.* (2006a) High-speed photography of the development of microdamage in trabecular bone during compression. *J. Mat. Res.* **21**, 1093–100.

Thurner PJ, Wyss P, Voide R, Stauber M, Stampanoni M, Sennhauser U and Muller R (2006b) Time-lapsed investigation of three-dimensional failure and damage accumulation in trabecular bone using synchrotron light. *Bone* **39**, 289–99.

Townsend PR, Rose RM and Radin EL (1975a) Buckling studies of single human trabeculae. *J. Biomech.* **8**, 199–201.

Townsend PR, Raux P, Rose RM, Miegel RE and Radin EL (1975b) Distribution and anisotropy of the stiffness of cancellous bone in the human patella. *J. Biomech.*, **8**, 363–7.

Turner CH and Cowin SC (1988) Errors introduced by off-axis measurements of the elastic properties of bone. *J. Biomech.* **110**, 213–14.

Turner CH (1989) Yield behavior of bovine cancellous bone. *J. Biomech. Eng.* **111**, 256–60.

Turner CH, Cowin SC, Rho JY, Ashman RB and Rice JC (1990) The fabric dependence of the orthotropic elastic constants of cancellous bone. *J. Biomech.* **23**, 549–61.

Turner CH, Rho J, Takano Y, Tsui TY, Pharr GM (1999) The elastic properties of trabecular and cortical bone tissue are similar: results from two microscopic measurement techniques. *J. Biomech.* **32**, 437–41.

Underwood EE (1970) *Quantitative Stereology*. Addison-Wesley.

Vajjhala S, Kraynik AM and Gibson LJ (2000) A cellular solid model for modulus reduction due to resorption of trabeculae in bone. *J. Biomech. Eng.* **122**, 511–15.

Van Reitbergen B, Weinans H, Huiskes R and Odgaard A (1995) A new method to determine trabecular bone elastic properties and loading using micromechanical finite-element models. *J. Biomech.* **28**, 69–81.

Weaver JK and Chalmers J (1966) Cancellous bone: its strength and changes with aging and an evaluation of some methods for measuring its mineral content. *J. Bone Joint Surg.* **48A**, 289–308.

Whitehouse WJ (1974) The quantitative morphology of anisotropic trabecular bone. *J. Microscopy* **2**, 153–68.

Williams JL and Lewis JL (1982) Properties and an anisotropic model of cancellous bone from the proximal tibial epiphysis. *J. Biomech. Eng.* **104**, 50–6.

Yamamoto E, Crawford RP, Chan DD and Keaveny TM (2006) Development of residual strains in human vertebral trabecular bone after prolonged static and cyclic loading at low load levels. *J. Biomech.* **39**, 1812–18.

Yang G, Kabel J, van Rietbergen B, Odgaard A, Huiskes R and Cowin SC (1999) The anisotropic Hooke's law for cancellous bone and wood. *J. Elasticity* **53**, 125–46.

Zysset PK, Guo XE, Hoffler CE, Moore KE and Goldstein SA (1999) Elastic modulus and hardness of cortical and trabecular bone lamellae measured by nanoindentation in the human femur. *J. Biomech.* **32**, 1005–12.

Zysset PK (2003) A review of morphology-elasticity relationships in human trabecular bone: theories and experiments. *J. Biomech.* **36**, 1469–85.

Plant parenchyma

Georget DMR, Smith AC and Waldron DW (2003) Modelling of carrot tissue as a fluid-filled foam. *J. Mat. Sci.* **38**, 1933–8.

Hiller S and Jeronimidis G (1996) Fracture in potato tuber parenchyma. *J. Mat. Sci.* **31**, 2779–96.

Kamiya N, Tazawa M and Takata T (1963) The relation of turgor pressure to cell volume in Nitella with special reference to mechanical properties of the cell wall. *Protoplasma* **57**, 501–21.

Lin T-T and Pitt RE (1986) Rheology of apple and potato tissue as affected by cell turgor pressure. *J. Texture Studies* **17**, 291–313.

Niklas KJ (1989) Mechanical behavior of plant tissues as inferred from the theory of pressurized cellular solids. *Amer. J. Botany* **76**, 929–37.

Niklas KJ (1992) *Plant Biomechanics: An Engineering Approach to Plant Form and Function*. Chicago, IL: The University of Chicago Press.

Oye ML, Vanstreels E, De Baerdemaeker J, *et al.* (2007) Effect of turgor on micromechanical and structural properties of apple tissue: a quantitative analysis. *Postharvest Biol. Technol.* **44**, 240–7.

Scanlon MG, Pang CH and Biliaderis CG (1996) The effect of osmotic adjustment on the mechanical properties of potato parenchyma. *Food Res. Int.* **29**, 481–8.

Scanlon MG, Day AJ and Povey MJW (1998) Shear stiffness and density in potato parenchyma. *Int. J. Food Sci. Tech.* **33**, 461–4.

Vincent JFV (1989) Relationship between density and stiffness of apple flesh. *J. Sci. Food Agr.* **47**, 443–62.

Warner M, Thiel BL and Donald AM (2000) The elasticity and failure of fluid-filled cellular solids: theory and experiment. *Proc. Natl. Acad. Sci.* **97**, 1370–5.

Adipose tissue

Comley K and Fleck NA (2009) The high strain rate response of adipose tissue. *IUTAM Symposium on Mechanical Properties of Cellular Materials* **12**, 27–33.

Comley K and Fleck NA (2010) The mechanical response of porcine adipose tissue, *J. Biomech. Eng.*, in press.

Greenwood MRC and Johnson PR (1983) The adipose tissue. In *Histology, Cell and Tissue Biology,* 5th edn., ed. Weiss L. London: Macmillan.

Samani A (2003) Measuring the elastic modulus of ex-vivo small tissue samples. *Phys. Med. Biol.* **48**, 2183–98.

Coral

Alvarez K, Camero S, Alarcon ME, Rivas A, Gonzalez G (2002) Physical and mechanical properties evaluation of *Acropora palmata* coralline species for bone substitution applications. *J. Mat. Sci.: Materials in Medicine* **13**, 509–15.

Barnes RD (1987) *Invertebrate Zoology*, 5th edn. New York: CBS College Publishing.

Chamberlain JA (1978) Mechanical properties of coral skeleton: compressive strength and adaptive significance. *Paleobiology* **4**, 419–35.

Isdale P (1977) Variation in growth rate of hermatypic corals in a uniform environment. *Proc. 3rd Int. Coral Reef Symposium*, University of Miami, May 1977, pp. 403–8.

Levinton JS (2001) *Marine Biology: Function Biodiversity and Ecology*, 2nd edn. Oxford: Oxford University Press.

Sponge

Aizenberg J, Sundar VC, Yablon AD, Weaver JC and Chen G (2004) Biological glass fibers: correlation between optical and structural properties. *Proc. Nat. Acad. Sci.* **101**, 3358–63.

Aizenberg J, Weaver JC, Thanawala MS, Sundar VC, Morse DE and Fratzl P (2005) Skeleton of *Euplectella* sp: structural hierarchy from the nanoscale to the macroscale. *Science* **309**, 275–8.

Barnes RD (1987) *Invertebrate Zoology*, 5th edn. New York: CBS College Publishing.

Levinton JS (2001) *Marine Biology: Function Biodiversity and Ecology*, 2nd edn. Oxford: Oxford University Press.

Miserez A, Weaver JC, Thurner PJ *et al.* (2008) Effects of laminate architecture on fracture resistance of sponge biosilica: lessons from nature. *Adv. Functional Mat.* **18**, 1241–8.

Weaver JC, Aizenberg J, Fantner GE *et al.* (2007) Hierarchical assembly of the siliceous skeletal lattice of the hexactinellid sponge *Euplectella aspergillum*. *J. Struct. Biol.* **158**, 93–106.

Woesz A, Weaver JC, Kazanci M *et al.* (2006) Micromechanical properties of biological silica in skeletons of deep-sea sponges. *J. Mat. Res.* **21**, 2068–78.

6 Cellular structures in nature

6.1 Introduction

In nature, structural elements often combine cellular materials with a nearly fully dense component to deliver increased mechanical performance for a given mass. Examples include bird skulls, with a sandwich structure in which a low-density core of trabecular bone separates nearly fully dense faces of cortical bone; bamboo, with a radial density gradient; and bird feather quills, with a fully dense, roughly cylindrical shell filled with a foam-like core (Fig. 6.1). In some cases, the solid phase is the same throughout the structure (e.g. in bone sandwich structures), while in others the solid phase varies (e.g. palm stems, which have increasing cellulose content and higher solid Young's modulus towards the periphery). Here, we describe the structure and mechanical behavior of these natural structures. The arrangement of the cellular and solid components typically confers a mechanical advantage: for each configuration, we give an estimate of the increased mechanical performance, compared with a similar component with a uniform distribution of the identical mass.

6.2 Sandwich structures

6.2.1 Leaves of monocotyledon plants

Monocotyledon plants, such as irises, cattails (called bulrushes in the UK) and grasses, have long, narrow leaves with parallel veins. In the case of iris and cattails, the leaves grow directly out of the ground, rather than from a stem. The leaf must provide its own structural support as well as a large surface area for photosynthesis. The leaves of many monocotyledon plants achieve this with a sandwich-like structure, with stiff outer faces separated by a lightweight core, giving increased flexural rigidity with little increase in weight. The sandwich structures of leaves from several monocotyledon plants (iris, cattail and two grasses *Lolium perenne*, or ryegrass, and *Stipa gigantea*, or giant feather grass) are shown in Fig. 6.2. The iris leaf has regularly spaced, nearly fully dense ribs, made up of sclerenchyma cells, running along the length of the outer surfaces of the leaf; the ribs are separated by a foam-like core of parenchyma cells (Fig. 6.2a). The cattail has more uniformly dense faces separated by a core that combines parenchyma cells and vertical ribs (Fig. 6.2b). Grass leaves, too, often have a sandwich structure, although the leaf blades typically have a V-shaped cross-section: the cross-section of

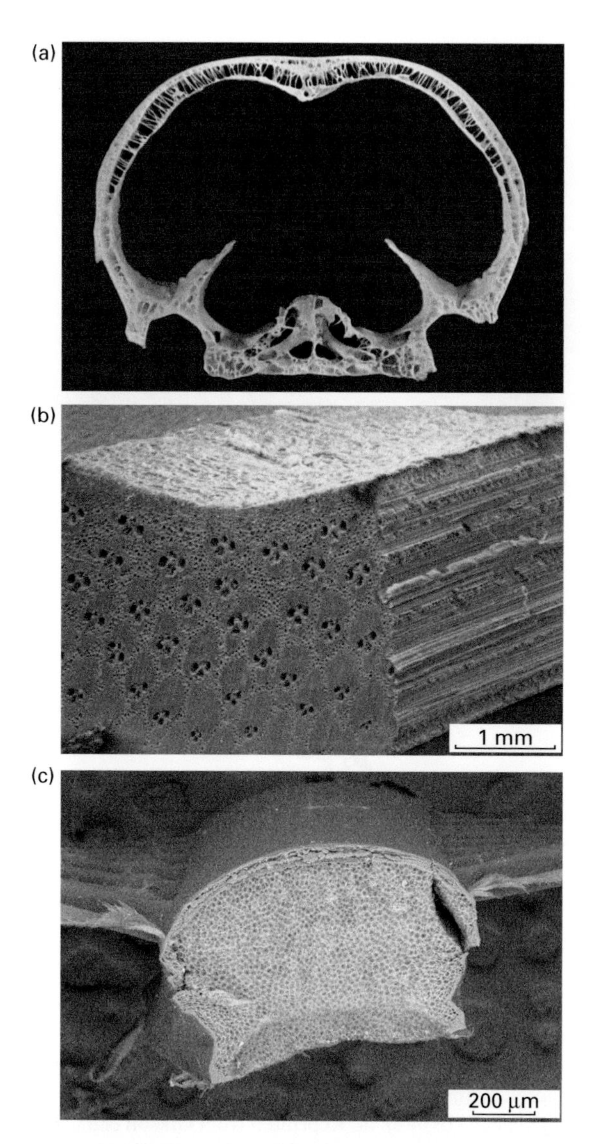

Fig. 6.1 (a) Carrion crow (*Corvus corone*) skull, showing the sandwich structure, with two faces of dense cortical bone separated by a lightweight core of trabecular bone. (b) Scanning electron micrograph of bamboo, showing the radial density gradient. (c) Scanning electron micrograph of blue jay feather rachis (or quill). (a, Reproduced from Buhler, 1972; b, reproduced from Gibson *et al.*, 1995).

the ryegrass, *Lolium perenne*, resembles that of the iris while that of the giant feather grass, *Stipa gigantea*, is more similar to that of the bulrush (Fig. 6.2c,d). The sandwich behavior of the leaves of iris (Gibson *et al.*, 1988) and of corn (*Zea mays L.*) (Moulia and Fournier, 1997) have been studied in detail, as we describe below.

Figure 6.3 shows longitudinal sections of the outer skin and the core of the iris leaf. The outer skin is made up of dense ribs connected by a single layer of cells of roughly square transverse section. Jointly they act like a fibre-reinforced sheet.

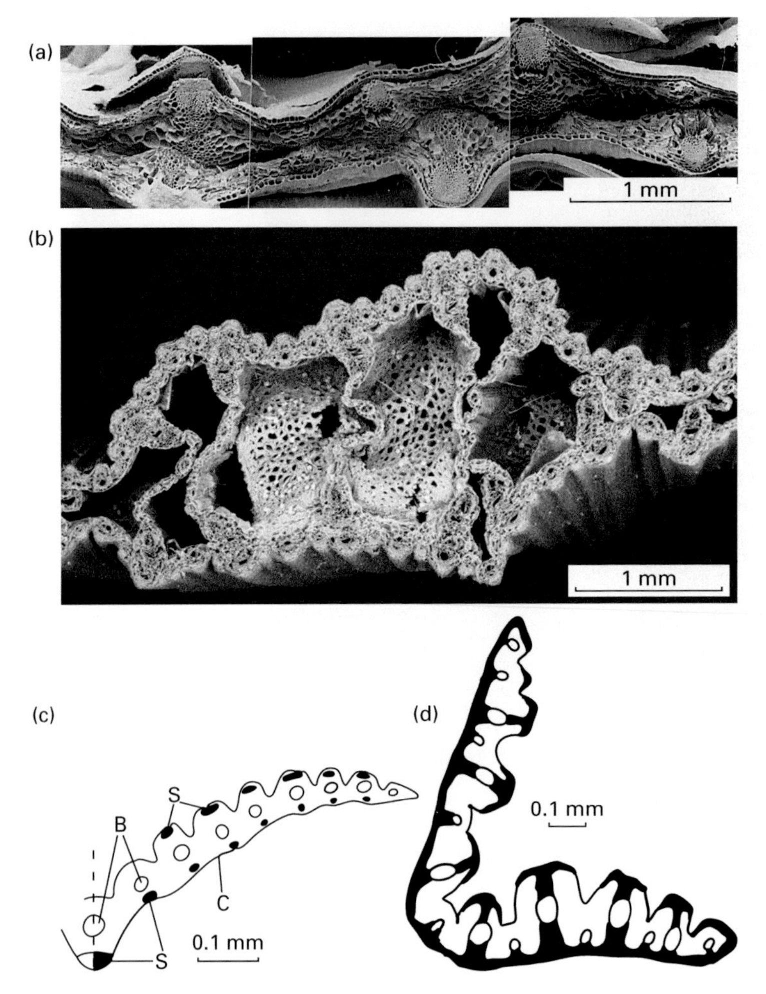

Fig. 6.2 Scanning electron micrographs of transverse sections of (a) an iris leaf and (b) a cattail leaf. Schematic drawings of transverse sections of (c) *Lolium perenne* and (d) *Stipa gigantea*, showing the stiff, dense sclerenchyma cells (S, black), the vascular bundles (B, ovals), the cuticle (C) and the parenchyma cells (white). (a, Reprinted from Gibson *et al.*, 1988, Fig. 2b, with kind permission from Springer Science+Business Media; c, reprinted from Vincent 1982, Fig. 1, with kind permission from Springer Science+Business Media; d, reprinted from Vincent 1991, Fig. 2, with kind permission from Springer Science+Business Media.)

The cells in the interior of the leaf differ, in that they are roughly equiaxed in the transverse section (Fig. 6.2a) and somewhat elongated in the longitudinal section (Fig. 6.3c). They have thin cell walls and form the low-density foam-like "core" of the leaf. A schematic drawing of a transverse section of the leaf is shown in Fig. 6.4. Typical dimensions are listed in Table 6.1. Most remain roughly constant along the length of the leaf; only the core thickness changes, tapering from about 6 mm at the base to 0.5 mm at the tip.

If the iris leaf, at constant, normal turgor pressure, derives its resistance to drooping from its sandwich-like construction it should be possible to calculate its stiffness, at

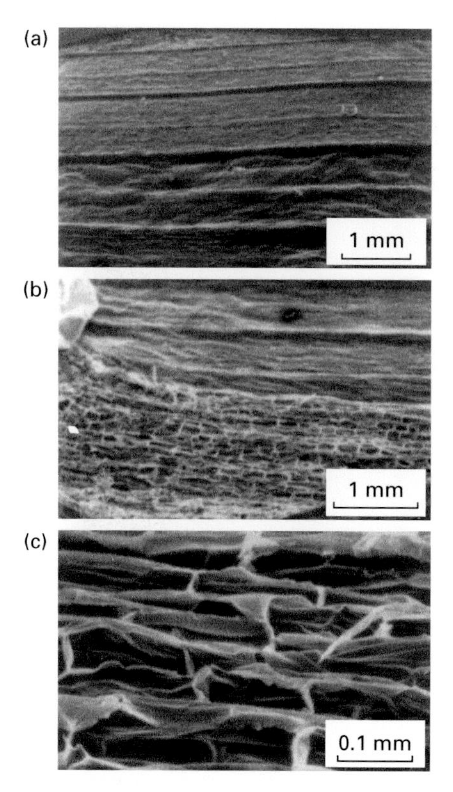

Fig. 6.3 Scanning electron micrographs of longitudinal sections of (a) the outer face, and (b,c) the inner core of an iris leaf. (Reprinted from Gibson *et al.*, 1988, Fig. 3, with kind permission from Springer Science+Business Media.)

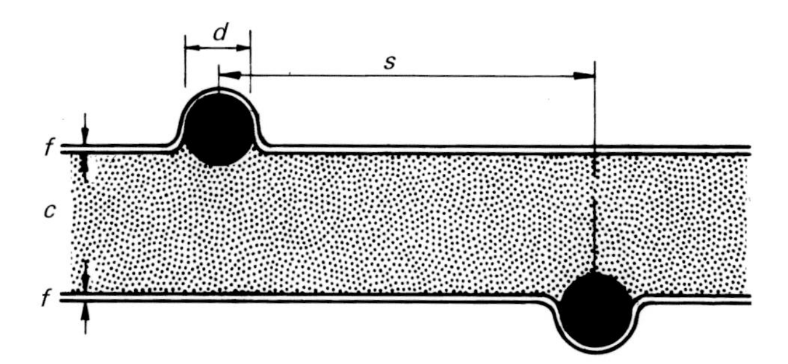

Fig. 6.4 A schematic drawing of a transverse section of an iris leaf, showing the parameters characterizing its structure. (Reprinted from Gibson *et al.*, 1988, Fig. 4, with kind permission from Springer Science+Business Media.)

least approximately, from (3.45). Gibson *et al.* (1988) measured the flexural stiffness of freshly picked iris leaves by cutting cantilever beams with their length parallel to the long dimensions of the leaf and bending them by hanging small weights from the free end. The bending stiffness was calculated from the load–deflection curve. The results are given in Table 6.2.

Table 6.1 Dimensions of the iris leaf

	Mean (mm)	Standard deviation (mm)
At thin end of leaf		
Depth of layer of "face" cells, f	0.03	0.0043
Thickness of square "face" cell wall t_f	0.0014	–
Length of square "face" cells, l_f	0.04	0.0058
Depth of "core" layer, c	0.5	–
Thickness of a "core" cell, t_c	0.0014	–
Length of "core" cells, l_c	0.05	0.023
Diameter of rib, d	0.13	0.04
Spacing of rib, s	1.2	0.46
Volume fraction of solid in rib, v_f	0.8	–
At mid-length of leaf		
Depth of layer of "face" cells, f	0.03	0.0054
Thickness of square "face" cell wall, t_f	0.0014	–
Length of square "face" cells, l_f	0.03	0.005
Depth of "core" layer, c	3.0	–
Thickness of "core" cell wall, t_c	0.0014	–
Length of "core" cells, l_c	0.07	0.025
Diameter of rib, d	0.19	0.058
Spacing of rib, s	0.92	0.32
Volume of fraction of solid in rib, v_r	0.8	–

The flexural stiffness of the iris leaf can be estimated using (3.45), once the Young's modulus of the face, E_f, and core, E_c, are known. Using tensile tests on the leaves, along with the solid area fraction of each component (ribs, face cells and parenchyma core cells) as well as the volume fraction of each component in the leaf (using the cell dimensions in Table 6.1), Gibson *et al.* (1988) estimated E_f to be 1.6 GPa and E_c to be 0.21 GPa, assuming that the value for the modulus of the solid cell-wall material is constant throughout the leaf (E_s = 4.4 GPa). This approach gave estimates of the beam stiffnesses that were 0.6 to 1.3 times the measured stiffnesses.

The extensive measurements of Young's modulus for the cell walls of plants are summarized in Table 6.3. Several early studies measured the cell-wall modulus of algae, in part because their cells are relatively large, simplifying the measurements; their moduli range from 0.1 to 4 GPa. Measurements of the cell-wall modulus of parenchyma in potato, a vascular plant, also give a value within that range. The modulus of the cell wall in sclerenchyma in bamboo, measured using nanoindentation, gives values between 4 and 26 GPa, depending on the orientation of the loading (Yu *et al.*, 2007); as discussed in Chapter 4, some caution is needed in interpreting these values

Table 6.2 Beam bending results

Specimen	1	2	3	4
Measured beam stiffness, P/δ (N/mm)	0.66	0.54	0.41	0.25
Beam length, l (mm)	35	35	35	35
Face thickness, f (mm)	0.03	0.03	0.03	0.03
Maximum core thickness, c (mm)	4.63	3.31	2.49	1.51
Width, b (mm)	18	18	18	18
Flexural rigidity, D (Nm^2)	0.027	0.016	0.0096	0.0051
Bending compliance, $(\delta/P)_b$ (m/N)	5.29×10^{-4}	8.98×10^{-4}	1.49×10^{-3}	2.83×10^{-3}
Shear compliance, $(\delta/P)_s$ (m/N)	2.99×10^{-4}	3.83×10^{-4}	4.83×10^{-4}	6.3×10^{-4}
Calculated beam stiffness, P/δ (N/mm)	1.21	0.78	0.51	0.29
Calculated/measured beam stiffness	1.83	1.44	1.24	1.16

since the equations for estimating the modulus in a nanoindentation test assume that the material is isotropic. Values of the modulus of fibers taken from leaves of various grasses range between 2 and 23 GPa; typically, the porosity of the sclerenchyma cells making up the fibers is relatively low, so that these measurements also give an indication of the modulus of the cell wall in sclerenchyma. The modulus of plant fibers with a high cellulose content range from 3 to 110 GPa, depending on the volume fraction of cellulose and the moisture content. Collectively, the data of Table 6.3 suggest that the Young's moduli of sclerenchyma cell walls are a factor of ten or more greater than those of parenchyma cell walls.

Several factors affect the cell-wall modulus: the most significant are the direction of loading (because the cell wall is made up of aligned cellulose microfibrils in a matrix) and the moisture content of the cell wall. The rate or the duration of loading has a small effect on the modulus; Sellen (1979), for example, measures a decrease in the relaxation modulus of about 20% when the duration of loading is increased by four orders of magnitude. As discussed in Section 5.3, at sufficiently high turgor pressures (Kamiya *et al.*, 1963, and Lin and Pitt, 1986, indicate about 0.4 MPa) the stiffness of parenchyma is independent of the pressure. We assume that for freshly picked leaves the turgor pressure is roughly constant at a relatively high level (around 0.6 MPa) and that the longitudinal stiffness of the leaf is independent of the pressure. The water content of the leaf also affects the measured longitudinal stiffness. Vincent (1982) reports that the longitudinal stiffness of two grasses was constant at water contents of between 100 and 400% of the dry weight, but rose sharply as the water content was reduced from 100% to 10%. Fresh leaves are always in a range in which the stiffness is independent of water content.

Here, we recalculate the bending stiffness of the iris leaves tested by Gibson and co-workers, using improved estimates for E_f and E_c. Data for the Young's moduli of fresh parenchyma (Table 5.3) suggest that they lie in the range of about 0.5 to 6 MPa, considerably smaller than the value of 210 MPa used by Gibson *et al.*, 1988. We will use a value of the Young's modulus of the parenchymal core of the iris sandwich of

Table 6.3 Young's modulus of plant cell walls

Plant material	Condition	E (GN/m²)	References
Cell wall: algae			
Penicillus dumetosus	Wet	0.16–0.22	a
Acetabularia crenulata	Wet	0.12–0.14	b
Nitella translucens	Wet	1.8–2.2	a
Nitella opaca	Wet	0.5–4	c
Nitella flexilis	Wet	0.4–0.7	d
Chara corallina	Wet	0.4–0.7	e
Cell wall: vascular plant parenchyma			
Potato tuber	Wet	0.5	f
Cell wall: vascular plant sclerenchyma			
Phyllostachys edulis (bamboo)	65% RH[1]	12–26 (longitudinal)	g
Phyllostachys edulis (bamboo)	65% RH[1]	4–9 (transverse)	g
Sclerenchyma cells (from grass leaf fibers)			
Lolium perenne	Wet	23	h
Holcus lanatus		3.0	i
Dactylis glomerata		7.4	i
Bromis hordescens		7.5	i
Deschampsia caespitosa		3.6	i
Stipa gigantea		2.1	i
Cell-wall components (theoretical)			
Cellulose		130	j
Lignin		2	j
Fibers with high cellulose content			
Flax		110	k
Ramie		60	k
Hemp	Wet	35	l
	Dry	70	l
Sisal fiber	Dry	10–21	m
Cotton hair	Dry	8.5	m
	Wet	2.9	m

Sources: [a] Sellen (1979); [b] Haughton and Sellen (1969); [c] Probine and Preston (1962); [d] Kamiya *et al.* (1963); [e] Toole *et al.* (2001); [f] Nilsson *et al.* (1958); [g] Yu *et al.* (2007); [h] Vincent (1982); [i] Vincent (1991); [j] Bodig and Jayne (1982); [k] Treitel (1946); [l] Wainwright *et al.* (1976); [m] Preston (1974).
[1] RH = relative humidity

E_c = 4 MPa, reflecting the data for parenchyma in Table 5.3; we take the core shear modulus to be half this value, G_c = 2 MPa. We estimate the Young's moduli of the ribs in the faces of the iris sandwich using the tensile test data for the iris leaf reported by Gibson and co-workers: E_{rib} = 21 GPa, which is consistent with the values for sclerenchyma cells in Table 6.3. The modulus of the face of the iris sandwich is then 21 GPa multiplied by the volume fraction of the ribs in the faces (0.39), or $E_f = 8.2$ GPa (assuming that the square face cells have the same modulus as the parenchyma and can be neglected in calculating the face modulus).

Calculation of the flexural stiffness is complicated by the varying thickness of the core across the section. This can be dealt with by dividing the transverse section into sub-units, each of constant thickness, and summing their contributions to the overall stiffness. The flexural stiffness of the leaf is then found from (3.45), with the constants B_1 and B_2 replaced by 3 and 1, respectively, for a cantilever beam (Table 3.4). The comparisons between the calculated and measured beam stiffnesses, using the improved estimates for $E_f = 8.2$ GPa, $E_c = 4$ MPa and $G_c = 2$ MPa, are given in the bottom row of Table 6.2: the calculations overestimate the stiffness of the beams, by between 16–83%. The agreement is as good as could be expected given the approximations made in estimating the moduli of the face and core, and in modeling the irregular cross-section of the specimens. The results clearly support the view that the iris leaf behaves mechanically like a sandwich structure.

The strength of the iris leaf can be approached in a similar manner. As discussed in Chapter 3, there are three main modes of failure in sandwich beams: face tensile failure (e.g. by yielding); face wrinkling, a local buckling failure; and core shear failure. The equations for calculating the failure load for each mode are given in (3.47), (3.49) and (3.50). To evaluate the failure loads of the iris leaf, we need a little more information on the strengths of the face and core. Vincent (1991) has measured the tensile strength of a variety of grasses and found that it increases linearly with the volume fraction of sclerenchyma, $V_{\mathrm{f\,sclerenchyma}}$, in the leaf:

$$\sigma_f \left(\text{in MPa}\right) = 1.44\left(V_{\mathrm{f\,sclerenchyma}} \times 100\right) + 1.53 \tag{6.1}$$

In the iris, the ribs, which we assume to be entirely sclerenchyma, are 80% dense and make up 40% of the face, so that the strength of the face is $\sigma_{yf} = 47$ MPa. The parenchyma in the core carry shear loading. The compressive strengths of other parenchymal tissues, such as apple and potato, range from about 0.25 to 1.3 MPa (Table 5.3) and the tensile strength has been measured at half the compressive strength (Oye *et al.*, 2007). Using an average value for the compressive strength of 0.8 MPa, and taking the shear strength to be half the tensile strength, we estimate $\tau_c^* = 0.2$ MPa. Using these values, along with the Young's modulus of the core $E_c^* = 4$ MPa, we find the failure loads corresponding to each mode. We assume that the beam is loaded as a cantilever with a uniformly distributed load (as from the wind, for instance) (so that $B_3 = 2$ and $B_4 = 1$ in (3.47)–(3.50)) and calculate the loads at the base of the leaf (so that the face thickness $t = 0.03$mm and $\ell \sim 600$ mm) where the stresses are highest:

$$\frac{P_{\text{face yield}}}{bc} = 4.7 \text{ kPa} \tag{6.2}$$

$$\frac{P_{\text{face wrinkling}}}{bc} = 2.8 \text{ kPa} \tag{6.3}$$

and

$$\frac{P_{\text{core shear}}}{bc} = 200 \text{ kPa} \tag{6.4}$$

where b and c are the width and core thickness of the leaf, respectively. The leaf is expected to fail by wrinkling of the face.

Corn (or maize, *Zea mays L.*) is a member of the grass family. Its leaves are composed of a stiff, V-shaped midrib with thin lamina on either side. The depth of the V of the midrib tapers along the length of the leaf and the thin laminae are usually wavy (Fig. 6.5a,b). The midrib itself has a sandwich-like structure, with dense, stiff sclerenchyma on the outer skins, separated by a core of parenchyma tissue of much lower relative density (Fig. 6.5c,d). Vascular bundles, of intermediate relative density, lie just within the lower skin of sclerenchyma tissue. The flexural response of the maize leaf midrib, accounting for the variation in the depth of the rib along the length of the leaf, has been shown to behave as a composite sandwich beam (Moulia and Fornier, 1997).

The studies on the iris and maize leaves demonstrate that they act mechanically like efficient sandwich structures. The iris leaf has another feature that also increases its efficiency: the depth of the iris leaf increases from the tip to the base, corresponding to the increase in wind-induced bending moment in the cantilevered leaf. But are the leaves optimized? Minimum weight design for a foam-core sandwich beam of a given stiffness indicates that at the optimum design, the shear deflection is two-thirds of the total deflection (Gibson and Ashby, 1997): the shear deflection of the iris is somewhat less than this (Table 6.2). In nature, stiffness is not the only function of the leaves. Most natural materials and structures must fulfill several functions simultaneously: the iris leaf, for example, must also be strong enough not to fail and must provide sufficient surface area for photosynthesis for the plant. Quantifying the relative importance of stiffness, strength and surface area is difficult, so that an overall engineering optimization analysis is not possible.

Sandwich structures have also been observed in marine plants – for instance, in the leaf-like fronds of *Durvillaea antarctica*, a seaweed that inhabits the temperate coasts of New Zealand and Chile (Stevens *et al.*, 2002). It is the largest intertidal seaweed in the world, with fronds up to 12 m long. Unlike other types of seaweeds, its fronds have a gas-filled, honeycomb-like core that provides buoyancy as well as flexural rigidity. The fronds float on top of the water, maximizing the surface area exposed to the sun for photosynthesis and minimizing the self-shading that occurs in some seaweeds (Harder *et al.*, 2005).

6.2.2 Sandwich structures in bones and shells

Plate- or shell-like bones in the body often have a sandwich structure, with outer faces of nearly fully dense cortical bone separated by a core of much lower density trabecular bone. Examples include the pelvis, the scapula and the skull (Figs. 6.1a and

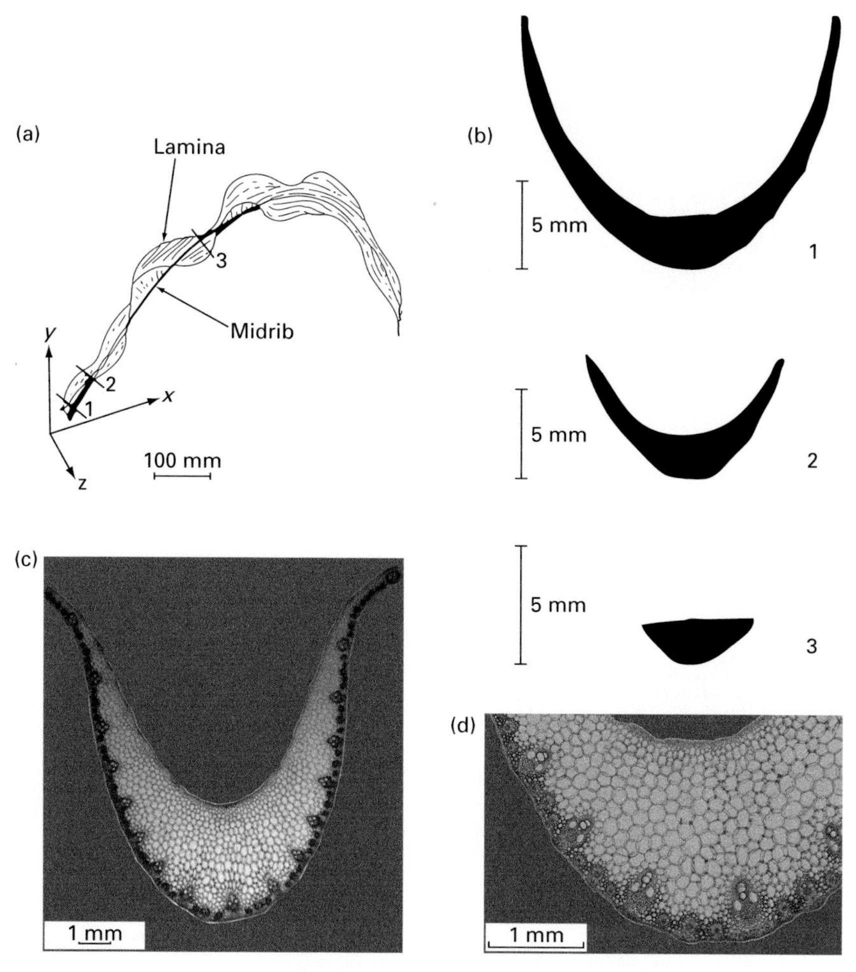

Fig. 6.5 (a) The geometry of the maize leaf, showing the central midrib and thin lamina on either side. (b) Outlines of three cross-sections of the midrib, corresponding to sections 1, 2 and 3 in (a). (c) Macrophotograph of a cross-section in the maize midrib. (d) Higher magnification image of a cross-section, showing the dense sclerenchyma on the outer skins, the low relative density parenchyma core and the vascular bundles just inside the lower skin. (a, b, Reprinted from Moulia and Fournier (1997), Fig. 1, with kind permission from Springer Science+Business Media; c,d courtesy of B. Moulia.)

6.6). The primary role of the skull is to protect the brain from impact. Loading of a hemispherical shell, like the skull, over a small area induces both membrane and bending stresses; for typical human skull dimensions and loading over a circular area 25 mm in diameter, the bending stresses are about three times the membrane stresses (Roark and Young, 1975), so that a sandwich structure is an effective means of resisting loading at low weight.

Minimizing weight is critical for flight in birds. Their skulls have a variety of sandwich-like structures (Fig. 6.7) (Buhler, 1972). In many birds, in regions of the skull where the inner and outer cortical faces are not concentric (e.g. at the base of the skull,

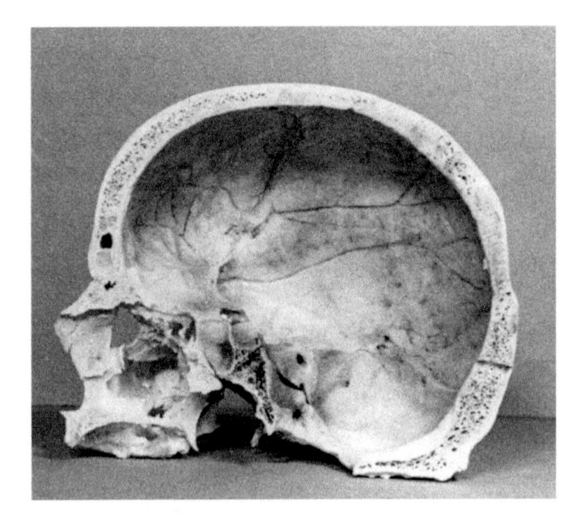

Fig. 6.6 Cross-section of a human skull, showing the sandwich structure. (Reprinted from Hodgson (1973) Fig. 1b, with kind permission from Springer Science+Business Media.)

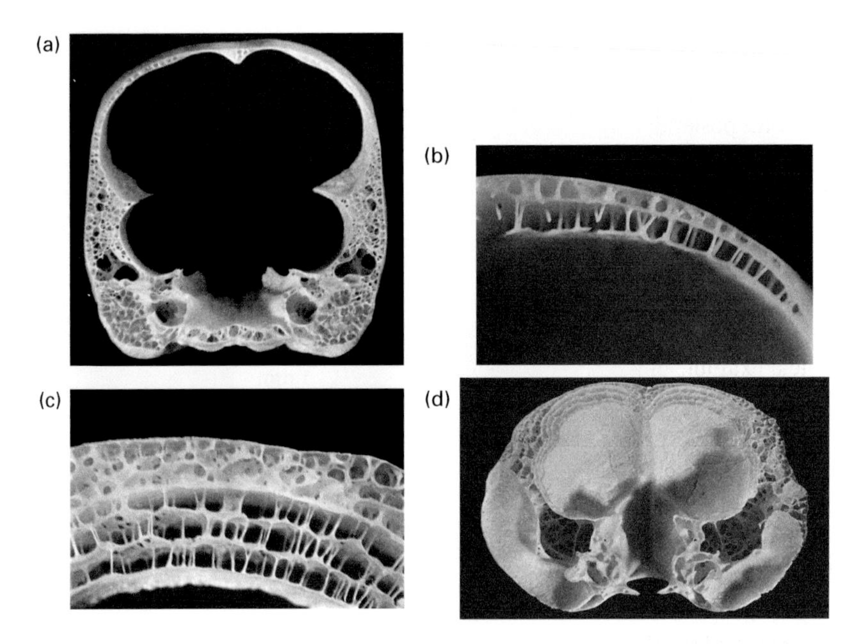

Fig. 6.7 Photographs of cross-sections of bird skulls showing their sandwich structure: (a) random, foam-like trabecular core (house pigeon); (b) double sandwich structure with trabeculae oriented normal to the outer shell (magpie, *Pica pica*); (c) multiple sandwich structure, with trabeculae oriented normal to the faces (long-eared owl, *Asio otus*); and (d) posterior view of the skull of a long-eared owl. The upper central chamber encases the brain while the lower, outer kidney-shaped bones are the posterior bone walls of the outer auditory canals. (Reproduced from Buhler, 1972).

Fig. 6.7a), the trabecular core has a random, foam-like stucture. In regions of the skull where the inner and outer cortical faces are concentric, the trabeculae are oriented normal to the faces; in small songbirds there is a single sandwich structure, while in larger birds, especially owls, there are multiple sandwiches (Fig. 6.7b,c,d). The reason for these multiple sandwich structures is not clear. We do know that honeycomb-like cores, with their walls aligned normal to the faces of a sandwich, are more efficient than foam-like cores, as their out-of-plane stiffness and strength vary linearly with relative density (Table 3.1) rather than with the square (Table 3.3). But the trabeculae in multiple sandwich structures in bird skulls appear to be rod-like, rather than honeycomb-like, so that they would be expected to bend, rather than stretch, under the shear loading expected in the core of a sandwich structure, making them more foam-like than honeycomb-like. The flat sheets making up the faces of the multiple sandwich structures are anchored at their ends in the much stiffer cortical bone, increasing their resistance to shear. The length of trabeculae in animals ranging in size from 4 gram shrews to 40 000 kg whales are similar (about 200 to 1500μm, Swartz *et al.*, 1998), so it may be that multiple layers allow increased thickness of the skull, and corresponding increase in moment of inertia, while maintaining the trabeculae normal to the cortical shells.

In owls, the skull both encases the brain and provides enlarged bone walls for the outer auditory canals that assist in owls' exceptional hearing: owls are known to catch small rodents rustling beneath the snow that they cannot see. The skull of a long-eared owl is shown in Fig. 6.7d: the central chamber in the top part of the skull encases the brain while the two large, kidney-shaped bones on either side of the skull in the lower part of the image are the posterior bone walls of the outer auditory canals that are thought to assist in focusing sound (Buhler, 1972). Notice that these bones are asymmetrical: in some species of owls, one ear opening is higher than the other, assisting in pinpointing the location from which sound emanates. The multiple sandwich structure provides a lightweight means of protecting the brain as well as supporting the auditory bones.

The performance of a bone sandwich structure can be compared with that of an equivalent cortical bone structure. To simplify the calculation, we consider a hypothetical example of a circular sandwich panel of radius R, simply supported around its circumference, and subjected to a uniformly distributed load, q. The central deflection of the plate is w; the sandwich panel has a given stiffness, qR/w. We take the density and Young's modulus of the solid in the trabecular bone to be the same as those for cortical bone, as discussed in Chapter 5. The values of the face and core thicknesses and of the core density that minimize the mass of the sandwich panel for the given stiffness are given by Gibson and Ashby (1997). Substituting these values into the equation for the mass, m, of the panel, we find that the mass depends only on the given panel stiffness, qR/w, and the density, ρ_s, and modulus, E_s, of cortical bone:

$$\frac{m_{\text{sandwich}}}{\pi R^2} = 1.49\left(\frac{qR/w}{E_s}\right)^{3/5} \rho_s R \qquad (6.5a)$$

We next consider an equivalent solid circular plate, of equal radius and stiffness, made from cortical bone. The depth of the plate can be related to the given stiffness, qR/w, and the modulus of cortical bone, E_s. The mass of an equivalent solid circular plate also depends only on the given stiffness and the solid density and modulus:

$$\frac{m_{\text{uniform}}}{\pi R^2} = 0.89\left(\frac{qR/w}{E_s}\right)^{1/3} \rho_s R \tag{6.5b}$$

Taking the ratio:

$$\frac{m_{\text{sandwich}}}{m_{\text{uniform}}} = 1.67\left(\frac{qR/w}{E_s}\right)^{0.27} \tag{6.5c}$$

For all practical values of the normalized stiffness, we find that the mass of the bone sandwich panel is substantially less than that of the cortical bone panel: for instance, for $(qR/w)/E_s = 10^{-4}$ (roughly corresponding to a 500 N load distributed over a circular area of 100 mm radius, producing a deflection of 1 mm, with $E_s = 18$ GPa) the mass of the optimized bone sandwich is 14% of that of the cortical bone panel.

The cuttlefish (Fig. 6.8), a member of the family Sepiidae, is related to the octopus and squid and is a mollusk rather than a fish. Like an octopus, it has tentacles that it uses to capture prey and can shoot ink (called sepia) into the water to avoid predators. It camouflages itself by changing its skin color (by contracting or expanding pigment cells in its skin) and texture (using muscles in the dermis). Here, we are interested in the cuttlebone, a hard, brittle structure within the cuttlefish, composed primarily of aragonite (see Chapter 2 for a description of this form of calcium carbonate). Its microscopic structure looks like a multi-layer sandwich with thin face sheets connected by numerous upright pillars. The cuttlefish uses the bone for buoyancy, by controlling the fraction of the chambers filled with water or air. Cuttlebones implode at depths of between 200 m and 600 m, depending on the species. Because of this limitation, most species of cuttlefish live in shallow waters. Sandwich structures are also observed in the shells of some arthropods, such as the horseshoe crab (Fig. 6.9).

We have seen that monocotyledon leaves such as the iris and maize behave as sandwich structures, making them efficient for resisting bending loads from their own weight and the wind. Bone sandwich shells reduce the mass of the bone relative to a fully dense cortical shell. Our calculations suggest that an optimally designed bone sandwich plate has a fraction of the mass of an equivalent cortical bone plate; we expect similar results for a bone sandwich shell. Finally, the sandwich structure of the cuttlebone provides buoyancy to the remarkable cuttlefish.

6.3 Circular sections with radial density gradients

6.3.1 Arborescent palm stems

Arborescent palms can grow to heights of 20 to 40 meters, similar to dicotyledon trees and conifers. Dicotyledon trees and conifers, as they grow over time, resist the additional loads by increasing their diameter, through the addition of new cells at the

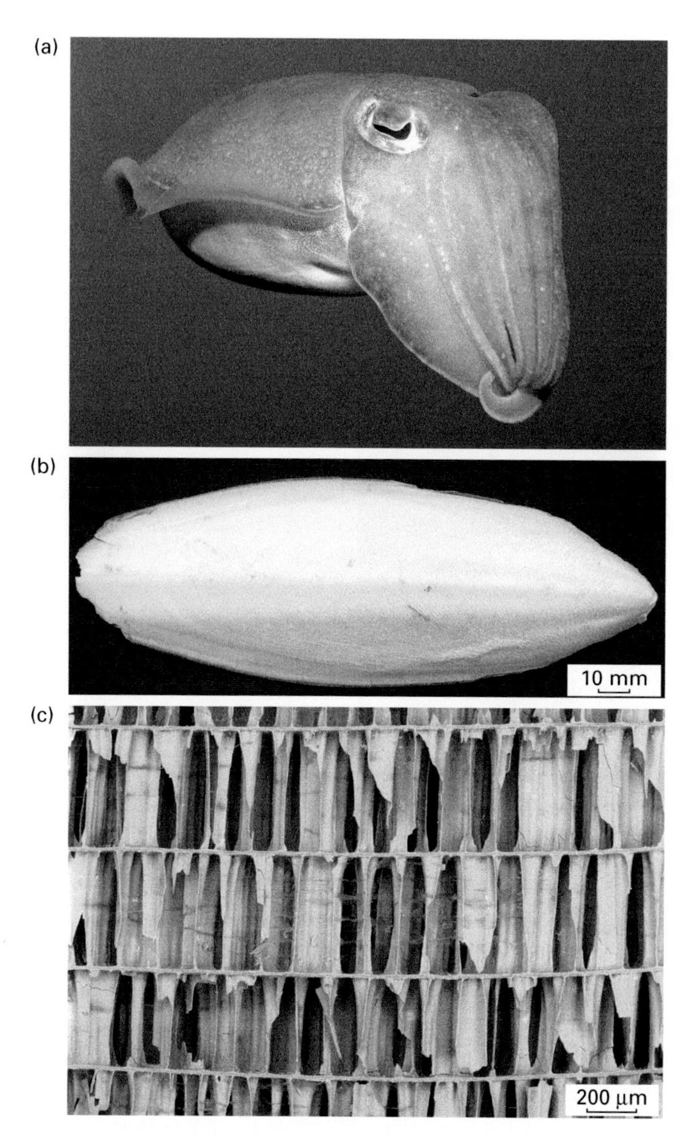

Fig. 6.8 (a) A cuttlefish, (b) the cuttlebone and (c) a scanning electron micrograph of the cuttlebone. (a, Reprinted from and copyright © Phillip Colla Photography.)

cambium at the periphery of the tree. Since palm stems lack a lateral cambium, they cannot add new cells so that the stem diameter typically remains roughly constant throughout its lifetime. Instead, as the stem grows taller, the increased load is resisted by an increase in the thickness of the cell walls. The largest stresses imposed on palm stems arise from hurricane winds, which cause bending of the stems.

The microstructure and mechanical properties of the stems of four species of palm have been studied in detail by Rich (1986, 1987a,b). Micrographs of the cross-section of *Iriartea gigantea* reveal parallel vascular bundles, with their cells aligned along the length of the stem, separated by ground tissue with more equiaxed parenchyma cells (Fig. 6.10a,b). The increased cell-wall thickness in older tissue can also be seen in the

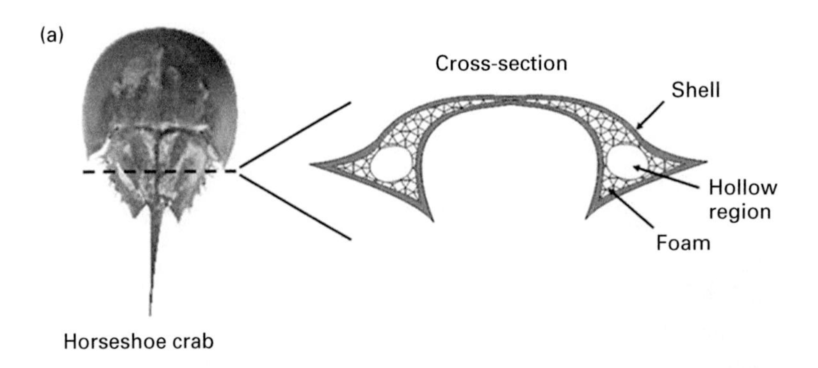

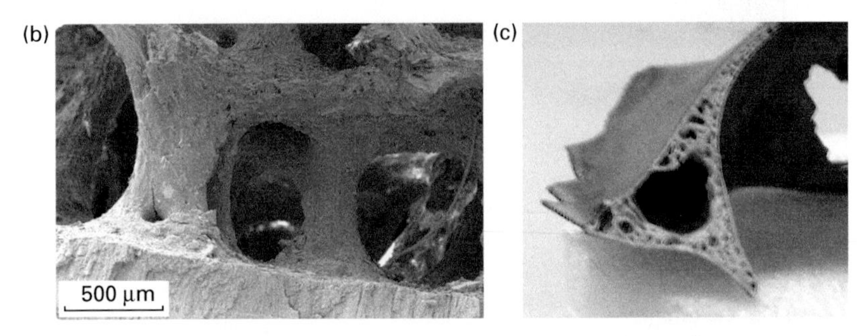

Fig. 6.9 (a) A horseshoe crab and a schematic drawing of a cross-section of its shell; (b) a scanning electron micrograph of the horseshoe crab shell; and (c) a photograph of the end section of the shell. (Reprinted from Meyers *et al.*, 2008, with permission from Elsevier.)

micrographs of Fig. 6.10a,b. Higher magnification scanning electron micrographs of a different palm, *Cocus nucifera*, indicate that the thicker walls have additional layers of secondary wall (Fig. 6.10c,d) (Kuo-Huang *et al.*, 2004). The density distribution in the young and old stems in Rich's studies is shown schematically in Fig. 6.11: the density is highest at the base and at the periphery of the stem. As the palm increases in height, the thickness of the outer cells increases higher up from the base. More quantitatively, the density distribution is plotted against radial position and against height above ground for two species of palms in Fig. 6.12. There is a remarkable range in the density within a single stem: the dry density of the *Iriartea gigantea* tissue varies from less than 100 kg/m^3 to about 1000 kg/m^3, spanning nearly the entire range of densities of woods, from balsa (100–200 kg/m^3) to lignum vitae (1300 kg/m^3). Common woods, such as spruce, pine and oak, have densities ranging from about 400 to 700 kg/m^3.

The Young's modulus of dry tissue from all four species of palm in Rich's (1987b) study varied with density according to (Fig. 6.13a)

$$E^* = C'\left(\rho^*\right)^{2.46} \tag{6.6}$$

For two individual species, E^* varied with density to the power 2.27 (*Welfia georgii*) and 2.47 (*Iriartea gigantea*). The increased density and modulus of tissue at the periphery of the palm stem is also confirmed by Kuo-Huang *et al.* (2004).

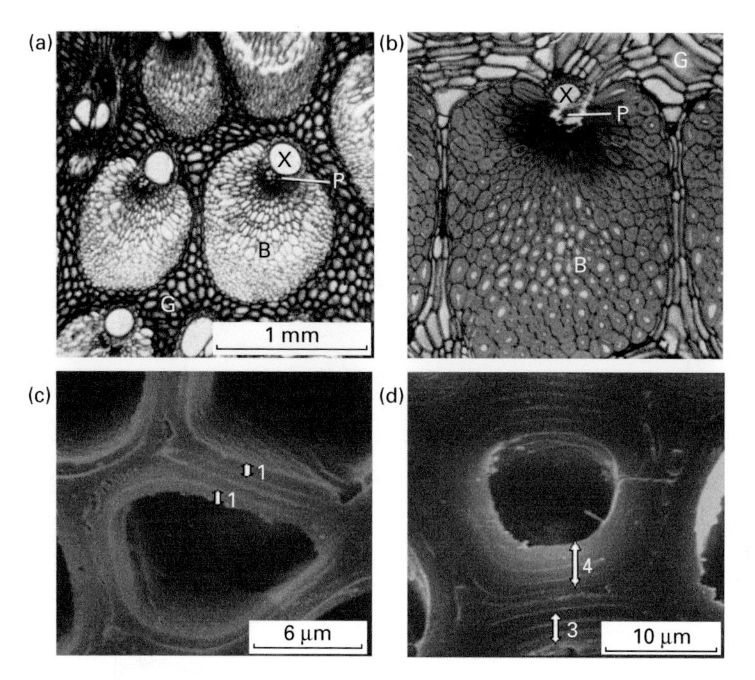

Fig. 6.10 (a,b) Optical micrographs of cross-sections of *Iriartea gigantea* showing (a) the peripheral stem tissue of a young individual and (b) the peripheral stem tissue of an older individual. B = vascular bundle containing xylem (X) and phloem (P), G = ground tissue. (c,d) Scanning electron micrographs of cross-sections of coconut palm *Cocus nucifera*, showing (c) a cell near the center of the stem, with a primary cell wall and one secondary layer and (d) a cell near the periphery of the stem, with a primary cell wall and three or four secondary layers. (a,b, Reprinted from Rich 1987a, Figs. 22, 23, with kind permission of *American Journal of Botany*; c,d, reprinted from Kuo-Huang *et al.*, 2004, Figs. 1e,f, with kind permission of *IAWA Journal*.)

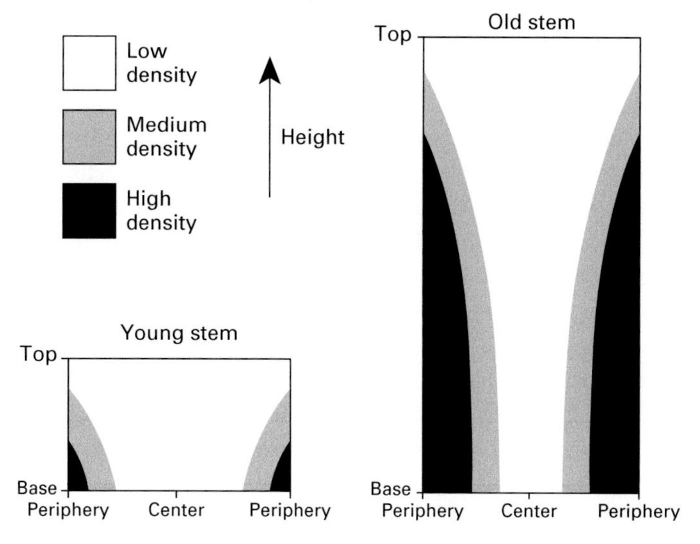

Fig. 6.11 Schematic showing the density distribution in young and old palm stems. This is typical of all palm species. (Reprinted from Rich, 1986, with permission from International Palm Society.)

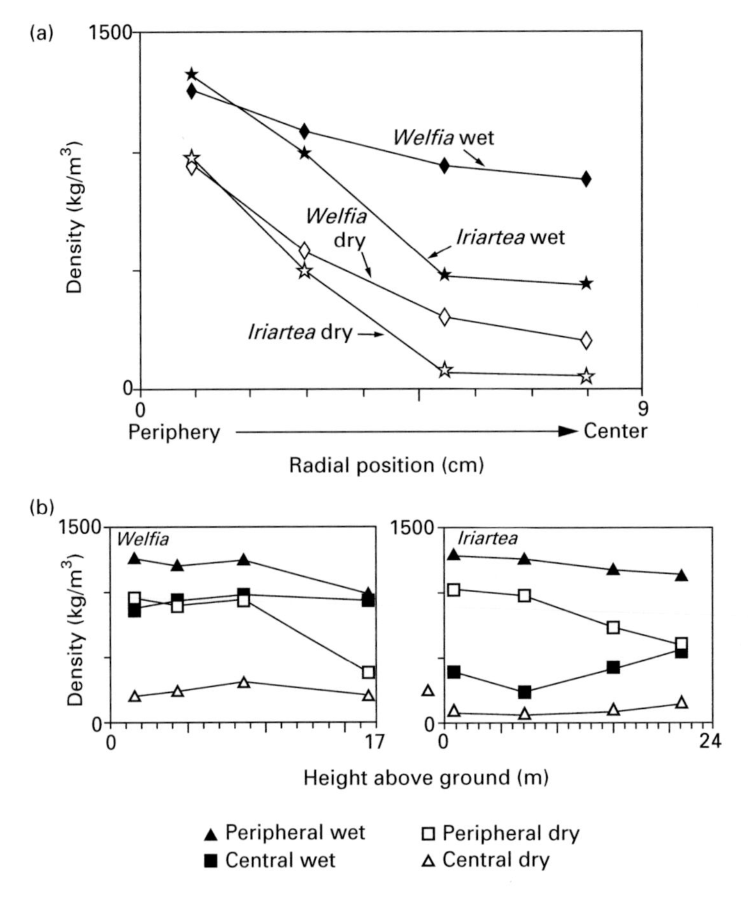

Fig. 6.12 (a) Density plotted against radial position (at breast height) in the stem for a 17 m tall *Iriartea gigantea* and for a 19 m tall *Welfia georgii*. (b) Density plotted against height in the stem for a 19 m tall *Welfia georgii* and for a 26 m tall *Iriartea gigantea*. (Reprinted from Rich, 1987b.)

Both the honeycomb-like vascular bundles as well as the more equiaxed, fluid-filled parenchyma cells undergo cell-wall stretching under loading in the axial direction, so that the Young's modulus in the axial direction would be expected to vary linearly with relative density ((3.13), Section 5.3), in contrast to the measured moduli (6.6). But unlike many other natural cellular materials, such as wood or trabecular bone, in which the cell-wall properties are constant, in palm the cell-wall properties themselves differ in cells of different relative densities. The additional secondary layers in the denser tissue probably increase the solid wall modulus, since secondary walls tend to have the cellulose fibers that are more aligned with the cell axis (Dinwoodie, 1981; Niklas, 1992; Kuo-Huang *et al.*, 2004). The cell walls of palms are also known to contain biosilica (Rich, 1987a; Tomlinson, 1990; Schmitt *et al.*, 1995), with a modulus of roughly 45 GPa (Table 2.2). Tensile measurements of the cell-wall modulus within a single vascular bundle of the Mexican fan palm, *Washingtonia robusta*, indicate that it increases by a factor of about 10 in cells with greater wall thickness (Ruggeberg *et al.*, 2008). Extrapolation of the Young's modulus data in Fig. 6.13a to the fully dense value

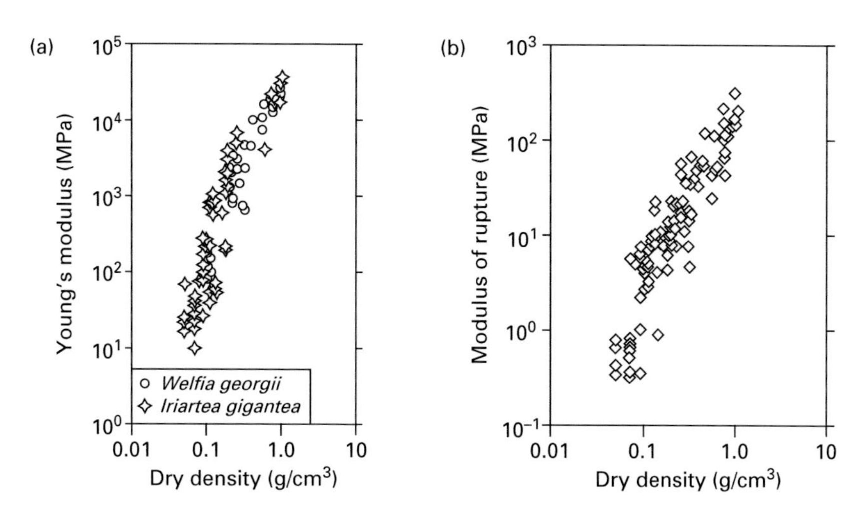

Fig. 6.13 (a) Young's modulus plotted against dry density for two species of palm. (b) Modulus of rupture plotted against dry density for six species of palm: *Welfia georgii, Iriartea gigantea, Socratea durissima, Euterpe macrospadix, Prestoea decurrens,* and *Cryosophila albida.* (Reprinted from Rich, 1987b.)

for the solid cell wall gives E_s = 150 GPa, assuming that the organic component of the cell wall has a density similar to that of wood (ρ_s=1500 kg/m^3) and that the fully dense wall is 10% biosilica (ρ_s= 2600 kg/m^3). Remarkably, the Young's modulus for the solid cell wall for palm is similar to the value for cellulose (E = 110–165 GPa, Table 2.2) and much higher than that for wood cell wall (E_s = 35 GPa, Table 4.2).

The modulus of rupture of dry tissue from all four species of palm (Fig. 6.13b, Rich, 1987b) varies as

$$\sigma^* = C''\left(\frac{\rho}{\rho_{\max}}\right)^{2.05} \tag{6.7}$$

Extrapolation of the modulus of rupture data to the fully dense value for the solid cell wall gives σ_{ys} = 700 MPa, again similar to the tensile strength of cellulose (σ_y = 750–1010 MPa, Table 2.2) and about double that for wood cell wall (σ_y = 350 MPa, Table 4.2).

The position of the denser, stiffer tissue at the periphery of the stem increases its flexural rigidity. Here, we compare the flexural rigidity of the palm, with its radially varying density, with that of an equivalent uniform density cross-section of the same diameter and mass. For both cases, the outer radius is r_o. The radial variation in the dry density of the stem tissue of two palms, *Iriartea gigantea* and *Welfia georgii*, (Fig. 6.12a) can be represented (3.51) by

$$\frac{\rho - \rho_{\min}}{\rho_{\max} - \rho_{\min}} = \left(\frac{r}{r_o}\right)^n \tag{6.8}$$

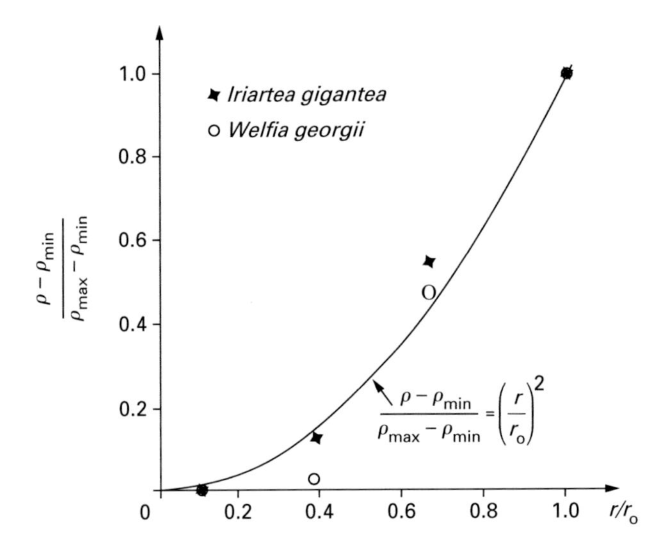

Fig. 6.14 Radial density distribution corresponding to (6.8), with $n = 2$, along with data for the dry density of *Welfia georgii* and *Iriartea gigantea*.

where ρ is the density at a radius r, ρ_{min} is the minimum density at the center of the cross-section, ρ_{max} is the maximum density at the periphery and $n = 2$ (Fig. 6.14). Rewriting (6.6) for the Young's modulus of the palm tissue gives

$$E = C\left(\frac{\rho}{\rho_{max}}\right)^m \tag{6.9}$$

The flexural rigidity for the equivalent section of uniform density (3.56) is

$$(EI)_{uniform} = C\left(\frac{2 + nR}{2 + n}\right)^m \frac{\pi r_o^4}{4} \tag{6.10}$$

where R is the ratio of the minimum to maximum density in the cross-section.

The flexural rigidity for the section with the radial density gradient (3.57) is

$$(EI)_{rad\,grad} = \int_0^{r_o} C\left[(1 - R)\left(\frac{r}{r_o}\right)^n + R\right]^m \pi r^3 \mathrm{d}r \tag{6.11}$$

We consider two cases. First, we note that for *Iriartea gigantea*, the minimum density is essentially zero, so that R, too, is zero. The flexural rigidity in this case is then

$$(EI)_{rad\,grad} = \frac{C\pi r_o^4}{mn + 4} \tag{6.12}$$

Using the appropriate factors of $m = 2.5$ and $n = 2$, we find that the *Iriartea gigantea* stem is about 2.5 times as stiff in bending as the equivalent uniform density section.

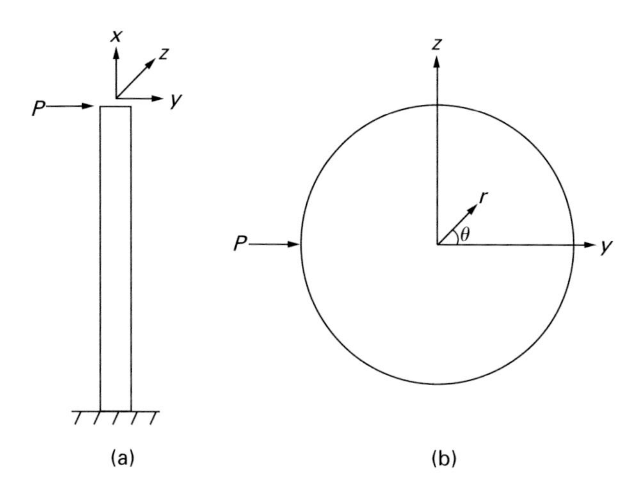

Fig. 6.15 (a) A palm stem in bending, showing the x-, y-, z-axis system and (b) the circular cross-section, with cylindrical coordinates of radius, r, and angle, θ.

Second, the data for *Welfia georgii* can be approximated by $m = 2$ and $n = 2$, which is analytically straightforward. We find that

$$(EI)_{\text{rad grad}} = C\pi r_o^4 \left[\frac{(1-R)^2}{8} + \frac{2R(1-R)}{6} + \frac{R^2}{4} \right] \tag{6.13}$$

For *Welfia georgii* the minimum dry density is about 200 kg/m^3 and the maximum is about 930 kg/m^3, giving $R = 0.22$ and an increase in flexural rigidity over that of the uniform density cross-section by a factor of 1.6.

We next consider the stress distribution within a palm stem loaded in bending (Fig. 6.15). From elementary bending theory, assuming that plane cross-sections remain plane, the normal stress within the circular cross-section varies according to:

$$\sigma(y) = E\varepsilon = E\kappa y \tag{6.14}$$

where κ is the curvature at the cross-section and y is the distance from the neutral axis at the center of the palm stem in the direction of loading (shown in Fig. 6.15). At a given cross-section, κ is a constant that depends on the bending moment at that section and the flexural rigidity of the cross-section. The Young's modulus of the palm stem is given by (6.9).

The *Iriartea gigantea* palm is the simplest case to analyze, as its minimum density is nearly zero so that the relative density varies with radial position according to

$$\frac{\rho}{\rho_{\text{max}}} = \left(\frac{r}{r_o} \right)^n \tag{6.15}$$

The normal stress in the palm stem varies with radial position and orientation θ (see Fig. 6.15):

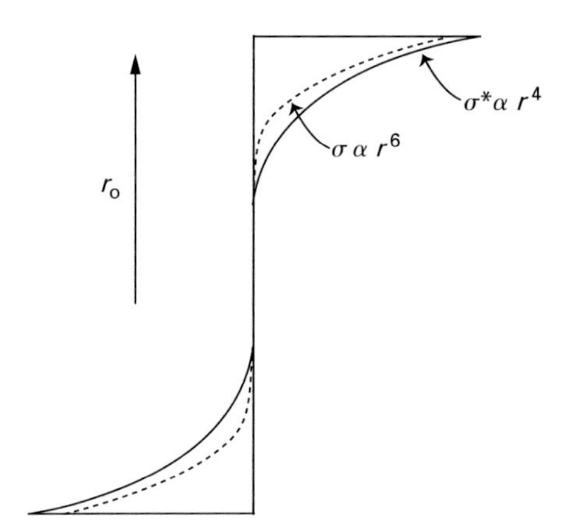

Fig. 6.16 Normal stress distribution on the palm cross-section (along a line with $\theta = 90°$) resulting from bending: $\sigma \propto r^6$. Modulus of rupture of the section, with $\sigma^* \propto r^4$. The maximum bending stress is set equal to the modulus of rupture at the extreme fiber of the cross-section. When the extreme fiber is near failure, the normal stress distribution closely follows the strength across the section.

$$\sigma(r,\theta) = C\left(\frac{\rho}{\rho_{max}}\right)^m \kappa y = C\left(\frac{r}{r_o}\right)^{mn} \kappa r \cos\theta \tag{6.16}$$

or, for a given angle, θ,

$$\sigma(r) \propto r^{mn+1} \tag{6.17}$$

For *Iriartea gigantea*, with $m = 2.5$ and $n = 2$, we find that the normal stress varies with r^6. The stress is maximum at $r = r_o$ and $\theta = 0°$. The stress distribution across the cross-section, for $\theta = 0°$, is plotted in Fig. 6.16: as expected, most of the normal stress is carried by the outer extremity of the stem.

Finally, we compare the normal stress across the palm stem with the bending strength of the stem. The modulus of rupture of palm varies as (Fig. 6.13b, Rich, 1987b):

$$\sigma^*(r) = C''\left(\frac{\rho}{\rho_{max}}\right)^q = C''\left(\frac{r}{r_o}\right)^{nq} \tag{6.18}$$

Rich gives $q = 2.052$; for simplicity, we take $q = 2$. For *Iriartea gigantea*, $n = 2$, so that the modulus of rupture varies with r^4. The variation in the modulus of rupture across the palm stem is also plotted in Fig. 6.16, with the maximum normal stress at $r = r_o$ and $\theta = 0°$ equal to the modulus of rupture. We see that, while the dependence of the bending stress (6.17) and the modulus of rupture (6.18) on radial position are not identical, in practice the bending strength of the palm stem closely follows the normal stress from bending, making it a highly efficient structure.

6.3.2 Bamboo

Photographs of cross- and longitudinal sections of bamboo show a tubular structure, with more or less evenly spaced solid diaphragms at the nodes from which the leaves

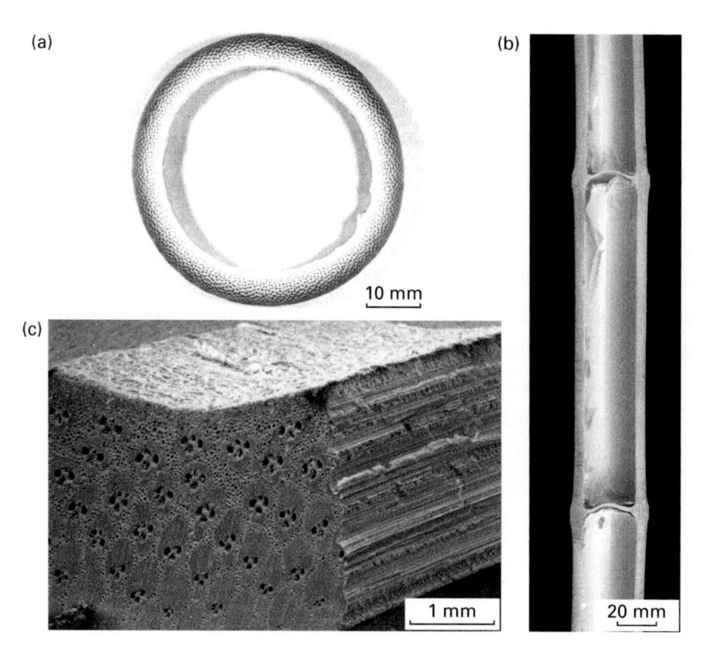

Fig. 6.17 (a) A cross-section of bamboo, showing the tubular structure and (b) a longitudinal section of bamboo, showing the more or less evenly spaced diaphragms. (c) Scanning electron micrograph of a cross-section of bamboo, showing the radial density gradient. (a,c, Reproduced from Gibson *et al.*, 1995; b, specimens provided by the Arnold Arboretum, Harvard University, Boston MA.)

grow (Fig. 6.17a,b). The higher magnification scanning electron micrograph reveals that the wall of the tube has a radial density gradient, with a higher volume fraction of stiff, longitudinal fibers towards the periphery of the wall (Fig. 6.17c). Hollow, thin-walled sections can be more efficient for carrying loads than thick-walled or solid ones. They use less material and are therefore lighter (more "economical") while resisting the same bending or torsional load; and this is true whether the design is based on stiffness or on strength. When the mode of loading is bending and the direction of loading is unknown, circular tubes are better than other shapes.

Both the tubular structure and the radial density gradient increase the bending stiffness of bamboo. The shape factor, ϕ, characterizes the moment of inertia, I, of a cross-section per unit area, A (Ashby, 2005):

$$\phi = \frac{4\pi I}{A^2} \tag{6.19}$$

Higher values of ϕ give higher bending stiffness for a given mass. The shape factor is non-dimensional; the factor 4π gives $\phi = 1$ for a solid circular section. For thin-walled tubes, $\phi = r/t$; for bamboo, $\phi = 2–6$. We can also apply the above analysis for a radial density gradient by setting the minimum density to zero, corresponding to the central void, n to a high value ($n = 6$ gives minimal density until $r/r_0 = 0.75$, corresponding to the cross-section in Fig. 6.17a) and $m = 1$, corresponding to a linear dependence of the Young's modulus on density for the aligned fibers and honeycomb-like cells of the bamboo. We note that nanoindentation tests have shown that the Young's modulus of the bamboo fibers, E_s, remains constant from the interior to the exterior of the cross-section (Yu *et al.*, 2007). These values give an increase in flexural rigidity over that of the uniform density cross-section of a factor of 1.6.

The aligned fibers and honeycomb-like cells of bamboo make the tube wall orthotropic, so that the moduli and strengths differ in the longitudinal, circumferential and radial directions. In orthotropic tubes, the dominant failure mode is controlled in part by the geometry of the cross-section, and in part by the ratio of the properties in the longitudinal and circumferential directions (Wegst and Ashby, 2007). If the ratio of the properties in the different directions is chosen properly, the orthotropic tube is both stiffer and stronger in bending than an equivalent isotropic one. Orthotropic tubes are exploited both by engineers (composites, highly drawn metals) and by nature (stems, bamboo culms), and are almost invariably structured so that the stiff, strong direction lies parallel to the axis of the tube.

Hollow tubes loaded in bending can fail in one of three ways. When the tube is bent, the cross-section ovalizes, reducing its moment of inertia and resistance to further bending. The modulus that determines ovalization is the circumferential one. Bent far enough, it fails by *local buckling*, the familiar kinking failure in a bent plastic straw. Local buckling depends both on the degree of ovalization (and therefore on the circumferential modulus) and on the longitudinal modulus. But bending also creates tensile and compressive stresses in the tube wall; if either of these exceeds the uniaxial strength of the tube wall in the longitudinal direction, it fails by *tensile yield or fracture*, or by *compressive collapse*. Finally, ovalization has another, subtler, consequence: it creates circumferential stresses in the tube wall, which, if they exceed the circumferential strength, cause *longitudinal splitting*.

The degree of anisotropy has consequences for the way the tube fails when it is bent. If the circumferential modulus is too low, it fails by ovalization and kinking, like a bent drinking straw. If the longitudinal strength is too low it fails by tensile or compressive failure of the tube wall in a plane parallel to the axis of bending, like a stick of celery bent until it snaps. And if the circumferential strength is too low, it fails by longitudinal splitting – like a stick of celery, pinched between the fingers.

Observations and analyses of the bending response of thin-walled orthotropic tubes appear in two quite separate bodies of literature. That relating to plant stems is largely experimental (e.g. Niklas 1992; Arce-Villalobos, 1993; Spatz *et al.*, 1998). That focusing on light-weight engineering structures, and particularly on polymer-composite tubes, is predominantly analytical, and generally treats only one aspect of what is a multi-faceted problem (Gerard, 1968; NASA, 1968; Cecchini and Weaver, 2002). Only two papers (Schulgasser and Witztum, 1992, and Wegst and Ashby, 2007) attempt a comprehensive survey of competing failure modes and explore how well plants – particularly bamboo – are structured to combat them.

Ovalization of the cross-section in bending plays a role in failure by both local buckling (or kinking) and longitudinal splitting. The diaphragms might be expected to act as ring stiffeners, suppressing ovalization. Calladine (1983) has shown that ovalization can be inhibited in an isotropic tube if the non-dimensional parameter

$$\Omega = \left(\frac{tL^2}{r^3}\right)^{1/2} < 0.5 \tag{6.20}$$

where t is the wall thickness of the tube, L is the spacing between the diaphragms and r is the tube radius. For an orthotropic tube, this becomes (Suo, 1990a,b):

$$\Omega = \left(\frac{E_{\|}}{E_{\perp}}\right)^{1/4} \left(\frac{tL^2}{r^3}\right)^{1/2} < 0.5 \tag{6.21}$$

Using typical values for bamboo, and noting that, in plants, the ratio of the moduli in the parallel and perpendicular directions is typically about 10, we find that the diaphragms in bamboo do not suppress ovalization.

The analyses presented by Schulgasser and Witzum (1992) and by Wegst and Ashby (2007) reach much the same conclusions: for a given geometry and ratio of base stiffness to strength (a ratio that differs greatly between living and dead plants), one of the three failure modes dominates. In bamboo, the dominant failure mode is longitudinal yield or fracture, resulting from the normal bending stresses along the length of the bamboo stem. Ideally, in engineering design, the greatest advantage is gained by adjusting the degree of anisotropy such that all three modes occur at approximately the same bending moment. It is a result that can be easily understood: if one failure mode occurs at a lower moment than the others there is advantage in adjusting the anisotropy to suppress it – until another mode intervenes.

6.4 Cylindrical shells with compliant cores: animal quills and plant stems

Plant stems and animal quills have dense cylindrical shells that are either partially or fully filled with foam- or honeycomb-like cores (Vincent and Owers, 1986; Karam and Gibson, 1994, 1995) (Fig. 6.18a,b,c). This architecture is seen so frequently in plant stems that botanists have a name for it: the "core-rind" structure (Niklas, 1992). In plant stems, the outer cylindrical shell is typically composed of dense sclerenchyma cells while the core is made up of thin-walled, foam-like parenchyma cells (see Section 5.3). The central region of plant stems is often hollow. Porcupine and hedgehog quills are composed of keratin that is solid in the outer shell and cellular in the compliant core. Particularly large bird beaks, such as those of the toucan and the hornbill, have a similar structure with a thin outer shell surrounding a foam-like core, although they are more elliptical in cross-section (Fig. 6.18d) (Seki *et al.*, 2005, 2006); here, as an approximation, we model them as cylindrical. Bird feather quills (called the rachis) have a foam-filled structure that is intermediate between that of a rectangular sandwich beam and a circular tube (Fig. 6.1c), probably because the loading on the feathers is primarily in bending in the dorso-ventral direction, as the wings beat during flight. Woodpecker tail feathers are known to be remarkably stiff relative to other types of feathers: when pecking, woodpeckers place their tails against the tree to brace themselves. The

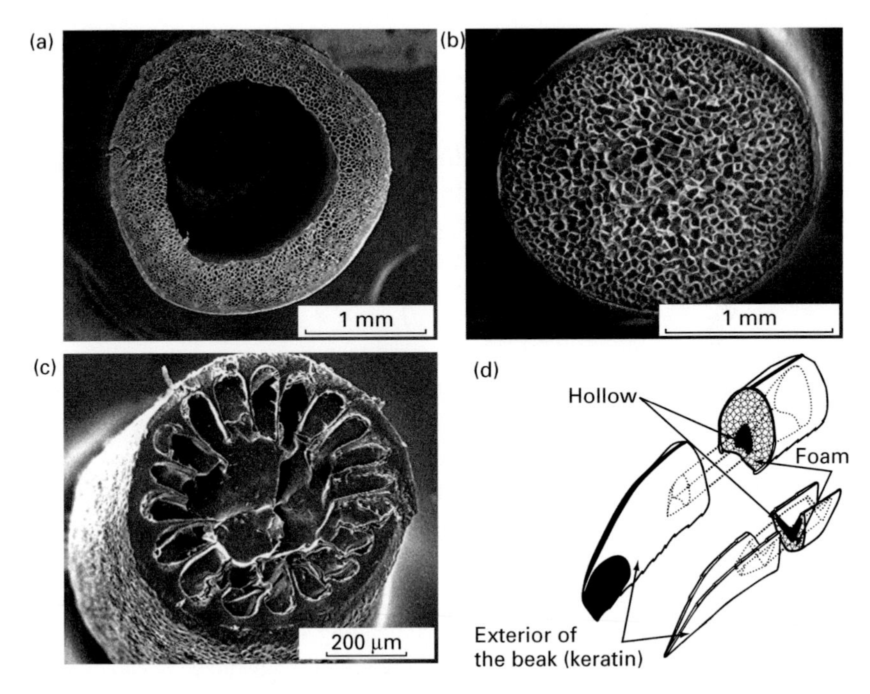

Fig. 6.18 Natural cylindrical shells with compliant cores: (a) grass, *Elytrigia repens*; (b) porcupine quill (*Erethizon dorsatum*); (c) hedgehog spine (*Erinaceus europaeus*); and (d) toco toucan beak (*Ramphastos toco*). (a,b,c, Reprinted from Karam and Gibson 1995, with permission from Elsevier; d, reprinted from Meyers *et al.*, 2008, with permission from Elsevier.)

asymmetrical geometry of bird feathers makes them more complicated to analyze than the circular cross-section of other animal quills and plant stems.

Plant stems are loaded in bending from the wind and from any flower or berries that the stem supports. The wind may also twist the stem, giving torsional loading. Although animal quills are loaded primarily in compression when they contact another animal, any eccentricity in the compressive loading also produces bending. Here, we focus our attention on the bending performance of filled cylindrical shells in nature.

The flexural rigidity of a hollow circular tube is maximized (for a given mass) by increasing the ratio of the tube radius, a, to the wall thickness, t (6.19). But at some point, with increasing a/t, the tube buckles locally, kinking like a bent drinking straw: for this reason, the maximum ratio of a/t in commercially available engineering alloy tubes is typically about 20 (Ashby, 2005). The compliant core of a natural cylindrical shell can act as an elastic foundation, supporting the outer shell, increasing its resistance to local buckling. In this section, we make use of results from Section 3.4.3 to compare the resistance of several natural cylindrical shells with compliant cellular cores to local buckling with that of an equivalent hollow shell without a compliant core. Cylindrical sections in bending can also fail by tensile or compressive failure along the outer edge of the cylinder, or, if the cylinder is hollow, by longitudinal splitting associated with ovalization of the section, as discussed in Section 6.3.2 on bamboo. We also consider failure by these additional modes.

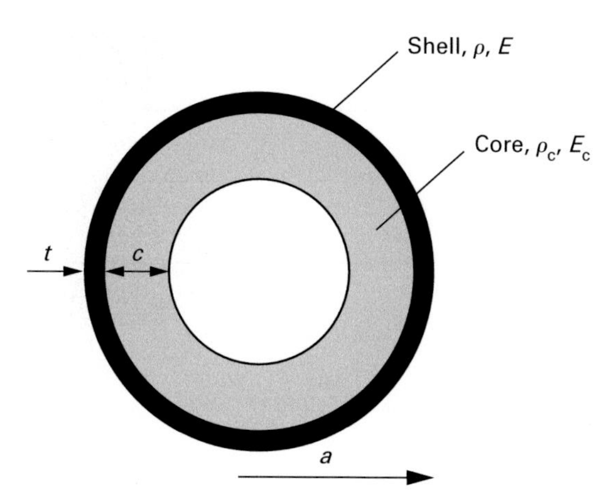

Fig. 6.19 Schematic cross-section of a natural cylinder with a fully dense shell and a cellular core.

A schematic drawing of a natural cylindrical shell is shown in Fig. 6.19. Values for the geometry and shell and core properties of a variety of natural cylindrical structures with compliant cores are listed in Table 6.4. The radius, a, shell thickness, t, and core thickness, c, were obtained from measurements on specimens. The density of the shell, ρ, was assumed to be the same as that of the solid making up the core, ρ_s. The relative density of the core, ρ_c/ρ, was estimated from the area fraction of solid in a plane section of the core. For the animal quills, the Young's modulus of the shell, E, was assumed to be the same as that of the solid making up the cellular core, E_s. For porcupine quills, with a foam-like core, $E_c/E = (\rho_c/\rho)^2$ (3.16), while for hedgehog quills, with a honeycomb-like core loaded radially in the quill, or in the out-of-plane direction of the honeycomb, $E_c/E = (\rho_c/\rho)$ (3.13). Plant stems are more complicated, as the Young's modulus of the sclerenchyma cells in the outer shell is greater than that of the solid making up the foam-like core of parenchyma cells. In addition, turgor pressure induces stretching, rather than bending in the cell walls of the foam-like parenchyma cells (see Section 5.3). Here we assume that the Young's modulus of the sclerenchyma cell wall is ten times that of the parenchyma cell wall, and that the Young's modulus of the core, E_c, relative to that of the solid it is made from, E_s, is linearly related to the relative density (as a result of wall stretching). We then find that the Young's modulus of the core, E_c, relative to that of the shell, E, is $E_c/E = 0.1\ (\rho_c/\rho)$. The ratio of E_c/E for the toucan beak was taken directly from Seki *et al.* (2005).

In Section 3.4.3 we saw that in a bent cylinder, local buckling occurs when the maximum compressive stress on the outer skin of the cylinder equals the uniaxial compressive buckling stress, σ_{cr} (3.61). If the core is sufficiently stiff, it acts as an elastic foundation and the uniaxial compressive buckling stress corresponds to wrinkling of the shell on an elastic foundation. At the other extreme, if the core is too compliant, it fails to support the outer shell, which then acts as a hollow cylinder. The criterion for the core to be sufficiently stiff to act as an elastic foundation can be identified by considering a plot of the buckling wavelength parameter, λ_{cr}, against a/t (Fig. 6.20). If the core is sufficiently stiff, λ_{cr} depends only on the ratio of E_c/E and is independent of a/t (the horizontal dashed lines in Fig. 6.20, (3.62a)). If the core fails to support the outer

Table 6.4 Geometry and local buckling resistance of natural cylindrical shells with compliant cores (based on Karam and Gibson, 1994)

Species	a/t	c/t	ρ_c/ρ	E_c/E	Core acts as an elastic foundation? (Fig. 6.20)	M_{lb}/M_{eq}	c/λ_{cr}
Quills							
Brazilian porcupine (*Coendou prehensilis*)	14.0	Fully filled	0.20	0.04	no	–	–
N. American porcupine (*Erethizon dorsatum*)	18.0	Fully filled	0.125	0.0156	no	–	–
Hedgehog (*Erinaceus europaeus*)	13.7	6.8	0.10	0.10	yes	3.36	4.56
Plant stems							
Tall blue lettuce (*Latuca biennis*)	58.6	12.1	0.10	0.01	yes	1.37	3.81
American pokeweed, wild-berry type (*Phytolacca americana*)	18.4	12.2	0.032	0.0032	no	–	–
Oat, rye grasses (*Avena, secale*)	50	12	0.10	0.01	yes	1.26	3.81
Sedge grass, quack grass, common barley (*Eleocharis, Elytrigia repens, Hordeum vulgare*)	25	8	0.22	0.022	yes	0.77	3.27
Beaks							
Toucan beak (*Ramphastos toco*)	35	7.2	0.1	0.034	yes	3.10	3.40

Source: Data for animal quills and plant stems from Karam and Gibson (1994). Oat and rye grass have similar ratios for a/t, c/t and ρ_c/ρ; typical values are given here. Sedge grass, grass, and barley have similar ratios for a/t, c/t and ρ_c/ρ; typical values are given here. Data for toucan beaks from Seki *et al.* (2005, 2006).

shell and acts as a hollow cylinder, then λ_{cr} depends only on a/t and is independent of E_c/E (the solid line on Fig. 6.20, (3.62b)). All but one of the partially filled cylinders have cores that act as an elastic foundation (Table 6.4); the exception is the American pokeweed, which has a particularly low value of E_c/E. Neither of the fully filled cylinders (the quills) has a core that acts as an elastic foundation. It is possible that their cores fulfill some other need: for instance, quills are modified hairs and the foam-filling may provide thermal insulation. Data for a/t and E_c/E for the natural cylinders with cores that act as elastic foundations are superimposed on Fig. 6.20. The data lie at the very left of the horizontal portion of the dashed lines: the cores are just stiff enough, and no stiffer, to act as an elastic foundation for the a/t of each shell.

We next compare the performance of the natural cylindrical shells with the compliant cores that act as elastic foundations to that of a hollow shell (with no core) of equal

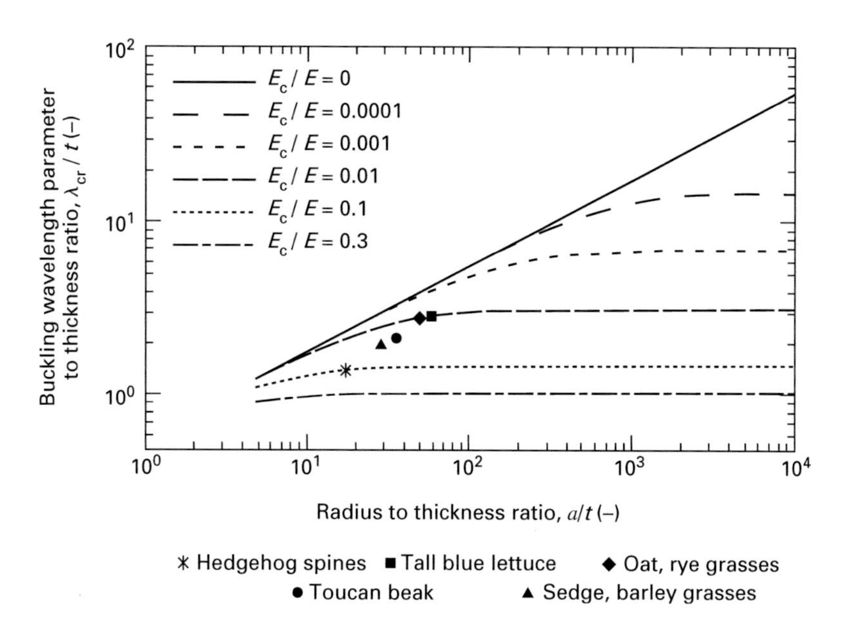

Fig. 6.20 Buckling wavelength parameter, λ_{cr}, normalized by the shell thickness, t, plotted against the shell radius, a, normalized by t. The horizontal dashed lines represent a cylindrical shell supported by an elastic foundation (3.62a), while the solid line represents the case of the hollow shell (3.62b). Values of a/t and E_c/E corresponding to several natural cylindrical shells are indicated on the plot. (Adapted from Karam and Gibson, 1995.)

radius and weight. The natural structures are primarily loaded in bending. We calculate the ratio of the local buckling moment of the shell with the compliant core to that of a hollow shell of equivalent radius and weight, using (3.73) for the partially filled shells. The local buckling moments for the partially filled shells are, in general, substantially higher than those of the hollow shells of equivalent radius and weight: they are up to a factor of about 3 higher for hedgehog spines and the toucan beak (Table 6.4).

In Section 3.4.3 we noted that the stresses within the core decay to about 5% of their maximum values at a radial distance $5\lambda_{cr}$ from the interface with the shell. Core material interior to this does not carry any significant load and can be removed. For the partially filled natural cylindrical structures with cores that act as elastic foundations, the ratio of the measured core thickness to the buckling wavelength parameter, c/λ_{cr} varies from 3.4 to 4.6 (Table 6.4): the core thickness matches the depth at which there are significant stresses within the core.

Finally, we examine other possible failure modes for the natural cylindrical shells that have cores that act as elastic foundations to resist local buckling (those plotted on Fig. 6.20). Ovalization of the cross-sections is negligible for these quills and stems: typically less than 1.5% of the cylinder radius (3.72), so that longitudinal splitting associated with ovalization is unlikely. Next, we compare the tensile stress in the shell when the local buckling moment is reached with the tensile strength of the shell material. The hedgehog quill and the toucan beak are both made of keratin which has a Young's modulus in the range 1–4 GPa and a tensile strength in the range 130–210 MPa (Table 2.2). Local buckling occurs when the compressive stress in the shell reaches the buckling stress in uniaxial compression: $\sigma_b = E(t/a)f_1$, (3.61). Quills are modified hairs: we assume that, like hair, they are made of relatively "soft" keratin, with a modulus in the lower range, $E = 1$ GPa. For the hedgehog quill, with $E = 1$ GPa, $\sigma_b = 138$ MPa, close to the lower range of tensile strengths. For the toucan beak, using $E = 1$ GPa gives a value of $\sigma_b = 65$ MPa; values of E up to 3.5 GPa give buckling stresses below the maximum tensile strength of keratin. It appears that in both cases, local buckling precedes tensile failure. For the plant stems, we compare the stress in the outer shell of the stem when local buckling occurs (3.61) with the tensile strength of the outer shell. The outer shells of the plant stems are made up largely of sclerenchyma cells. Leaf fibers are also largely sclerenchyma: their Young's moduli have been measured to be in the range 2–26 GPa (Table 6.3), with a value of 23 GPa for *Lolium perenne*, a rye grass. Tensile strengths of leaf fibers correlate with the Young's modulus: for leaf fibers with moduli of roughly 20 GPa, the tensile strengths range from 800 to 1500 MPa (McLaughlin and Tait, 1980). Using $E = 20$ GPa, we calculate that the stress in the shell when local buckling occurs is 690 MPa for tall blue lettuce and 720 MPa for oat and rye grasses, somewhat below the tensile strength of the shell, so that local buckling again precedes tensile failure. Local buckling is the dominant failure mode in each of these cases.

Overall, these findings indicate that the partially filled cores of the hedgehog spine, the toucan beak and some plant stems inhibit ovalization and increase their resistance to local buckling or kinking by acting as an elastic foundation supporting the dense outer shell.

6.5 Biomimicking of natural cellular structures

In this chapter, we have seen how nature makes use of sandwich structures, cylinders with radial density gradients and filled cylindrical tubes to increase the mechanical performance of leaves, skulls, palms, bamboo, animal quills and plant stems. Engineers, too, make use of the same structural arrangements, sometimes by using the principles of structural mechanics and sometimes by mimicking structures in nature.

The concept of a sandwich panel, with two stiff, strong faces separated by a lightweight core, was first developed in the 1800s by Duleau and Fairbairn, using basic mechanics. But it was not until the 1930s, when structural adhesives were developed, that engineering sandwich panels became practical (Timoshenko, 1953; Zenkert, 1995). The first major engineering use of sandwich panels was in World War II "Mosquito" aircraft: they had wood veneer faces and a balsa core. Today, structural sandwich panels are often used in aircraft components and in sporting equipment (Fig. 6.21). Typical face materials include aluminum and fiber-reinforced composites such as Kevlar. Honeycomb cores are preferred in applications where weight is at a premium; foam cores are used when the panel must provide thermal insulation, as well as structural support, as in buildings.

Engineers are also interested in fabricating cellular materials with a density gradient, similar to the structure of palms and bamboo. The cellular microstructure of rattan (*Calamus rotang*), a tropical climbing palm, has been replicated with remarkable fidelity

Fig. 6.21 Examples of engineering sandwich structures: (a) a helicopter rotor blade and (b) a downhill ski with a foam core.

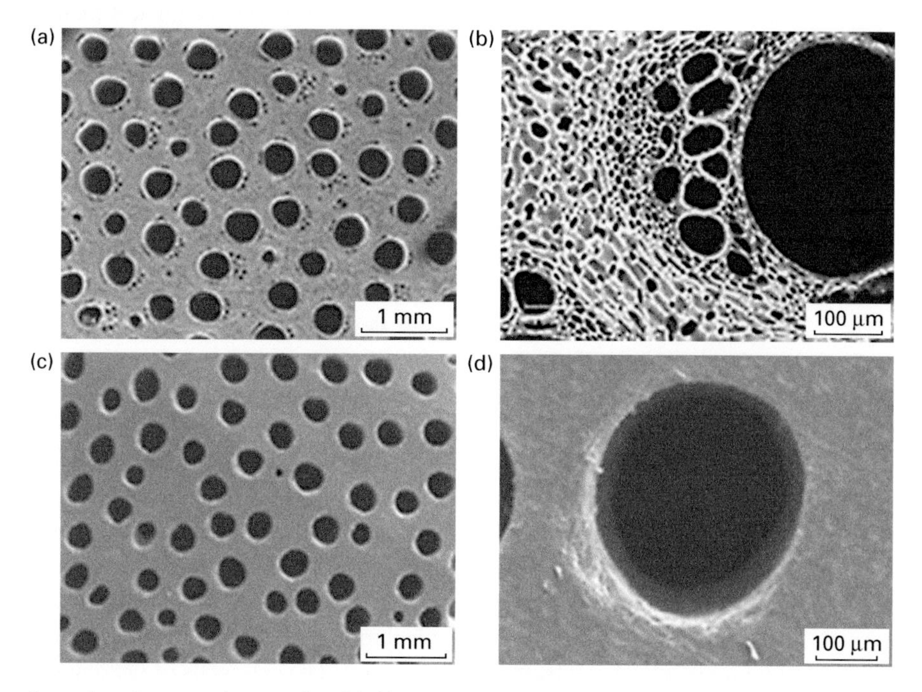

Fig. 6.22 Scanning electron micrographs of (a,b) rattan (*Calamus rotang*) template and (c,d) bioceramic Si-SiC composite made by pyrolysis of the rattan and infiltration with liquid Si. (Reprinted from Zampieri *et al.*, 2005, with permission from Wiley.)

by Greil and co-workers using techniques similar to those described in Section 4.2.6 for mimicking the structure of wood (Zampieri *et al.*, 2005). Rattan samples with cell sizes of the order of 450 μm were pyrolyzed in an inert atmosphere, producing an amorphous carbon char that preserved the cellular microstructure (although with some shrinkage). These biocarbon preforms were then infiltrated with liquid silicon that reacts with the carbon to form silicon carbide. Pores smaller than 50 μm filled with silicon, so that the resulting cellular material was a composite of Si-SiC (Fig. 6.22). While the rattan samples in this study were limited to about 40 mm in diameter, so that the radial density gradient was not significant, in principle, this technique could be used to replicate the density gradient in palm and bamboo. In this study, in a final processing step, zeolites were crystallized on the cell walls of the porous biocomposite. The final material had a hierarchy of pore sizes combined with sufficient mechanical and thermal stability for applications in catalytic and sorption/separation processing. In another study, the bioceramic replica was completely infiltrated with a metal, giving a fully dense ceramic-reinforced metal composite with superior mechanical properties (Zollfrank *et al.*, 2005). Using a larger palm specimen or bamboo as the preform would produce a ceramic-reinforced metal composite in which the volume fraction of the ceramic increased towards the periphery of the composite.

Ceramic and metal foams with density gradients have also been fabricated. Ceramic foams with a density gradient are of interest in applications ranging from biomedical implants to thermal barrier coatings and thermal shock resistant structures. Columbo and Hellman (2002) foamed silicone resin by direct blowing and then converted the polymer foam into a ceramic one by pyrolysis; graded porosity was achieved by varying the processing parameters during the foaming. Zeschky *et al.* (2005) used a similar technique, foaming a blend of poly(silsesquioxane), Si and SiC, and then pyrolyzing the resulting foam. Ceramic foams with a gradient in relative density, from 10% to 50% through the thickness of the specimen, were obtained through control of the temperature during the polymer foaming process.

Metallic foams with a density gradient are of interest for mechanically efficient structural components, for reducing stress concentrations at interfaces between dissimilar materials and in heat transfer devices. Metal foams with density gradients have been made by first compressing the precursor open-cell polymer foam to varying degrees (e.g. by compressing a conical foam sample in a cylindrical container) (Brothers and Dunand, 2006). The compressed polymer foam was then infiltrated with plaster and removed by pyrolysis. Infiltration with molten metal then replicated the polymer foam; the plaster mold was removed by immersion in water. This method has the disadvantage that the cell walls in the denser regions are highly deformed by the initial compression of the precursor polymer foam. In a second method, Dunand and co-workers selectively removed material from a 10% dense open-cell aluminum foam of initially uniform density by chemical milling in baths of either hydrochloric acid or sodium hydroxide (Matsumoto *et al.*, 2007). The cell walls of the foams immersed in HCl had large pits and cracks in them after processing, reducing their mechanical properties. Those immersed in NaOH were more uniform in thickness, with fewer defects. Gradient densities can be produced by gradually draining the dissolution bath in which the metal foam sample is placed; samples with relative densities varying from 5% to 10% across their thickness were produced in this way.

Thin-walled cylindrical tubes for engineering applications are often reinforced with longitudinal and circumferential stiffeners (as, for example, in the legs of offshore oil platforms). These stiffeners increase the resistance of the tubes to buckling, but do not act as a continuous elastic foundation, as the compliant cores in hedgehog spines, some plant stems and the toucan beak do. Recently, there have been attempts to mimic these natural cylindrical tubes with their compliant cores. Palumbo and Tempesti (2001), for instance, found that a foam core prevented local buckling in a circular,

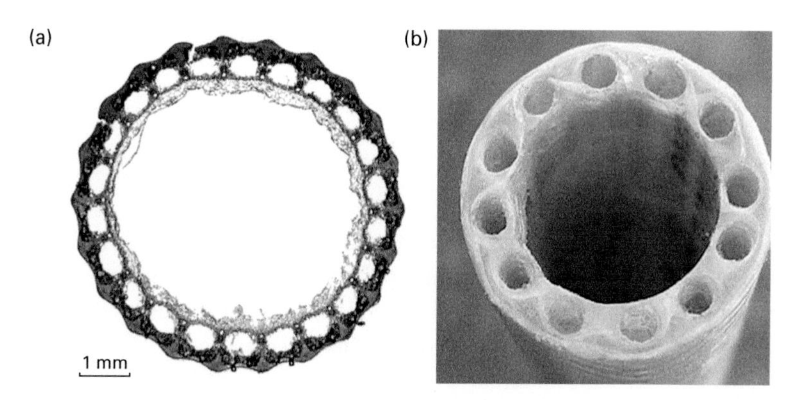

Fig. 6.23 (a) Cross-section of the stem of the horsetail (*Equisetum hyemale*) (diameter = 6 mm) and (b) the "technical plant stem" made with double braid textile construction. (Reprinted from Milwich *et al.*, 2006, with permission from the *American Journal of Botany*.)

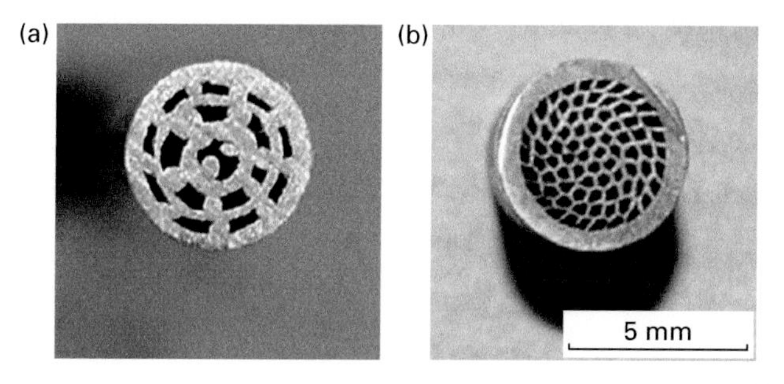

Fig. 6.24 Cylindrical shells with compliant cores, mimicking (a) lotus roots (*Nelumbo nucifera*) and (b) plant stems. (Reprinted from Utsunomiya *et al.*, 2008, with permission of Wiley.)

glass-fiber composite tube, allowing the tube to reach the full tensile strength of the glass-fiber composite. Milwich *et al.* (2006) examined the hierarchical structures of horsetail (*Equisetum hyemale*) and the giant reed (*Arundo donax*) and used them as models for developing the "technical plant stem," made by a fiber pultrusion process that uses a braiding machine to control the placement of diagonal fiber bundles into the structure (Fig. 6.23). Utsunomiya *et al.* (2008) placed a combination of copper and aluminum wires into a cylindrical copper tube, extruded the entire structure to densify it and then leached out the aluminum, leaving a dense cylindrical shell supported by a porous core, similar to the root of the lotus plant (*Nelumbo nucifera*). The relative density of the core was decreased by using copper-coated rather than uncoated aluminum wires (Fig. 6.24). Recent numerical studies of engineering cylindrical shells with compliant cores have indicated that, in addition to increasing the resistance to local buckling, the core also decreases the sensitivity of the cylinder to imperfections (Obrecht *et al.*, 2006, 2008). Knock-down factors associated with imperfection sensitivity in cylindrical shells are large: for instance, a shell with an imperfection amplitude of 0.4 times the shell thickness has a buckling strength of only 30% of that of the perfect shell. For comparison, Obrecht and co-workers find that for a cylindrical shell with a radius to wall thickness ratio of 100 this can be increased to 65% by adding an inner compliant core that is 1% as stiff as the shell and has a thickness of 10% of the shell radius.

6.6 Summary

In this chapter, we have seen that combining cellular and nearly fully dense components can give improved mechanical performance compared with a uniform material. The sandwich structure of a plate-like bone, such as the scapula, reduces the weight by a factor of the order of ten, compared with that of a dense cortical plate of the same span and stiffness; we expect similar weight reductions for shell-like sandwich bones, such as the skull. The radial density gradient in two palms, *Iriartea gigantea* and *Welfia georgii*, roughly doubles their flexural rigidity over that of an equivalent section of uniform density. The flexural rigidity of bamboo is increased both by its radial density gradient as well as its tubular cross-section. Hollow tubes like bamboo can fail by several modes (local buckling, longitudinal tensile or compressive failure, longitudinal splitting); the dominant mode depends both on the geometry of the cross-section and on the material properties in the longitudinal and circumferential directions. And the compliant cellular cores supporting dense outer cylindrical shells in hedgehog quills, some plant stems and the toucan beak increase their resistance to local buckling by a factor of up to 3.4. Engineers are increasingly interested in mimicking these natural cellular structures.

References

Introduction

Buhler P (1972) Sandwich structures in the skull capsules of various birds: the principle of lightweight structures in organisms. In *Information of the Institute for Lightweight Structures*, Vol. IL4. Stuttgart: University of Stuttgart.

Gibson LJ, Ashby MF, Karam GN, Wegst U and Shercliff HR (1995) The mechanical properties of natural materials II Microstructures for mechanical efficiency. *Proc. Roy. Soc. Lond.* **A450**, 141–62.

Sandwich structures – leaves

Bodig J and Jayne BA (1982) *Mechanics of Wood and Wood Composites*. New York: Van Norstrand Reinhold.

Gibson LJ, Ashby MF and Easterling KE (1988) Structure and mechanics of the iris leaf. *J. Mat. Sci.* **23**, 3041–8.

Gibson LJ and Ashby MF (1997) *Cellular Solids: Structure and Properties,* 2nd edn. Cambridge: Cambridge University Press.

Harder DL, Stevens CL, Speck T and Hurd CL (2005) The role of blade buoyancy and reconfiguration in the mechanical adaptation of the southern bullkelp Durvillaea. In *SEB Book on Biomechanics and Biology*, ed. Herrel A, Speck T and Rowe NP. Boca Raton, FL: CRC Press, pp. 61–84.

Haughton PM and Sellen DB (1969) Dynamic mechanical properties of the cell walls of some green algae. *J. Exp. Bot.* **20**, 516–35.

Kamiya N, Tazawa M and Takata T (1963) The relation of turgor pressure to cell volume in *Nitella* with special reference to mechanical properties of the cell wall. *Protoplasma* **57**, 501–21.

Lin T-T and Pitt RE (1986) Rheology of apple and potato tissue as affected by cell turgor pressure. *J. Texture Studies* **17**, 291–313.

Moulia B and Fournier M (1997) Mechanics of the maize leaf: a composite beam model of the midrib. *J. Mat. Sci.* **32**, 2771–80.

Nilsson SB, Hertz CH and Falk S (1958) On the relation between turgor pressure and tissue rigidity 2: theoretical calculations on model systems. *Physiol. Plant* **11**, 818–37.

Oye ML, Vanstreels E, De Baerdemaeker J *et al.* (2007) Effect of turgor on micromechanical and structural properties of apple tissue: a quantitative analysis. *Postharvest Biol. Technol.* **44**, 240–7.

Preston RD (1974) *The Physical Biology of Plant Cell Walls*. London: Chapman and Hall.

Probine MC and Preston RD (1962) Cell growth and the structure and mechanical properties of the wall in internodal cells of *Nitella opaca* II: mechanical properties of the walls. *J. Exp. Bot.* **13**, 111–27.

Sellen DB (1979) The mechanical properties of plant cell walls. In *The Symposia of the Society for Experimental Biology: The Mechanical Properties of Biological Materials*, Leeds University, September 4–6, pp. 315–29.

Stevens CL, Hurd CL and Smith MJ (2002) Field measurement of the dynamics of the bull kelp Durvillaea antarctica (Chamisso) Heriot. *J. Exp. Marine Biol. Ecol.* **268**, 147–71.

Toole GA, Gunning PA, Parker ML, Smith AC and Waldron KW (2001) Fracture mechanics of the cell wall of *Chara coralline*. *Planta* **212**, 606–11.

Treitel O (1946) Elasticity, plasticity and fine structure of plant cell walls. *J. Colloid Sci.* **1**, 327–70.

Vincent JFV (1982) The mechanical design of grass. *J. Mat. Sci.* **17**, 856–60.

Vincent JFV (1991) Strength and fracture of grasses *J. Mat. Sci.* **26**, 1947–50.

Wainwright SA, Gibbs WD, Currey JD and Gosline JM (1976) *Mechanical Design in Organisms*. Princeton, NJ: Princeton University Press.

Yu Y, Fei B, Zhang B and Yu X (2007) Cell wall mechanical properties of bamboo investigated by in-situ imaging nanoindentation. *Wood Fiber Sci.* **39**, 527–35.

Sandwich structures – bone

Buhler P (1972) Sandwich structures in the skull capsules of various birds: the principle of lightweight structures in organisms. In *Information of the Institute for Lightweight Structures*, Vol. IL4. Stuttgart: University of Stuttgart.

Hodgson VR (1973) Head model for impact tolerance. In *Human Impact Response: Measurement and Simulation*, ed. King WF and Mertz HJ. New York: Plenum Press, pp. 113–28.

Meyers MA, Chen P-Y, Lin AY-M and Seki Y (2008) Biological materials: structure and mechanical properties. *Prog. Mat. Sci.* **53**, 1–206.

Roark RJ and Young WC (1975) *Formulas for Stress and Strain*, 5th edition. New York: McGraw Hill.

Swartz SM, Parker A and Huo C (1998) Theoretical and empirical scaling patterns and topological homology in bone trabeculae. *J. Exp. Biol.* **201**, 573–90.

Palm and bamboo stems

Arce-Villalobos OA (1993) *Fundamentals of the design of bamboo structures*. Unpublished PhD thesis, Einhoven University of Technology, The Netherlands.

Ashby MF (2005) *Materials Selection in Mechanical Design,* 3rd edn. Oxford: Butterworth Heinemann.
Calladine CR (1983) *Theory of Shell Structures.* Cambridge: Cambridge University Press.
Cecchini LS and Weaver PM (2002) Optimal fiber angles to resist the Brazier effect in orthotropic tubes. *AIAA J.* **40**, 2136–8.
Dinwoodie JM (1981) *Timber: Its Nature and Behaviour.* New York: Van Nostrand Reinhold.
Gerard G (1968) *Minimum Weight Design of Compressive Structures.* New York: New York University Press/Interscience.
Gibson LJ, Ashby MF, Karam GN, Wegst U and Shercliff HR (1995) The mechanical properties of natural materials, II: microstructures for mechanical efficiency. *Proc. Roy. Soc. Lond.* **A450**, 141–62.
Kuo-Huang L-L, Huang Y-S, Chen S-S and Huang Y-R (2004) Growth stresses and related anatomical characteristics in coconut palm trees. *IAWA J.* **25**, 297–310.
NASA (1968) *Space Vehicles Design Criteria (Structures) – Buckling of Thin-walled Circular Cylinders.* Technical report, NASA SP-8007.
Niklas KJ (1992) *Plant Biomechanics: An Engineering Approach to Plant Form and Function.* Chicago, IL: University of Chicago Press.
Rich PM (1986) Mechanical architecture of arborescent rain forest palms. *Principes* **30**, 117–31.
Rich PM (1987a) Developmental anatomy of the stem of *Welfia geogii*, *Iriartea gigantica* and other arborescent palms: implications for mechanical support. *Amer. J. Botany* **74**, 792–802.
Rich PM (1987b) Mechanical structure of the stem of arborescent palms. *Botanical Gazette* **148**, 42–50.
Ruggeberg M, Speck T, Paris O *et al.* (2008) Stiffness gradients in vascular bundles of the palm *Washingtonia robusta. Proc. Roy. Soc.* **B275**, 2221–9.
Schmitt U, Weiner G and Liese W (1995) The fine structure of the stegmata in *Calamus axillaris* during maturation. *IAWA J.* **16**, 61–8.
Schulgasser K and Witztum (1992) On the strength, stiffness and stability of tubular plant stems and leaves. *J. Theor. Biol.* **155**, 497–515.
Spatz HC, Kohler L and Speck T (1998) Biomechanics and functional anatomy of hollow-stemmed sphenopsids. I: *Equisetum giganteum* (*Equisetaceae*). *Amer. J. Bot.* **85**, 305–14
Suo Z (1990a) Delamination specimens for orthotropic materials. *J. Appl. Mech.* **57**, 627–34.
Suo Z (1990b) Singularities, interfaces and cracks in dissimilar anisotropic media. *Proc. Roy. Soc. Lond.* **A427**, 331–58.
Tomlinson PB (1990) *The Structural Biology of Palms.* Oxford: Oxford University Press.
Wegst UGK and Ashby MF (2007) The structural efficiency of orthotropic stalks, stems and tubes. *J. Mat. Sci.* **42**, 9005–14.
Yu Y, Fei B, Zhang B and Yu X (2007) Cell wall mechanical properties of bamboo investigated by in-situ imaging nanoindentation. *Wood Fiber Sci.* **39**, 527–35.

Shells with compliant cores

Ashby MF (2005) *Materials Selection in Mechanical Design,* 3rd edn. Oxford: Butterworth Heinemann.
Karam GN and Gibson LJ (1994) Biomimicking of animal quills and plant stems: natural cylindrical shells with foam cores. *Mat. Sci. Eng.* **C2**, 113–32.
Karam GN and Gibson LJ (1995) Elastic buckling of cylindrical shells with elastic cores I: Analysis. *Int. J. Solids Struct.* **32**, 1259–83.
McLaughlin EC and Tait TA (1980) Fracture mechanism of plant fibres. *J. Mat. Sci.* **15**, 89–95.

Niklas KJ (1992) *Plant Biomechanics: An Engineering Approach to Plant Form and Function.* Chicago, IL: University of Chicago Press.

Seki Y, Schneider MS and Meyers MA (2005) Structure and mechanical behavior of a toucan beak. *Acta Mat.* **53**, 5281–96.

Seki Y, Kad B, Benson D and Meyers MA (2006) The toucan beak: structure and mechanical response *Mat. Sci. Eng.* **C26**, 1412–20.

Vincent JFV and Owers P (1986) Mechanical design of hedgehog spines and porcupine quills. *J. Zool. Lond.* **A210**, 55–75.

Biomimicking

Brothers AH and Dunand DC (2006) Density-graded cellular aluminum. *Adv. Eng. Mat.* **8**, 805–9.

Columbo P and Hellmann JR (2002) Ceramic foams from preceramic polymers. *Mat. Res. Innov.* **6**, 260–72.

Matsumoto Y, Brothers AH, Stock SR and Dunand DC (2007) Uniform and graded chemical milling of aluminum foams. *Mat. Sci. Eng.* **A447**, 150–7.

Milwich M, Speck T, Speck O, Stegmaier T and Planck H (2006) Biomimetics and technical textiles: solving engineering problems with the help of nature's wisdom. *American Journal of Botany* **93**, 1455–65.

Obrecht H, Rosenthal B, Fuchs P, Lange S and Marusczyk C (2006) Buckling, postbuckling and imperfection-sensitivity: old questions and some new answers. *Comput. Mech.* **37**, 498–506.

Obrecht H, Fuchs P, Reinicke U, Rosenthal B and Walkowiak (2008) Influence of wall constructions on the load-carrying capability of light-weight structures. *Int. J. Solids Struct.* **45**, 1513–35.

Palumbo M and Tempesti E (2001) Fiber-reinforced syntactic foams as a new lightweight structural three-phase composite. *Appl. Comp. Mat.* **8**, 343–59.

Timoshenko SP (1953) *A History of Strength of Materials*. New York: McGraw Hill.

Utsunomiya H, Koh H, Miyamoto J and Sakai T (2008) High strength porous copper by cold-extrusion. *Adv. Eng. Mat.* **10**, 826–9.

Zampieri A, Sieber H, Selvam T *et al.* (2005) Biomorphic cellular SiSiC/Zeolite ceramic composites: from rattan palm to bioinspired structured monoliths for catalysis and sorption. *Adv. Mat.* **17**, 344–9.

Zenkert D (1995) *An Introduction to Sandwich Construction*. Warrington, UK: Engineering Materials Advisory Services Ltd.

Zeschky J, Hofner T, Arnold C *et al.* (2005) Polysilsesquioxane derived ceramic foams with gradient porosity. *Acta Mat.* **53**, 927–37.

Zollfrank C, Travitzky N, Sieber H, Selchert T and Greil P (2005) Biomorphous SiSiC/Al-Si ceramic composites manufactured by squeeze casting: microstructure and mechanical properties. *Adv. Eng. Mat.* **7**, 743–6.

7 Property charts for natural cellular materials and their uses

7.1 Introduction

The properties of engineering materials are usefully displayed as *material property charts*. The charts give perspective, showing the range and relationship of material attributes. Material indices, developed in Chapter 3, can be plotted on them, allowing the selection of materials meeting given constraints and objectives (Ashby, 2005; Ashby *et al.*, 2007). We use the design of a light, stiff beam as an example; the other material indices are used in a similar way.

Figure 7.1 shows a chart of the Young's modulus, E, plotted against density, ρ, for engineering materials. The scales are logarithmic, allowing the material indices E/ρ, $E^{1/2}/\rho$ and $E^{1/3}/\rho$, derived in Chapter 3, to be plotted onto the chart. The condition

$$\frac{E}{\rho} = C \tag{7.1}$$

or, taking logs,

$$\log E = \log \rho + \log C \tag{7.2}$$

describes a family of straight parallel lines of slope 1 on a plot of $\log E$ against $\log \rho$, where each line corresponds to a value of the constant C. The condition

$$\frac{E^{1/2}}{\rho} = C \tag{7.3}$$

or, taking logs again,

$$\log E = 2\log \rho + 2\log C \tag{7.4}$$

gives another set, this time with a slope of 2. That for the third index gives yet another set, with slope 3. These *selection guidelines* give the slope of the family of parallel lines belonging to that index. The charts for natural material that follow in this chapter have these guidelines plotted on them.

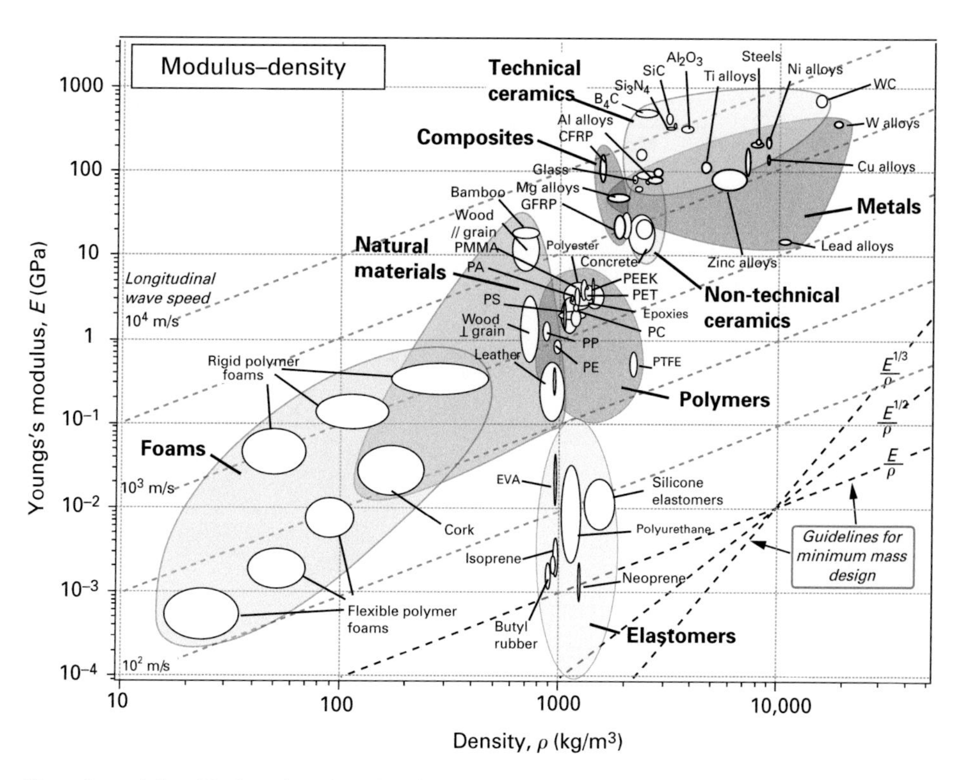

Fig. 7.1 Young's modulus, E, plotted against density, ρ. The heavy envelopes enclose data for a given class of material. The diagonal contours show the longitudinal wave velocity. The guidelines of constant E/ρ, $E^{1/2}/\rho$, and $E^{1/3}/\rho$ allow selection of materials for minimum weight, deflection-limited design. See plate section for color version.

It is now easy to read off the chart the subset of materials that optimally maximize performance for each loading geometry. All the materials that lie on a line of constant $E^{1/2}/\rho$ perform equally well as a light, stiff beam; those above the line are better, those below, worse. Figure 7.2 shows a schematic of the chart with a grid of lines corresponding to values of $E^{1/2}/\rho$ from 0.1 to 10 in units of GPa$^{1/2}$/(Mg.m^{-3}). A material with an index value of 3 gives a beam that has one-tenth the weight of one with a value of 0.3 that has the same bending stiffness.

Property charts for natural materials are equally useful (Wegst and Ashby, 2004). In this chapter we examine five charts that display density, modulus, strength and toughness of natural cellular materials (Figs. 7.3–7.7), allowing a comparison of their properties and an exploration of the possible constraints that led to their evolution. Natural cellular materials all have low densities (ρ = 0.1–1.2 Mg/m^3) because of the high volume fractions of voids they contain. Most are anisotropic as a result of the shape and orientation of the fibers within their cell walls and of the shape of the cells themselves; the prismatic cells of wood, for instance, give a much greater stiffness and strength along the grain than across it. When anisotropy is great, the modulus parallel (symbol ||) and perpendicular (symbol ⊥) to the fiber orientation or grain have been plotted separately. Classes of natural materials are circumscribed by large balloons; class members are shown as smaller white bubbles within

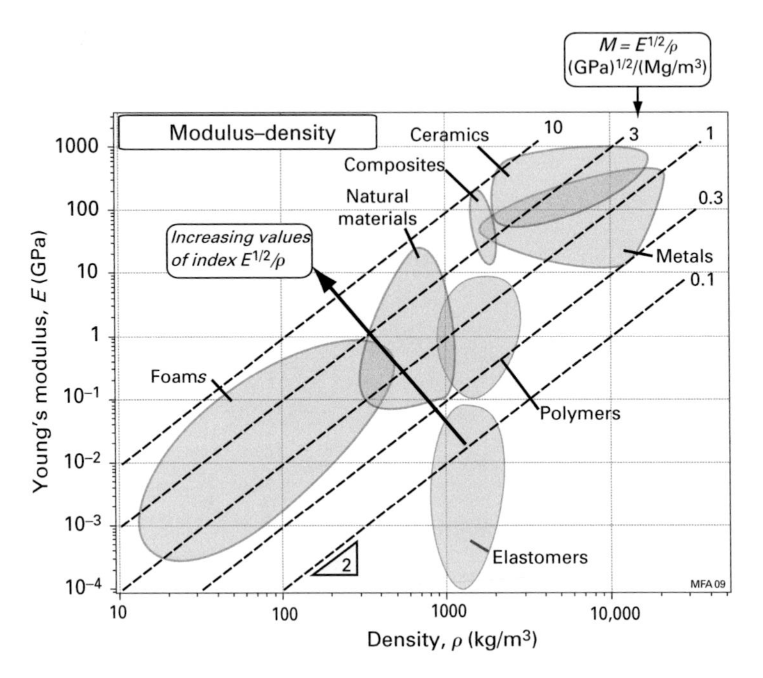

Fig. 7.2 A schematic E–ρ chart showing a grid of lines for the material index $M = E^{1/2}/\rho$.

them. Where the range of density is large, as it is in trabecular bone, the data sets have been broken up into three ranges: low density (LD), medium density (MD) and high density (HD). Fluid-filled cells (here, plant parenchyma) are differentiated from those that, when tested, were not filled. The basic building blocks (minerals, polysaccharides and proteins) are included where appropriate. To distinguish palms from woods, data and envelopes for palm are plotted with a green broken line. To distinguish trabecular bone from other mineralised cellular materials such as corals, the bone data and envelopes are enclosed in blue dash-dot lines.

The charts can be used to explore some of the mechanical functions that natural cellular materials have evolved to fill, and to examine how structural shape can be combined with material to enhance performance further.

7.2 Using property charts for natural cellular materials

Chapter 3, Section 3.5 explained how engineering materials are chosen to perform desired *functions* while meeting a number of *constraints*. Among those that meet the constraints the best choices are those that minimize one or more *objectives*, which, in engineering design, are frequently those of minimizing cost, mass or volume. In making the choice, designers have at their disposal a number of *free variables*, parameters that can be varied to best meet the objectives. Function, constraints, objectives and free variables (Table 3.6) define the selection problem. The choice is made by *screening out* all materials that fail to meet the constraints and then *ranking* the survivors by their ability to minimize the objective.

Applying methods like these might be termed "forward engineering." Reverse engineering – dissecting successful designs to find out how the designer overcame constraints and met objectives – is a useful tool of education and one that is also practiced in industry to explore the methods used by competitors.

Nature, through the evolutionary process, develops her own materials. They, too, perform functions. The matrix of Table 7.1 shows the multiple functions that some examples of natural cellular materials and structures perform. The material must perform its functions subject to constraints, among them the ability to tolerate the environment in which the organism lives: the climate, the acidity of the soil, the available diet. Among the materials that meet the constraints, the best are those that minimize one or more objectives, of which the most obvious is to minimize the resources required to allow the organism to grow as large (or larger) or as strong (or stronger) than those with which it competes. The matrix brings out the multi-functional role that many of them fill. Table 7.2 lists common constraints and objectives for natural materials, in a similar way to Table 3.7 for engineering materials.

The constraints and objectives that have most influenced the evolutionary development of natural materials are not always obvious. It is interesting to ask: could the methods of reverse engineering be applied to natural materials to infer the constraints and objectives that shaped their attributes? Caution is needed here. Some constraints may not be obvious (a tolerance of some aspect of soil chemistry, a defense against local predators or reaction to local climate, for example). There are the constraints associated with the ability to change in size as an organism matures, to thrive on locally available nutrients and to survive extremes of climate. And some attributes of natural materials may not have evolved to meet constraints that exist today, but are a result of evolutionary inheritance instead. None of these are easily identified from a study of mechanical properties. But others – such as the provision of stiffness, strength or toughness at minimum "cost" (here meaning demand on nutrients) – are more easily identified. So it is worth a try.

Chapter 3 also introduced the idea of a *material index*: the property or property-group that is an indicator of performance or "measure of excellence" in performing a particular function. Table 3.8 listed indices for stiffness and strength-limited design for three generic components,a tie, a beam and a panel (Fig. 3.29), for each of four objectives. Selecting materials with the objective of minimizing mass uses as little material as possible, conserving material resources. It also minimizes the energy of locomotion because, in nature as in engineering, fuel consumption scales with weight. The charts allow those that excel to be identified.

The modulus–density chart

Figure 7.3 shows data for the Young's modulus, E, and density, ρ, for natural cellular materials and the materials of which they are made. Three stiffness guidelines are shown, each representing the material index for a particular mode of loading:

$$M_1 = E/\rho \quad \text{(stiff tie in tension)} \tag{7.5}$$

$$M_2 = E^{1/2}/\rho \quad \text{(stiff beam in flexure)} \tag{7.6}$$

Table 7.1 Natural cellular materials and structures and the multiple functions they perform

			Mechanical				*Thermal*		*Transport*		*Other*			
			Stiffness	*Strength*	*Toughness*	*Energy absorption*	*Insulation*	*Fire protection*	*Exchange within organism*	*Exchange with environment*	*Buoyancy*	*Create large surface area*	*Nutrient storage*	*Bio-protection*
Materials	*Plant*	*Wood*	■	■	■				■					
		Cork					■	■		■				■
		Plant parenchyma	■	■					■				■	
	Animal	*Trabecular bone*	■	■	■	■			■			■		
		Adipose tissue		■	■								■	
		Coral	■	■								■		
		Sponge	■	■	■							■		
Structures	*Sandwich structures*	*Monocot leaves*	■	■	■							■		
		Sandwich bones	■	■	■	■								
		Sandwich shells	■	■	■									
		Cuttlefish bone	■	■							■			
	Gradient cylinders	*Palm*	■	■										
		Bamboo	■	■										
	Filled cylinders	*Plant stem*	■	■										
		Animal quill	■	■			■							■

Table 7.2 Some constraints and objectives governing material choice in nature

Common constraints	Common objectives	Free variables
Meet a target value of	**Minimize:**	**Evolutionary design of**:
Stiffness	Resource input for given performance	Certain dimensions
Strength	Mass for given strength	Shape and configuration at every level from the macro to the nano
Toughness	Volume for given strength	
Thermal conductivity		
Service temperature		
Must be		
Compatible with the environment		
Scalable		
Be made from available resources		
Self repairing		

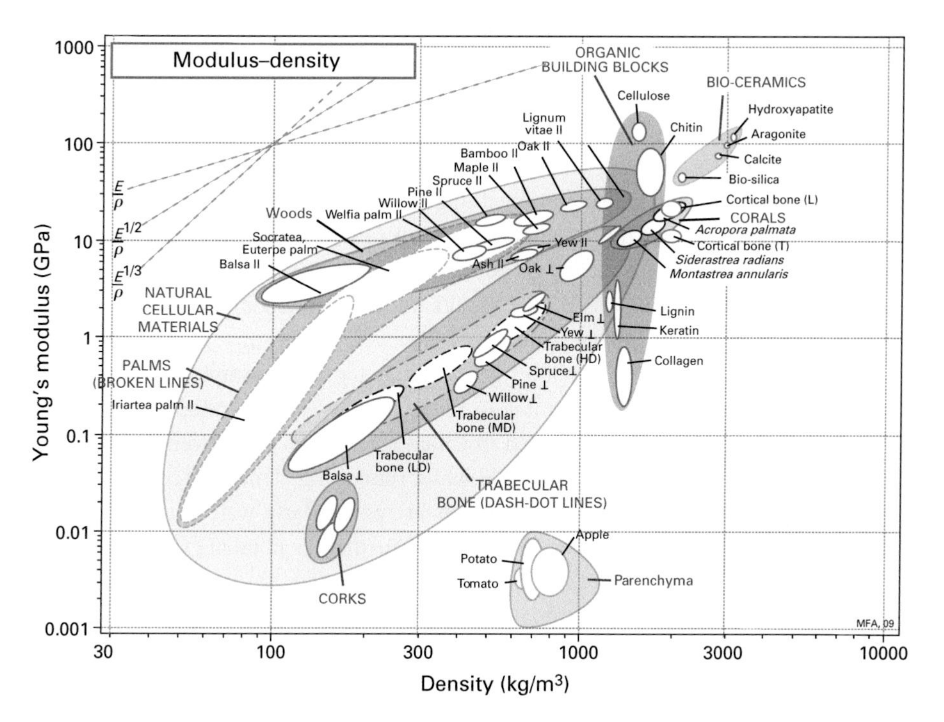

Fig. 7.3 The modulus–density chart for natural cellular materials. Materials that are stiff and light have high values of the indices indicated at the upper left. See plate section for color version.

$$M_3 = E^{1/3}/\rho \quad \text{(stiff plate in flexure)} \tag{7.7}$$

The natural polymer with the highest efficiency in tension, measured by the index E/ρ, is cellulose (it exceeds that of steel by a factor of about 3.2), giving it exceptional stiffness performance in the form of fibers. Wood, palm and bamboo have high values of

Table 7.3 The efficiency of natural cellular materials

Material	$E^{1/2}/\rho$ $\left[\frac{GPa^{1/2}}{Mg/m^3}\right]$	$\sigma_f^{2/3}/\rho$ $\left[\frac{MPa^{2/3}}{Mg/m^3}\right]$	σ_f^2/E $\left[\frac{MJ}{m^3}\right]$	σ_f/E [–]	J_c/ρ $\left[\frac{kJ/m^2}{Mg/m^3}\right]$
Steel	1.8	7	10	0.006	2.0
Aluminum	3.0	17	4.0	0.002	3.7
CFRP (quasi isotropic)	6.5	54	7.0	0.007	0.8
Butyl rubber	0.04	4.0	35	4.0	5.4
Balsa, II	7.5–15.0	29–46	0.12–0.14	0.005–0.0055	0.26–0.4
Spruce, II	8.5–8.9	35–40	0.44–0.48	0.003–0.0035	1.8–2.3
Oak, II	5.0–5.4	22–24	0.9–1.1	0.005–0.006	3.5–4.5
Palm, II	5.3–6.0	50–55	3.0–7.0	0.01–0.03	3.7–6.4
Parenchyma	0.06–0.28	0.3–1.0	0.05–0.12	0.11–0.12	0.08–0.2
Cork	0.5–0.7	6.5–7.0	0.06–0.07	0.04–0.05	0.5–2.0
Trabecular bone	1.4–2.0	5.5–7.5	0.04–0.16	0.01–0.05	0.25–0.6
Coral	2.1–2.3	5.1– 6.7	0.04–0.09	0.003–0.005	0.002–0.004

the flexure index $E^{1/2}/\rho$ when loaded parallel to the grain; they are particularly efficient in bending and resistant to buckling. Balsa wood, for example, can be five times stiffer in bending, per unit weight, than steel (Table 7.3). This is no surprise: the principal loads woods must carry in performing their natural function are bending (branches under their own weight, and trunks under wind loads) and axial compression leading to buckling (of the trunk under its own weight and snow loads). The moduli and densities for palm and bamboo reflect values for small samples cut from the stems. They have values of $E^{1/2}/\rho$ similar to those of woods. The radial and longitudinal density gradients in palm, discussed in Chapter 6, makes palm even more efficient, allowing palms to grow to great heights while remaining slender. Bamboo, too, is highly efficient, because the fibers it contains are particularly well oriented along the stem; and in the plant, the efficiency is increased further (as in most grasses) by the tubular shape and radial gradient of modulus across the tube.

Structural efficiency in nature, at least in part, means making the stiffest, strongest structure from a limited allocation of material. As an example, think of making the branch of a tree from a limited, fixed mass, m, of material. The longest branch allows the leaves it bears to capture the most sunlight, so the objective is to maximize its length, L, while providing enough stiffness that it does not droop too much. Thus there are two constraints: the fixed mass, which we approximate by

$$m = \pi r^2 L \rho \tag{7.8}$$

where r is the (mean) radius of the branch and ρ is the density of the material of which it is made, and the limit on the droop per unit length is δ/L where δ is the end-deflection of the branch under its own weight. The deflection of a cantilever beam under a total load F, uniformly distributed along its length, is

$$\frac{\delta}{L} = \frac{FL^2}{8EI} = \frac{mgL^2}{2\pi r^4 E} \tag{7.9}$$

where F has been replaced by mg (g being the acceleration due to gravity) and the second moment of area I has been set equal to $\pi r^4/4$. Substituting for the radius r from equation (7.8) and rearranging gives an expression for the longest branch that can be made from the fixed allocation m of material:

$$L = \left(\frac{2}{\pi}\frac{m}{g}\frac{\delta}{L}\right)^{1/4}\left(\frac{E}{\rho^2}\right)^{1/4} \tag{7.10}$$

All the terms in the first bracket are fixed. The longest branch, for a given mass m and droop δ/L, is that made from the material with the largest value of $(E/\rho^2)^{1/4}$ or, equivalently of $E^{1/2}/\rho$.

Figure 7.3 shows that, by this criterion, woods are remarkable efficient. Table 7.3 reinforces this finding, showing that the bending stiffness per unit weight of many woods is greater than that of steel, aluminum and, in some cases, even carbon-fiber reinforced polymer composites (CFRP).

The strength–density chart

Data for the tensile strength, σ_f, and density, ρ, for natural cellular materials appears in Fig. 7.4. Strength guidelines are shown for the material indices:

$$M_4 = \sigma_f / \rho \quad \text{(strong tie under uniaxial load)} \tag{7.11}$$

$$M_5 = \sigma_f^{2/3} / \rho \quad \text{(strong beam in flexure)} \tag{7.12}$$

$$M_6 = \sigma_f^{1/2} / \rho \quad \text{(strong plate in flexure)} \tag{7.13}$$

Cellulose emerges as the material with the highest value of σ_f/ρ, confirming its excellence as a fiber. Where strength in bending or buckling are required we expect to find materials with high $\sigma_f^{2/3}/\rho$. Woods have high values of the index when loaded parallel to the grain, giving resistance to flexural failure; among engineering materials only CFRP can compete with them. Certain palms, particularly *Welfia*, have even higher strength to weight values than woods. Trabecular bone, mineralized with hydoxyapatite, is mechanically more efficient than corals, mineralized with calcite with almost the same strength and density as hydroxyapatite.

The strength–modulus chart

Data for the strength, σ_f, and Young's modulus, E, for natural cellular materials are shown in Fig. 7.5. Guidelines are shown for the material indices:

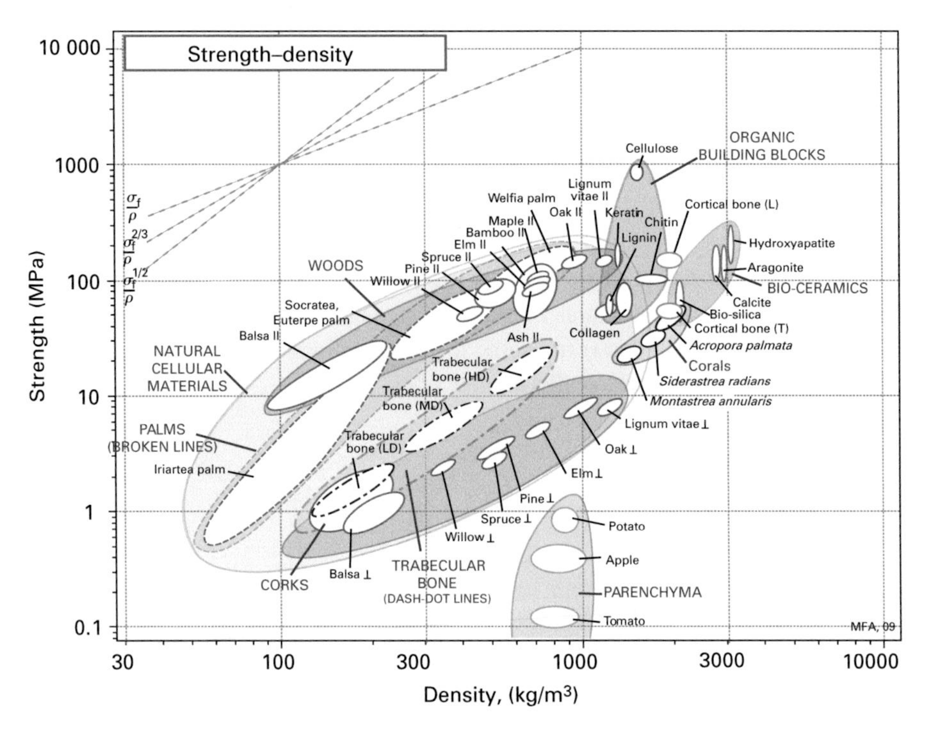

Fig. 7.4 The strength–density chart for natural cellular materials. Materials that are strong and light have high values of the indices indicated at the upper left. See plate section for color version.

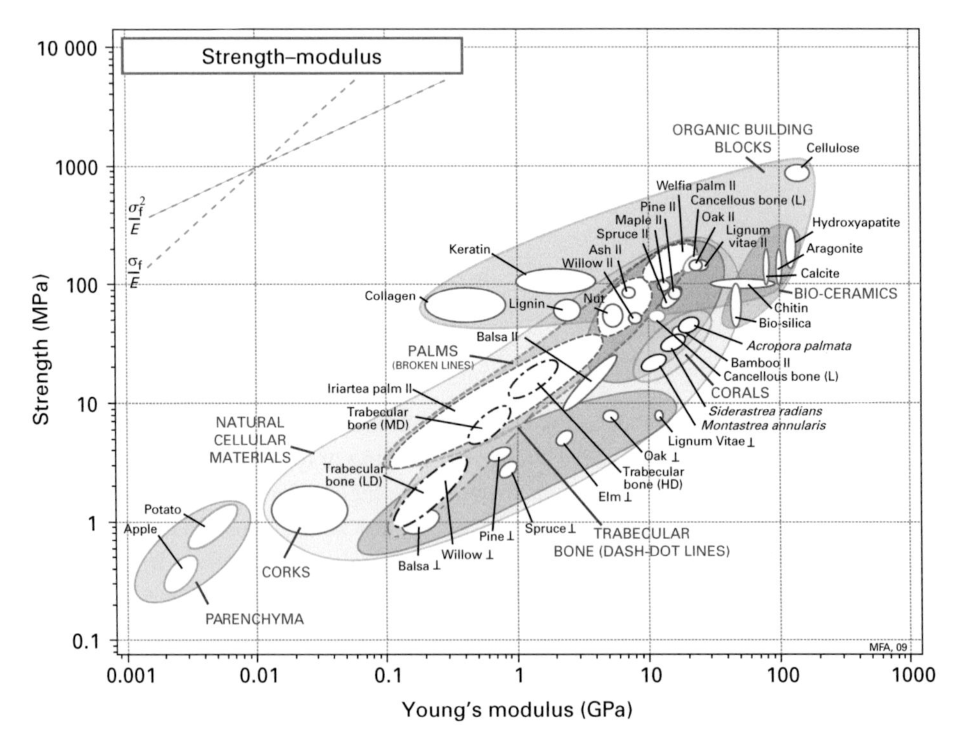

Fig. 7.5 The strength–modulus chart for natural cellular materials. Materials that excel at storing elastic strain energy per unit volume or are resilient have high values of the indices indicated at the upper left. See plate section for color version.

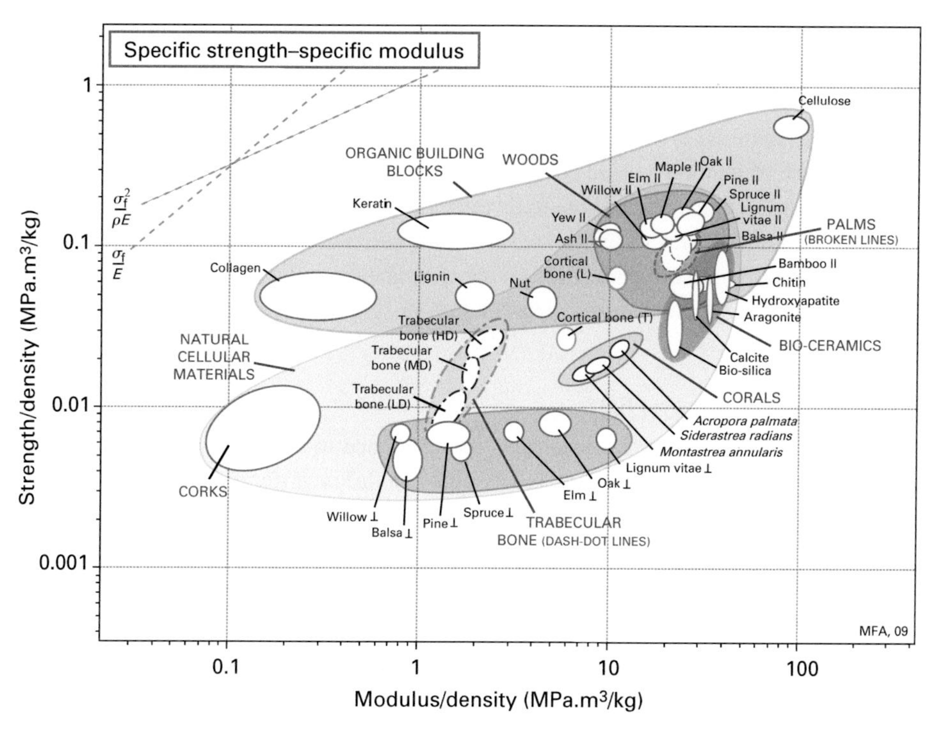

Fig. 7.6 The specific strength–specific modulus chart for natural cellular materials. Materials that excel at storing elastic strain energy per unit weight or that are resilient have high values of the indices indicated at the upper left. See plate section for color version.

$$M_7 = \sigma_f^2 / E \quad \text{(maximum elastic strain energy per unit volume)} \tag{7.14}$$

$$M_8 = \sigma_f / E \quad \text{(allow large, recoverable deformations; elastic hinges)} \tag{7.15}$$

Materials with large values of $\sigma_f^{\,2}/E$ store elastic energy and make good springs; and those with large values of σ_f/E have exceptional resilience – the ability to bend without breaking. Many palms have particularly high values of both indices, allowing them to flex in a high wind; the drag on the palm is reduced when the trunk bends into an arc, allowing the fronds to form a streamlined shape.

The strength/density–modulus/density chart

Data for the specific strength, σ_f/ρ, and specific modulus, E/ρ, for natural cellular materials are shown in Fig. 7.6. Guidelines are shown for the material indices:

$$M_9 = \sigma_f^2 / \rho E \quad \text{(maximum elastic strain energy per unit weight)} \tag{7.16}$$

$$M_8 = \sigma_f / E \quad \text{(allow large, reoverable deformations; elastic hinges)} \tag{7.17}$$

Materials with large values of $\sigma_f^2/\rho E$ store the most elastic energy per unit weight and make good springs, and, as already said, those with large values of σ_f/E have exceptional resilience – the ability to bend without breaking. This chart can be thought of as a map of material efficiency – the ability to provide stiffness and strength with the minimal use of material; those with the largest E/ρ and σ_f/ρ in any one family are the most efficient. Woods and palms, with values of E and ρ that span a factor of 30–100 on Figs. 7.3 and 7.4, cluster tightly on this chart, indicating that although their range of modulus and strength are large, they all use the material at their disposal with about the same efficiency. Among woods, spruce stands out for its efficiency.

The toughness–modulus chart

The toughness of a material measures its resistance to fracture under impact. The limited data for the toughness, J_c, and Young's modulus, E, for natural cellular materials are shown in Fig. 7.7. Data for the fibrous (across-grain) and splitting (along grain) fracture modes of woods are enclosed in separate envelopes.

The material index for fracture-safe design depends on the design goal. When the component is required to absorb a given *impact energy* with minimum use of resources, the best materials are those with the largest value of

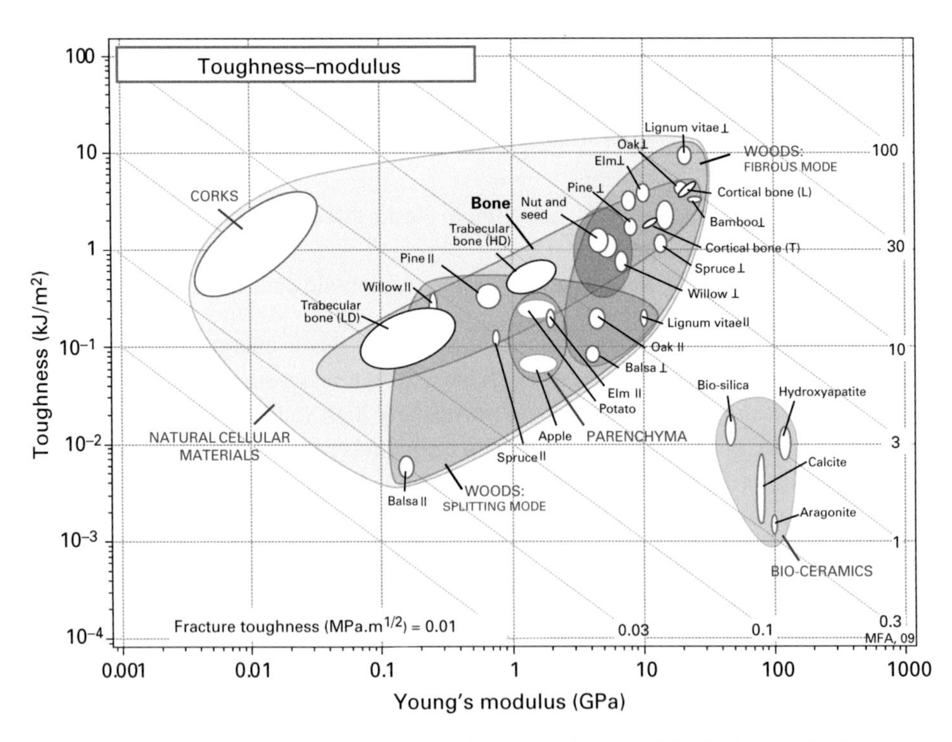

Fig. 7.7 The toughness–modulus chart for natural cellular materials. Materials that best resist fracture have high values of toughness or of fracture toughness, shown as diagonal contours. See plate section for color version.

$$M_{10} = J_c \tag{7.18}$$

These materials lie at the top of Fig. 7.7: bamboo and oak stand out. When instead a component containing a crack must carry a given *load* without failing, the safest choice of material is that with the largest values of the fracture toughness K_{1c}:

$$M_{11} = K_{1c} \approx (EJ_c)^{1/2} \tag{7.19}$$

Diagonal contours sloping from the upper left to lower right on Fig. 7.7 show values of this property. Among natural cellular materials, bamboo, high-density woods and dense trabecular bone strongly resist the propagation of cracks that traverse the grain (the fibrous mode). The resistance to splitting is much less dependent on wood density.

Table 7.3 compares the performance of natural cellular materials with four representative materials of engineering: steel (for basic structures), aluminum (representing light alloys), CFRP (high performance structures) and butyl rubber (typifying elastomers). In all but one case (where it is irrelevant) the performance is compared per unit mass since this is a measure of the efficiency in the use of material. It is striking that the materials of engineering, homogeneous except for CFRP, are so frequently outperformed by the hierarchically structured materials of nature.

7.3 Case studies: Selection of natural materials using the property charts

For tens of thousands of years people have used natural materials to craft tools, weapons and structures. Among natural materials, wood, bamboo and palm are some of the most adaptable and easy to shape, and include a large number of species offering a wide range of mechanical properties. Experience has led to certain of these species becoming the preferred choice for particular applications. We conclude this chapter by using the charts to examine three of these, seeking an explanation for the choice.

7.3.1 Materials for oars

Boats with oars appear in carved relief on monuments built in Egypt about 3300 BC. Wooden oars became the dominant mode of propulsion for small boats of all types and remained so for the next 3000 years. What is the best wood for an oar (Fig. 7.8)?

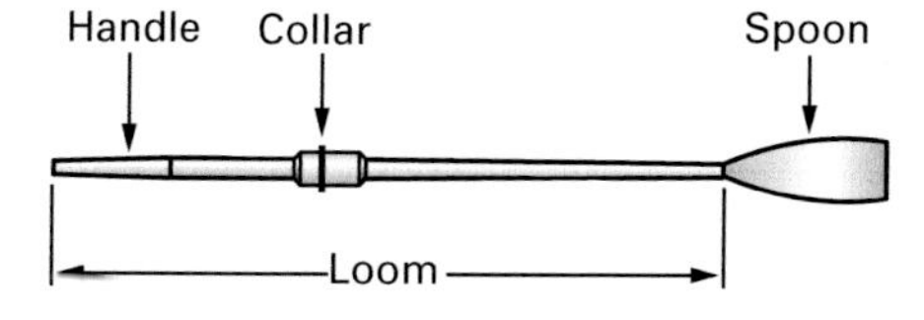

Fig. 7.8 An oar. Oars are designed for stiffness and they must be light.

Mechanically speaking, an oar is a beam, loaded in bending. It must be strong enough to carry, without breaking, the bending moment exerted by the rower, and it must have a stiffness to match the rower's own characteristics and give the right "feel." Meeting the strength constraint is easy. Competition oars are designed on *stiffness*, that is, to give a specified elastic deflection under a given load.

The oar must also be light; extra weight increases the wetted area of the hull and the drag that goes with it. Thus an oar is a beam of specified stiffness and minimum weight. The material index for material selection is that for a light, stiff beam:

$$M_2 = \frac{E^{1/2}}{\rho} \tag{7.20}$$

where E is the Young's modulus and ρ is the density.

Figure 7.9 shows the appropriate chart: that in which the Young's modulus, E, is plotted against density, ρ. The thick selection line for the index M_2 is positioned so that a small group of woods, and wood-like materials, is left above it. They are the materials with the largest values of M_2, and it is these that are the best choice provided they meet any other requirements of the design. Six materials stand out: balsa, spruce, willow, bamboo, and *Welfia* and *Socratea* palms. Balsa has an exceptionally high value of M_2, which is why it is the preferred material for model aircraft, but it is too soft to withstand

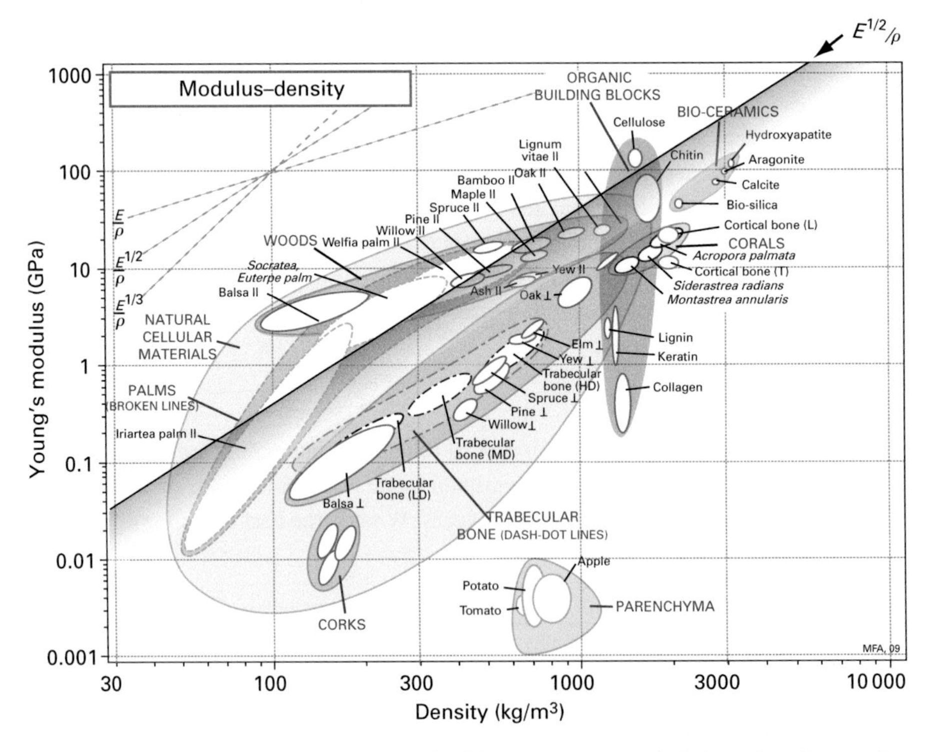

Fig. 7.9 The selection of woods for oars. Spruce and willow emerge as good choices. See plate section for color version.

indentation and damage when used as an oar. The next best is spruce. Willow comes close – it is the preferred material for cricket bats, but that is another story.

High-performance wooden oars are made the same way today as they were 100 years ago, by craftsmen working largely by hand. The shaft and blade are of Sitka spruce from the northern USA or Canada, the further north the better because the short growing season gives a finer grain. The wood is cut into strips, four of which are laminated together to average the stiffness and the blade is glued to the shaft. The rough oar is then shelved for some weeks to settle down, and finished by hand cutting and polishing. When finished, a spruce oar weighs between 4 and 4.3 kg.

Composite blades are a little lighter than wood for the same stiffness. The component parts are fabricated from a mixture of carbon and glass fibers in an epoxy matrix, assembled and glued. The advantage of composites lies partly in the saving of weight (typical weight: 3.9 kg) and partly in the greater control of performance: the shaft is molded to give the stiffness specified by the purchaser. Until recently a CFRP oar cost more than a wooden one, but the price of carbon fibers has fallen sufficiently that the two now cost about the same.

7.3.2 Archery bows

The evidence for the use of archery dates back about 50 000 years. Bows today have evolved into sophisticated structures using high-tech materials, but over most of their history they were made from wood, bamboo or palm. Here we explore the choice of wood species for bows (Fig. 7.10).

The primary mechanical function of a bow is to store elastic energy when drawn, which is transformed into kinetic energy of the bow and the arrow when released. The kinetic energy is distributed to the components in proportion to their mass, so that

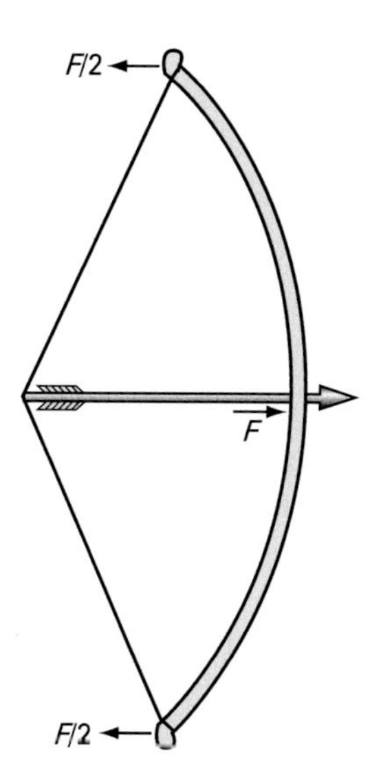

Fig. 7.10 An archery bow.

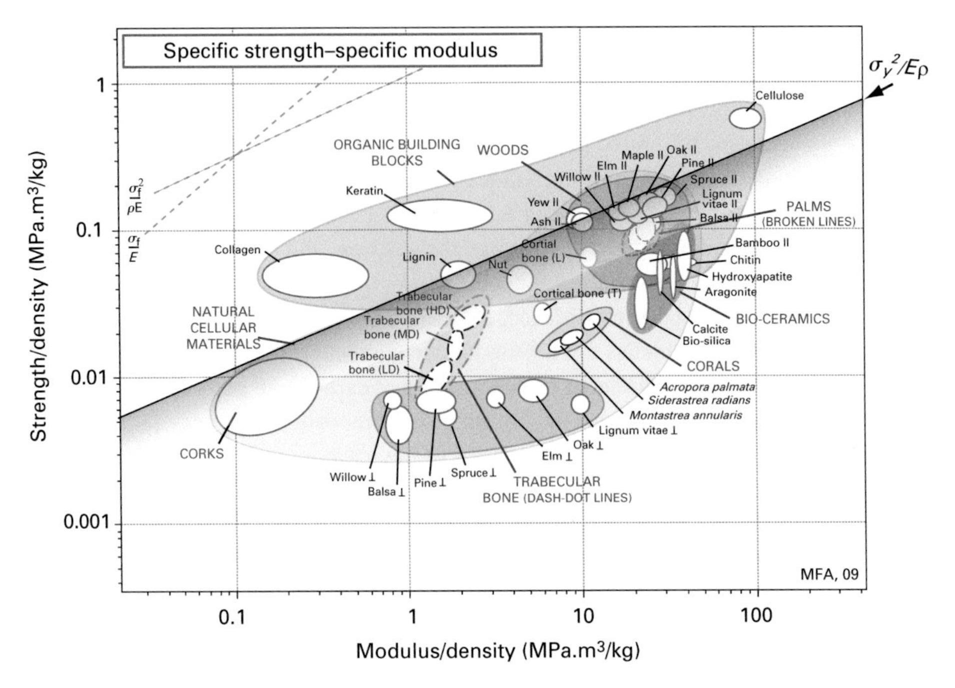

Fig. 7.11 Materials used for archery bows. As noted in the text, yew and ash stand out. See plate section for color version.

making the bow light releases more energy to the arrow. Thus the best material for a bow is one that stores the most elastic energy per unit mass.

The elastic energy per unit mass U_m stored in a bent beam under constant moment is

$$U_m \propto \frac{\sigma^2}{E\rho} \tag{7.21}$$

where σ is the stress in the outermost element of the beam, E is Young's modulus and ρ is the density. The bow will suffer damage if the stress exceeds the yield or failure strength σ_f of the material, setting an upper limit to it. Thus the best material for a bow, storing the most energy per unit mass, is one with the largest value of

$$M_9 = \frac{\sigma_f^2}{E\rho} \tag{7.22}$$

Figure 7.11 shows the chart of specific strength, σ_f/ρ, and specific modulus, E/ρ, with a selection line (slope 1/2) corresponding to the index M_9 plotted on it, positioned so that a small number of woods lie above it. Two wood species, yew and ash, stand out. This is because of their relatively low modulus (see Fig. 7.3) which, because E appears on the bottom of the equation for M_9, increases the elastic energy it can store.

The traditional wood used for bows in Europe is yew. The sapwood of yew has a higher tensile strength and lower modulus than the heartwood, which performs better in compression. Medieval bows were made with a D-shaped cross-section with sapwood on the tensile side, maximizing performance. Traditional Japanese archery, kyudo, uses a bamboo bow. Indigenous New Guinean warriors used bows made from palm with bamboo string. Figure 7.11 indicates that bamboo and palm are somewhat less efficient than yew and ash as materials for bows.

7.3.3 Wood for sound boards

It seems possible that instruments used for making music – sound, at any rate – are as old as language itself. Wegst (2006, 2008) analyzes the behavior required by the structural materials of musical instruments. Three properties are crucial in determining acoustic performance: the modulus, E, the density, ρ, and the loss coefficient, η. Modulus and density determine the natural vibration frequencies and speed of sound, c, in the material:

$$c = \sqrt{\frac{E}{\rho}} \tag{7.23}$$

The same two properties determine the acoustic impedance, z:

$$z = \sqrt{E\rho} \tag{7.24}$$

And the sound radiation coefficient:

$$R = \frac{c}{\rho} = \sqrt{\frac{E}{\rho^3}} \tag{7.25}$$

The loss coefficient measures the dissipation of vibrational energy in the material: low damping gives a bright sound, high damping a dull one.

Consider the specific case of materials for the sound board of stringed instruments (Fig. 7.12). Wegst points out that the sound board acts as a resonator, coupling the

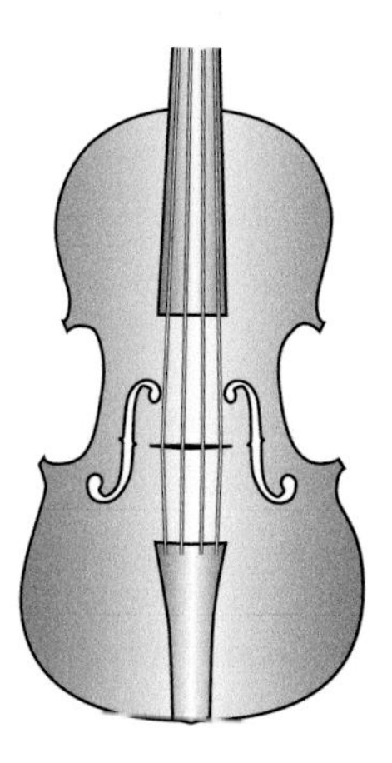

Fig. 7.12 A sound board.

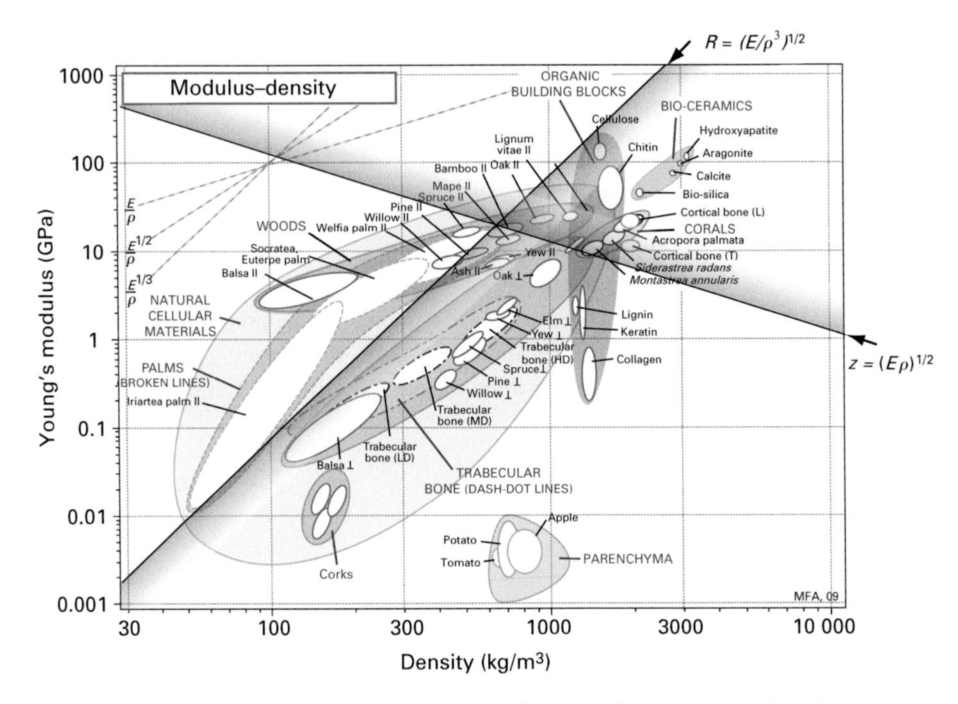

Fig. 7.13 The selection of woods for sound boards. Spruce, pine and willow stand out. See plate section for color version.

vibration of the string (which, alone, radiates very little energy) to the surrounding air. To do this well, the sound board material should have a high radiation coefficient, and, to match it to the strings, a low impedance.

Both properties are plotted on the modulus–density chart of Fig. 7.13. Materials with a high radiation coefficient, R, lie above the R-line of slope 3. Those with low impedance, z, lie below the z-line of slope –1. Those most distant from the two lines maximize R and minimize z. Spruce, pine, willow, balsa and palm radiate well. Sound boards are under stress because of the tension in the strings: a secondary requirement is that the material of the sound board should be strong enough to support this, ruling out the woods and palms of lowest density. Historically, spruce and pine are the traditional choices.

7.4 Summary

Natural cellular materials perform many functions. One, commonly, is mechanical – providing stiffness, strength and toughness, allowing stretch and flexure, or acting as a spring. They do this using a limited palette of materials, mainly proteins, polysaccharides, hydroxyapatite, calcites and aragonites, often arranged in elaborate structures. The mechanical efficiency of natural materials and of structures made from them can be explored by creating material property charts, of which five are presented here. The charts condense large bodies of data into single maps, showing the relationship between the properties of differing classes of materials, and allowing ranking by a set of material indices that measure performance. The results presented here confirm the high efficiency of natural materials and suggest that a number of them have evolved to meet specific mechanical requirements.

References

Introduction

Ashby MF (2005) *Materials Selection in Mechanical Design,* 3rd edn. Oxford: Butterworth Heinemann, chapter 4.

Ashby MF, Shercliff HR and Cebon D (2007) *Materials: Engineering, Science, Processing and Design.* Oxford: Butterworth Heinemann.

Wegst UGK and Ashby MF (2004) The mechanical efficiency of natural materials. *Phil. Mag.* **84**, 21, 2167–86.

Case study: materials for oars

Redgrave S (1992) *Complete Book of Rowing.* London: Partridge Press.

Wegst UGK and Ashby MF (2004) The mechanical efficiency of natural materials. *Phil. Mag.* **84**, 21, 2167–86.

Case study: archery bows

Gillelan GH (1974) *Complete Book of the Bow and Arrow.* Galahad Books.

Longman CJ, Waldron H and Legh AB (1901) *Archery.* Longmans, Green.

Case study: wood for sound boards

Wegst UGK (2006) Wood for sound. *Amer. J. Bot.* **93**, 1439–48.

Wegst UGK (2008) Bamboo and wood in musical instruments. *Ann. Rev. Mat. Res.* **38**, 323–49.

Proteins (collagen, keratin, chitin, elastin, resilin, abductin)

Fraser R and Macrae T (1980) Molecular structure and mechanical properties of keratins. In *The Mechanical Properties of Biological Materials (Proceedings of the Symposia of the Society for Experimental Biology; no. 34)*, ed. Vincent JFV and Currey JD. Cambridge: Cambridge University Press, pp. 211–46.

Fung YC (1993) *Biomechanics: Mechanical Properties of Living Tissues.* Berlin: Springer.

Geddes LA and Baker LE (1967) The specific resistance of biological material – a compendium of data for the biomedical engineer and physiologist. *Med. Biol. Eng.* **5**, 271–93.

Gentleman E, Lay AN, Dickerson DA, Nauman EA, Livesay GA and Dee KC (2003) Mechanical characterization of collagen fibers and scaffolds for tissue engineering. *Biomaterials*, **24**, 3805–13.

Gosline JM and French CJ (1979) Dynamic properties of elastin. *Biopolymers* **18**, 2091–103.

Park JB (1979) *Biomaterials, an Introduction.* New York and London: Plenum Press.

Tao XM and Postle R (1989) A viscoeleastic analysis of keratin. *Textile Res. J.* **59**, 5, 300–6.

Vincent JFV and Currey JD (editors) (1980) *The Mechanical Properties of Biological Materials (Proceedings of the Symposia of the Society for Experimental Biology; no. 34).* Cambridge: Cambridge University Press for Society for Experimental Biology.

Vincent JFV (1990) *Structural Biomaterials*, revised edn. Princeton, NJ: Princeton University Press.
Vogel S (2003) *Comparative Biomechanics – Life's Physical World.* Princeton, NJ: Princeton University Press.
Wainwright S (1980) Adaptive materials: a view from the organism. In *The Mechanical Properties of Biological Materials (Proceedings of the Symposia of the Society for Experimental Biology; no. 34)*, ed. Vincent JFV and Currey JD. Cambridge: Cambridge University Press, pp. 438–53.
Wainwright SA, Biggs WD, Currey JD and Gosline JM (1976) *Mechanical Design in Organisms.* London: Edward Arnold.

Polysaccharides (cellulose, lignin)

Cousins WJ, Armstrong RW and Robinson WH (1975) Young's modulus of lignin from a continuous indentation test. *J. Mat. Sci.* **10**, 1655–8.
Ferranti L Jr., Armstrong RW and Thadhani NN (2004) Elastic/plastic indentation of hard materials. *Mat. Sci. Eng. A* **371**, 251–5.
Mwaikambo LY and Ansell MP (2001) The determination of porosity and cellulose content of plant fibers by density methods. *J. Mat. Sci. Lett.* **20**, 23, 2095–6.
Vincent JFV and Currey JD (editors) (1980) *The Mechanical Properties of Biological Materials (Proceedings of the Symposia of the Society for Experimental Biology; no. 34).* Cambridge: Cambridge University Press for Society for Experimental Biology.

Minerals (calcite, aragonite, bio-silica, hydroxyapatite)

Broz ME, Cook RF and Whitney DL (2006) Micro-hardness, toughness and modulus of Mohs scale minerals. *Amer. Mineralogist* **91**, 135–42.
Evis Z and Doremus RH (2005) Coatings of hydroxyapatite. *Mat. Lett.* **59**, 3824–928.
Khalil KA, Kim SW, Dharmaraj N, Kim KW and Kim HY (2006) Novel mechanisms to improve toughness on hydroxyapatite bioceramics. *J. Mat. Process. Technol.* **187–188**, 417–20.
Park JB (1979) *Biomaterials, an Introduction*. New York and London: Plenum Press.
Skinner HCW and Jahren AH (2004) Biomineralization. In *Treatise on Geochemistry*, Vol. 8, Section 8.04. Oxford and Burlington, MA: Elsevier, USA, pp. 117–84.
Vincent JFV and Currey JD (editors) (1980) *The Mechanical Properties of Biological Materials (Proceedings of the Symposia of the Society for Experimental Biology; no. 34).* Cambridge: Cambridge University Press for Society for Experimental Biology.

Wood and wood-like materials (wood-cell-wall, wood, cork, bamboo, palm, rattan)

Amada S, Munekata T, Nagase Y, Ichikawa Y, Kirigai A and Yang ZF (1996) The mechanical structures of bamboos in viewpoint of functionally gradient and composite materials. *J. Composite Mat.* **30**, 7, 800–17.
Bhat K and Mathew A (1995) Structural basis of rattan biomechanics. *Biomimetics* **3**, 2, 67–80.
Frühwald A, Peek R-D and Schulte M (1992) *Nutzung von Kokospalmenholz am Beispiel von Nordsulawesi, Indonesien*. Technical report, Mitteilungen der Bundesforschungsanstalt für Forst- und Holzwirtschaft, Hamburg, Germany, Vol. 171, Kommissionsverlag Max Wiederbusch.
Gibson LJ, Easterling KE and Ashby MF (1981) The structure and mechanics of cork. *Proc. Roy. Soc. Lond.* **A377**, 1769, 99–117.
Godbole VS and Lakkad SC (1986) Effect of water-absorption on the mechanical properties of bamboo. *J. Mat. Sci. Lett.* **5**, 3, 303–4.

Janssen JJA (1991) *Mechanical Properties of Bamboo.* Dordrecht: Kluwer Academic Publishers.

Killmann W (1983) Some physical-properties of the coconut palm stem. *Wood Sci. Tech.* **17**, 3, 167–85.

Killmann W (1993) *Struktur, Eigenschaften and Nutzung von Stämmen wirtschaftlich wichtiger Palmen.* Unpublished PhD thesis, Universität Hamburg, Germany.

Kloot N (1952) Mechanical and physical properties of coconut palm. *Austr. J. Appl. Sci.*, **3**, 4, 293–322.

Lakkad SC and Patel JM (1981) Mechanical properties of bamboo, a natural composite. *Fibre Sci. Tech.*, **14**, 4, 319–22.

Mark RE (1967) *Cell Wall Mechanics of Tracheids.* New Haven, CT: Yale University Press.

Rich PM (1987) Mechanical structure of the stem of arborescent palms. *Botanical Gazette* **148**, 1, 42–50.

Rosa ME and Fortes MA (1988a) Stress-relaxation and creep of cork. *J. Mat. Sci.* **23**, 1, 35–42.

Rosa ME and Fortes MA (1988b) Rate effects on the compression and recovery of dimensions of cork. *J. Mat. Sci.* **23**, 3, 879–85.

Rosa ME and Fortes MA (1991) Deformation and fracture of cork in tension. *J. Mat. Sci.* **26**, 2, 341–8.

Plant tissue (apple and potato parenchyma, fruit skins, seaweed, nuts)

Blahovec J (1988) Mechanical properties of some plant materials. *J. Mat. Sci.* **23**, 10, 3588–93.

Gibson LJ and Ashby MF (1997) *Cellular Solids: Structure and Properties*, 2nd edn. Cambridge: Cambridge University Press.

Greenberg AR, Mehling A, Lee M and Bock JH (1989) Tensile behavior of grass. *J. Mat. Sci.* **24**, 7, 2549–54.

Jennings JS and Macmillan NH (1986) A tough nut to crack. *J. Mat. Sci.* **21**, 5, 1517–24.

Venkataswamy MA, Pillai CKS, Prasad VS and Satyanarayana KG (1987) Effect of weathering on the mechanical properties of midribs of coconut leaves. *J. Mat. Sci.* **22**, 9, 3167–72.

Vincent JFV (1989) Relationship between density and stiffness of apple flesh. *J. Sci. Food. Agr.* **47**, 4, 443–62.

Vincent JFV (1990) Fracture properties of plants. *Adv. Bot. Res.* **17**, 235–87.

Wang CH, Zhang LC and Mai YW (1994a) Deformation and fracture of macadamia nuts. 1. Deformation analysis of nut-in-shell. *Int. J. Fracture* **69**, 1, 51–65.

Wang CH and Mai YW (1994b) Deformation and fracture of macadamia nuts. 2. Microstructure and fracture mechanics analysis of nutshell. *Int. J. Fracture* **69**, 1, 67–85.

Bone

Ashman RB, Corin JD and Turner CH (1987) Elastic properties of cancellous bone – measurement by an ultrasonic technique. *J. Biomech.*, **20**, 10, 979–86.

Ashman RB and Rho JY (1988) Elastic modulus of trabecular bone material. *J. Biomech.* **21**, 3, 177–81.

Brear K, Currey JD, Raines S and Smith KJ (1988) Density and temperature effects on some mechanical properties of cancellous bone. *Eng. Medicine* **17**, 163–7.

Currey JD (1984) *The Mechanical Adaptions of Bones.* Princeton, NJ: Princeton University Press.

Currey JD (1988) The effects of drying and re-wetting on some mechanical properties of cortical bone. *J. Biomech.* **21**, 5, 439–41.

Currey JD (1990) Physical characteristics affecting the tensile failure properties of compact bone. *J. Biomech.* **23**, 8, 837–44.

Dickenson RP, Hutton WC and Stott JRR (1981) The mechanical properties of bone in osteoporosis. *J. Bone Joint Surg. – Brit.* **63**, 2, 233–8.

Gibson LJ (1985) The mechanical behavior of cancellous bone. *J. Biomech.* **18**, 5, 317–28.

Goldstein SA (1987) The mechanical properties of trabecular bone – dependence on anatomic location and function. *J. Biomech.* **20**, 11–12, 1055–61.

Kitchener A (1991) The evolution and mechanical design of horns and antlers. In *Biomechanics in Evolution (Society for Experimental Biology seminar series; no.36)*, ed. Rayner J and Wootton R. Cambridge: Cambridge University Press, pp. 229–53.

Kopperdahl DL and Keaveny TM (1998) Yield strain behavior of trabecular bone. *J. Biomech.* **31**, 601–8.

Linde F and Hvid I (1987) Stiffness behavior of trabecular bone specimens. *J. Biomech.* **20**, 1, 83–7.

Lotz JC, Gerhart TN and Hayes WC (1991) Mechanical properties of metaphyseal bone in the proximal femur. *J. Biomech.* **24**, 5, 317–27.

Moyle DD and Gavens AJ (1986) Fracture properties of bovine tibial bone. *J. Biomech.* **19**, 11, 919–27.

Moyle DD and Walker MW (1986) The effects of a calcium deficient diet on the mechanical – properties and morphology of goose bone. *J. Biomech.* **19**, 8, 613–25.

Sharp DJ, Tanner KE and Bonfield W (1990) Measurement of the density of trabecular bone. *J. Biomech.* **23**, 8, 853–7.

Sun JJ and Geng J (1987) A study of Haversian systems. *J. Biomech.* **20**, 8, 815.

Swanson S (1980) The elastic properties of rubber-like protein and highly extensible tissue. In *The Mechanical Properties of Biological Materials (Proceedings of the Symposia of the Society for Experimental Biology; no. 34)*, ed. Vincent JFV and Currey JD. Cambridge: Cambridge University Press, pp. 377–95.

Watkins M (1987) The development of a tough artificial composite based on antler bone. Unpublished PhD thesis, University of Reading, UK.

Coral

Alvarez K, Camero S, Alcaron ME, Rivas A, Gonzalez G (2002) Physical and mechanical properties evalation of *Acropora palmata* coralline species for bone substitution applications. *J. Mat. Sci. Mat. Med.* **13**, 509–15.

Chamberlain J (1978) Mechanical properties of coral skeleton: compressive strength and its adaptive significance. *Paleobiol.* **4**, 419–35.

Kim K, Goldberg WM and Taylor GT (1992) Architectural and mechanical properties of the black coral skeleton (*Coelenterata, Antipatharia*) – a comparison of 2 species. *Biol. Bull.* **182**, 2, 195–207.

Scott PJB and Risk MJ (1988) The effect of *Lithophaga* (*Bivalvia, Mytilidae*) boreholes on the strength of the coral *Porites lobata*. *Coral Reefs* **7**, 3, 145–51.

Part III

Cellular materials in medicine

8 Cellular solids as biomedical materials

8.1 Introduction

Engineered cellular materials are increasingly being used either to substitute for tissue in the body or to provide an environment for the regeneration of tissue. Such engineering materials are sometimes referred to as "biomaterials."

Materials used for hip and knee implants are examples of bone substitute materials. The femoral ball and stem of a hip implant, for instance, is typically made from a fully dense metal (a titanium, stainless steel or cobalt-chromium alloy), sometimes coated with a porous sintered metal or ceramic to increase bone ingrowth into the implant. The acetabular cup in the pelvis, into which the metal ball fits, is often made from ultra high molecular weight polyethylene. With the increase in research into metallic foams over the last decade, open-cell titanium and tantalum foams are finding application as bone substitute materials in, for instance, vertebral cages for spinal fusion and as coatings for hip and knee implants. As bone substitute materials, open-cell metal foams offer a number of advantages over fully dense metals: their Young's moduli more closely match that of the bone they replace, reducing stress shielding and subsequent loosening of the implant; their interconnected porosity allows tissue ingrowth; and their high specific surface area promotes the delivery of biological factors such as cells, genes, proteins and growth factors. Section 8.2 describes the range of processing techniques now available for making metallic foams for biomedical applications, as well as their microstructures and mechanical properties and their use in clinical applications.

Rather than substitute an engineering material for a particular tissue in the body, the goal of tissue engineering is to regenerate diseased or damaged tissues. In native tissue, cells attach to the extracellular matrix, a porous, fibrillar network typically made up of structural proteins, such as collagen, and a wide variety of proteoglycans. Tissue engineering scaffolds are designed to mimic the extracellular matrix, to provide an environment for biological cells to function: to attach, proliferate, migrate, differentiate, express proteins and form tissue. They are highly porous and often have a microstructure like an open-cell foam or a network of fibers. The interconnected porosity allows cells to migrate into the scaffold and the surface chemistry of the scaffold struts or fibers is designed to promote cell attachment. Over time, as biological cells form their own extracellular matrix, the scaffold resorbs, degrading into products that can be safely eliminated from the body.

Tissue engineering scaffolds are typically made from either natural polymers, such as collagen, glycosaminoglycans and chitosan (described in Chapter 2), or synthetic polymers such as the aliphatic polyesters used in resorbable sutures. Those for regenerating bone usually incorporate calcium phosphate or a bioactive glass or glass-ceramic. In this chapter, we summarize the materials, processing and microstructures of a range of tissue engineering scaffolds. We describe the application of the models for cellular solids (Chapter 3) to a collagen-GAG scaffold used for the regeneration of skin in burn patients, a mineralized collagen-GAG scaffold designed for the regeneration of bone and a number of honeycomb-like scaffolds.

8.2 Metal foams as bone substitute materials

For decades, fully dense alloys of cobalt-chromium, titanium, tantalum and stainless steel have been used to replace bone in prosthetic implants: in hip and knee joint replacements, in dental implants, and as plates for maxillofacial reconstruction and cranioplasty. All have a demonstrated history of biocompatibility and corrosion resistance but their Young's moduli are substantially higher than that of the bone they replace (Table 8.1), leading to stress shielding: a reduction in stress in the remaining bone, which can lead to bone resorption and loosening of the implant. This is especially critical in load-bearing applications such as implants for joint replacement. To improve the mechanical interaction between implant and bone, porous sintered metal beads, with their reduced moduli, were introduced as coatings on solid metal implants (see, for instance, Spector, 1987). Such coatings have been successful in achieving bone ingrowth as well as sufficient mechanical strength. Similarly, wire mesh coatings have also been developed, primarily for bone growth into flat implant surfaces (Ryan *et al.*, 2006).

Recently, there has been increasing interest in the use of metal foams in orthopedic implants (Ryan *et al.*, 2006; Levine, 2008). Nearly all the work to date has focused on open-cell foams made from tantalum and alloys and intermetallics of titanium, as they have already been demonstrated to have excellent biocompatibility and corrosion resistance; in addition, titanium has a modulus about half that of cobalt-chromium and stainless steel alloys. The initial application has been as a coating for prosthetic devices, similar to porous sintered beads, but there is also interest in their potential for broader use, for instance, as a replacement for vertebral bodies. Metal foams offer a number of advantages over solid metals. A variety of processing techniques allow control over the relative density, cell size and the fraction of open and closed cells. Control over the relative density of the foam allows matching of the foam and bone moduli. In open-cell foams, the interconnected porosity allows tissue ingrowth and the high specific surface area promotes the delivery of biological factors such as cells, genes, proteins and growth factors.

Once the tissue ingrowth is complete, however, then an open-cell metal foam is no longer porous, again potentially producing some stress shielding, although significantly less than with fully dense implants. And unlike tissue engineering scaffolds, discussed

Table 8.1 Properties of solid metals used in biomedical implants, compared with those of bone

Metal	Young's modulus, E (GPa)	Yield strength, σ (MPa)	Reference
Current implant metals			
Co-28Cr-6Mo	210		Levine *et al.*, 2006
Titanium alloys	106–115		Levine *et al.*, 2006
Tantalum	186		Levine *et al.*, 2006
316L stainless steel	193		Matweb.com
Metals used in foamed biomaterials			
Titanium	110	730	Cachinho and Correia, 2008
Ti-6Al-4V	114	924	eFunda.com
Ni-Ti	82	500	Ryan *et al.*, 2006
Tantalum	186	170	Eaglealloys.com
Nickel	207	59	Gauthier *et al.*, 2004
Bone			
Cortical bone (human)	18	182	Currey, 2002 Martin *et al.*, 1998
Trabecular bone	0.01–2	0.3–70	see Fig. 5.7, 5.8

in Section 8.3, metal foams are a permanent implant that do not resorb into the body over time. We next describe the processing, microstructure, mechanical properties and clinical applications of open-cell tantalum and titanium-based metal foams.

8.2.1 Tantalum foams

Tantalum foams are made by replicating the microstructure of an open-cell polyurethane foam. The polyurethane foam is first pyrolyzed to give a 2% dense vitreous carbon skeleton which is then coated with tantalum by chemical vapor deposition. The microstructure of the tantalum foam replicates that of the polyurethane preform (Fig. 8.1a). Tantalum foams are typically made with cell sizes between 400 and 600 μm, consistent with the observation that bone ingrowth is optimized for pore sizes in the range 50–400 μm (Ryan *et al.*, 2006). The thickness of the tantalum coating typically ranges from 40 to 60 μm, giving relative densities in the range 0.15–0.25. The solid struts within the foam are 99% by weight pure tantalum and 1% vitreous carbon (Levine *et al.*, 2006).

Tantalum forms a surface oxide, Ta_2O_5, which does not bond to bone. Tantalum foams can, however, readily be coated with an apatite layer that does bond to bone. Treatment of the foam in dilute NaOH, followed by heat treatment at 300 °C, induces the formation of a surface layer of amorphous tantalite. Submersion of the foam in

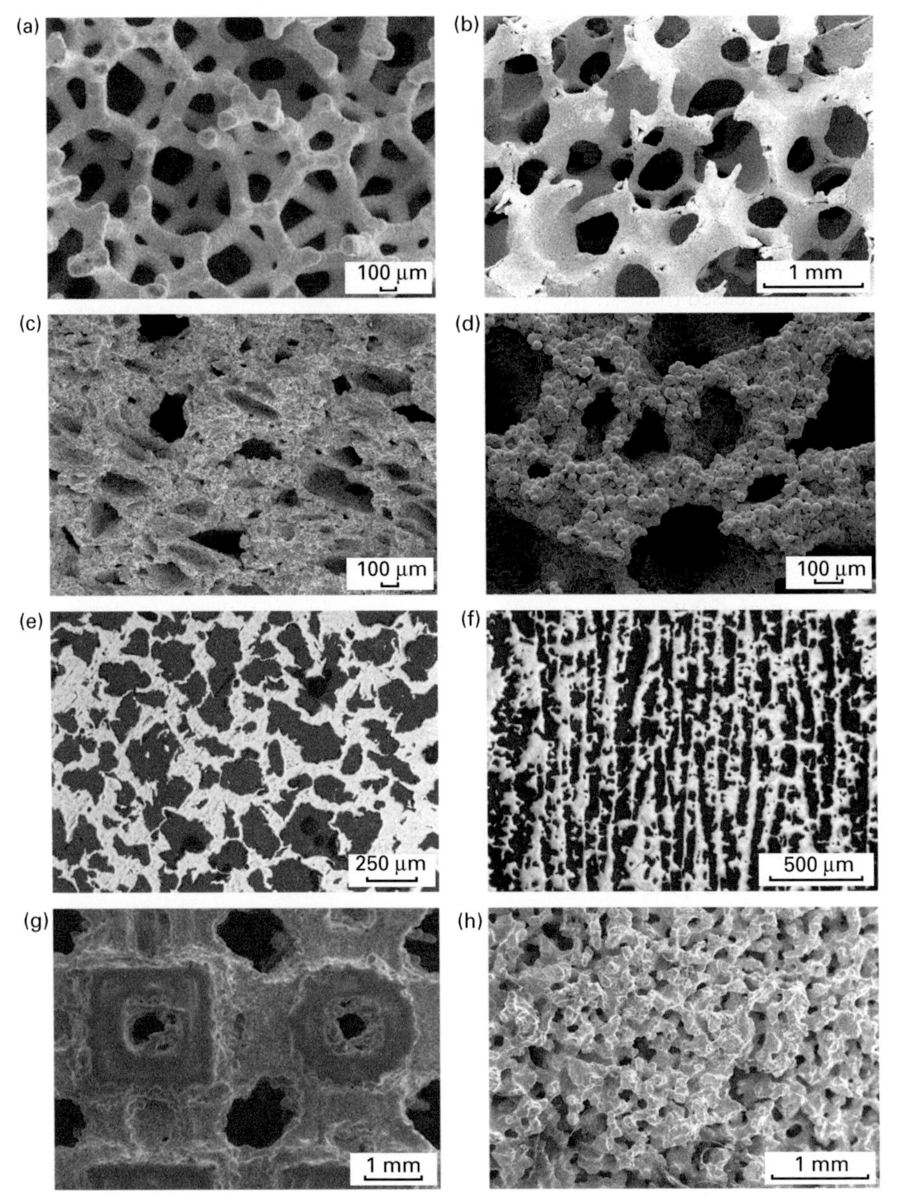

Fig. 8.1 Scanning electron micrographs of metal foams for implants: (a) tantalum foam replicating a polyurethane foam using CVD; (b) titanium foam made by replication of a polyurethane foam using slurry infiltration and sintering; (c) titanium foam made by the fugitive phase method; (d) titanium foam made with a foaming agent; (e) titanium foam made by expansion of argon gas; (f) titanium foam made by freeze-casting; (g) titanium foam made by selective laser melting; (h) nickel-titanium foam made by self-propagating high temperature synthesis. (a, Reproduced with permission and copyright © of the British Editorial Society of Bone and Joint Surgery from Bobyn *et al.*, 1999; b, reproduced with kind permission from Springer Science + Business Media, from Cachinho and Correia, 2008, Fig. 2; c, reproduced with kind permission from Springer Science + Business Media, from Hong *et al.*, 2008, Fig. 4; d, reproduced courtesy of L-P Lefebvre, National Research Council of Canada; e, reproduced from Li *et al.*, 2004 with kind permission from The Japan Institute of Metals; f, reproduced from Fife *et al.*, 2009, with kind permission from the Materials Research Society; g, reproduced from Lin *et al.*, 2007 with kind permission of John Wiley and Sons; h, reproduced from Arciniegas *et al.*, 2007, with permission of Elsevier.)

simulated body fluid, which has ion concentrations matching those of human blood plasma, induces an apatite coating on the strut surfaces of the foam (Levine *et al.*, 2006).

Bone ingrowth into the tantalum foam has been measured by implanting cylindrical specimens into holes drilled in the femoral diaphysis of mongrel dogs and examining the histology at several time points up to 52 weeks (Bobyn *et al.*, 1999). Bone ingrowth increased from 41.5% of the void area at 4 weeks to 79.7% at 52 weeks in tantalum foam implants with a cell size of 450 μm. Over time, the bony trabeculae penetrated the void and increased in thickness, until at 52 weeks, there was nearly solid bone impregnating the implant.

Tantalum foam components, marketed under the name Trabecular Metal™ (Zimmer, Parsippany, NJ), are available for clinical use in a range of orthopedic applications, including acetabular components for primary and revision total hip arthroplasty and tibial components for primary and revision total knee arthroplasty. Additional potential applications include cages for spinal fusion and implants for osteonecrosis (Levine *et al.*, 2006). Data on the long-term behavior of these components in vivo is not yet available. In particular, no studies have been done of the long-term effects of tantalum particles or debris within a joint (Levine *et al.*, 2006). Widespread use of the material is also limited by its high cost.

8.2.2 Titanium foams

A variety of techniques have been developed for fabricating open-cell titanium foams: replication of an open-cell polymer foam; removal of a fugitive phase (also known as a space filler); expansion of a foaming agent; directional freeze casting; rapid prototyping and self-propagating high temperature synthesis (SHS).

Titanium foams can be made by replicating an open-cell polyurethane foam in a manner similar to that used for tantalum foams. In this case, the titanium is deposited on the polyurethane foam by low-temperature arc vapor deposition (LTAVD) rather than CVD (Levine, 2008). As with the tantalum foam, polyurethane preforms of varying cell size are available and the thickness of the coating can be controlled to give titanium foams of varying cell size and relative density. Titanium foams made by this technique are currently available for clinical use as a coating for acetabular components for total hip arthroplasty (Tritanium™, Stryker, Mahwah, NJ).

A second replication method for producing titanium foams involves infiltration of a polymer sponge with a slurry containing titanium hydride (TiH_2) powder, removal of excess slurry by slight compression of the sponge, and then heat treatment to decompose the hydride, sinter the resulting titanium and remove the sponge (Cachinho and Correia, 2008). The resulting titanium foam has a pore size in the range 100–600 μm and a relative density of 0.25 (Fig. 8.1b).

Fugitive filler methods involve mixing titanium powder with the fugitive phase, compressing the mixture into a green compact and then heating the mixture in two stages, first to decompose the filler at a lower temperature (typically about 200 °C) and then to sinter the titanium powder at a higher temperature (typically about 1200 °C) (Fig. 8.1c). The filler is generally chosen to decompose at relatively low temperatures to avoid reaction with the titanium. A variety of fugitive phases have been used: ammonium hydrogen carbonate and carbamide (Bram *et al.*, 2000; Wen *et al.*, 2001, 2002), para-formaldehyde (Mullner *et al.*, 2007) and polypropylene carbonate (Hong *et al.*, 2008). The cell size can be controlled by the particle size of the fugitive phase and the relative density can be controlled by the volume fraction of the fugitive phase. Typical cell sizes are in the range 200–500 μm and relative densities are in the range 0.2–0.65 (Bram *et al.*, 2000; Wen *et al.*, 2001, 2002; Muller *et al.*, 2006; Hong *et al.*, 2008). Schematics of this process and several others, described below, are shown in Fig. 8.2.

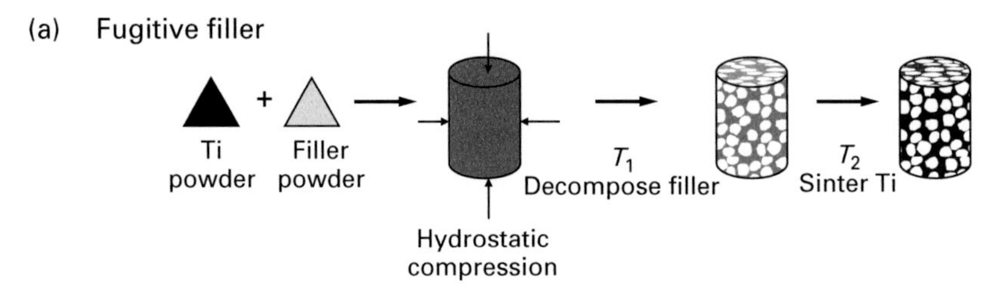

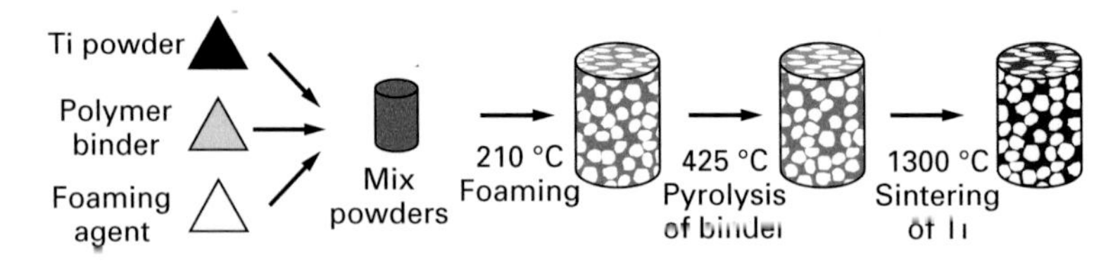

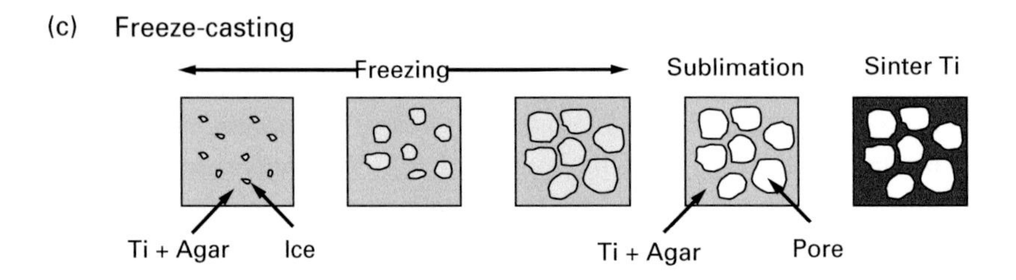

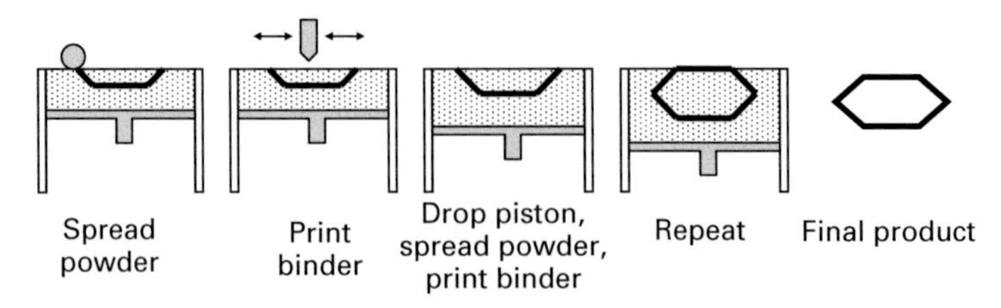

Fig. 8.2 Schematics of several processes used in making metal foams for biomedical applications: (a) fugitive filler method; (b) expansion of a foaming agent; (c) freeze-casting; and (d) rapid prototyping by three-dimensional printing.

Titanium foams have also been made by expansion of a foaming agent. In one process, titanium powder is mixed with a polymer binder and a foaming agent that evolves a gas at temperatures at which the binder is liquid (Gauthier *et al.*, 2003; Cheung *et al.*, 2007; Lefebvre and Baril, 2008). The mixture is subjected to a three-stage heat treatment: foaming at 210 °C, pyrolysis of the binder at 425 °C and final sintering of the remaining titanium at 1300 °C. The resulting foam is shown in Fig. 8.1d. The cell sizes are typically in the range 435–730 μm and the relative densities are in the range 20–35%. This material is currently being developed for use in dental implants by Opencell Biomedical (Toronto, Canada); it is not yet available for clinical use. A schematic of this process is shown in Fig. 8.2b.

Titanium foams can also be made by direct expansion of an inert gas, so that the gas acts as a physical foaming agent. Titanium powder is packed in a can, which is then evacuated and backfilled with argon gas and hot isostatically pressed. Specimens of compacts of the titanium, containing the pressurized gas (removed from the can), are then cyclically heated between 840 and 980 °C, allowing the gas to expand as the titanium superplastically deforms around the resulting pores (Li *et al.*, 2004; Spoerke *et al.*, 2005). The cell size is typically 150–300 μm with a maximum porosity of the titanium of 50%. A micrograph of a foam made using this method is shown in Fig. 8.1e. Titanium with elongated pores, with aspect ratios from 2 to 25, can be made by a similar method, using titanium wires instead of a powder. The maximum porosity achieved is low, however, about 12% (Spoerke *et al.*, 2007).

Freeze-casting can also be used to make titanium foams. A slurry of titanium powder and agar in water is frozen. As the ice nucleates and grows, the powder is forced into the regions between the ice grains and is bound together by the agar. Once the slurry has completely solidified, the ice is removed by sublimation and the remaining solid is sintered to densify the titanium (Fife *et al.*, 2009). The microstructure of a freeze-cast titanium foam is shown in Fig. 8.1f. The pores are elongated as a result of the directional freezing front. The relative density is typically 0.55 (depending on the concentration of the titanium powder in the solution) and the pore size is about 50 μm, which is somewhat small for bone ingrowth. A schematic of this process is shown in Fig. 8.2c.

Cellular metals with more controlled cell geometries can be made by rapid prototyping. Methods such as three-dimensional (3D) printing (Curodeau *et al.*, 2000) and selective laser melting (Lin *et al.*, 2007) have been used to fabricate orthopedic implants with porous surfaces designed for bone ingrowth. In both methods, the material is built up one two-dimensional layer at a time. In 3D printing, a layer of powder is rolled onto a powder delivery bed that is supported by a piston. A computer-controlled inkjet printer head is used to deliver a liquid adhesive to the areas of the layer that are to remain solid. Once all the adhesive is applied to a layer, the powder bed is lowered by the piston, another layer of powder is rolled onto the surface and the process is repeated until the entire three-dimensional object has been bonded. The loose powder is then removed and the remainder is heat treated to consolidate the bonded material. Minimum pore sizes of roughly 200 μm are achievable with this method. A schematic of this process is shown in Fig. 8.2d.

Selective laser melting involves a similar process, with the powder in each layer selectively bonded together by melting using a laser, rather than by an adhesive deposited by an inkjet head and subsequent sintering. The advantage of selective laser melting is that it gives an improved bond between the powder particles compared with that obtained by 3D printing. Lin *et al.* (2007) produced a porous titanium material with a cubic array of pores designed for use in a spinal interbody cage using selective laser melting. Their samples had relative densities of 0.52, with pore sizes of 700 μm (Fig. 8.1g).

Intermetallic NiTi foams can be made using self-propagating high-temperature synthesis (SHS). Ni and Ti powders of equiatomic composition are mixed, pressed into a green compact and then ignited (for instance, by a tungsten coil heated by an electrical current) (Li *et al.*, 2000; Zhang *et al.*, 2001; Arciniegas *et al.*, 2007; Sevilla *et al.*, 2007). On heating, there is a highly exothermic reaction between the Ni and Ti that results in self-ignition similar to an explosion. This method is also sometimes called combustion or ignition synthesis. NiTi foams with relative densities in the range 0.35–0.65 and cell sizes in the range 230–510 μm can be made using this technique (Fig. 8.1h). Intermetallic NiTi foams are currently used clinically for vertebral interbody devices marketed under the name Actipore™ (Biorthex, Laval, QC, Canada). Possible future applications of this material for surface coatings of hip and knee implants, for a lumbar vertebral fusion device and for replacement for entire vertebral bodies are under study.

Like tantalum, titanium forms a surface oxide, TiO_2, that can be coated with an apatite layer by immersion in NaOH, followed by drying, heat treatment and immersion in simulated body fluid (Kokubo *et al.*, 1990; Kim *et al.*, 1996; Wen *et al.*, 2002). The surfaces of the foam can also be coated with a hydroxyapatite layer by infiltrating the foam with a mixture of calcium nitrate tetrahydrate ($Ca(NO_3)_2.4H_2O$) and diethyl phosphite ($C_4H_{11}O_3P$) diluted in water, drying after infiltration at 110 °C and then annealing at 700 °C (Cachinho and Correia, 2008). A third method of coating foam with an organoapatite is to pretreat the surfaces with poly(L-lysine) followed by poly(L-glutamic acid) and to then immerse the foam in a solution of calcium hydroxide, phosphoric acid and poly(L-lysine), (Spoerke *et al.*, 2005). In another variation, the pores of an open-cell titanium foam can be filled with a self-assembly of peptide amphiphile nanofibers that are highly bioactive and can nucleate calcium phosphate from cells encapsulated in the fibers (Sargeant *et al.*, 2008).

The compressive stress–strain curve of a titanium foam is shown in Fig. 8.3. It has the three regimes characteristic of cellular solids: linear elasticity followed by a stress plateau and, at high strains, densification. Data for the relative Young's modulus, E^*/E_s, and relative compressive plateau stress, σ^*/σ_{ys}, of a number of metal foams being proposed as bone substitute materials are plotted against relative density in Figs. 8.4 and 8.5; values of the normalizing parameters are given in Table 8.1. Most of the data is for foams made from titanium (and its alloys and intermetallics) at relative densities over 0.3. The Young's moduli of these materials range from 1 to 100 GPa while their compressive strengths range from 10 to nearly 1000 MPa. The scatter in the data is large, reflecting the varied microstructures obtained from the different processes used to make the materials. For instance, the modulus and strength of the titanium foam shown in Fig. 8.1d, made by the use of a foaming agent, would be expected to depend

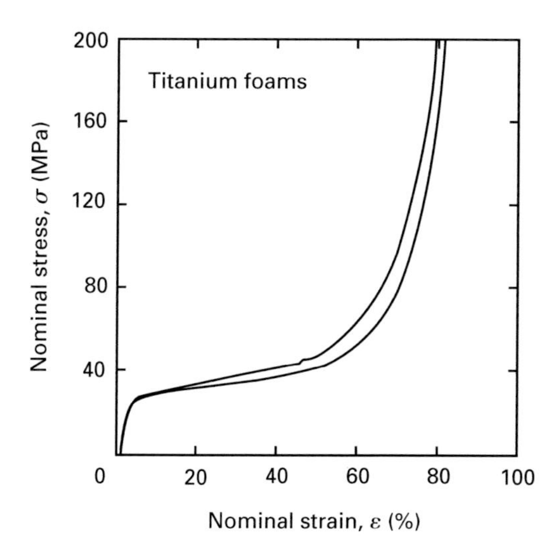

Fig. 8.3 Compressive stress–strain curve for a titanium foam ($\rho^*/\rho_s = 0.22$). The curve has the three regimes of linear elasticity, stress plateau and densification typical of cellular solids. (Reproduced from Wen *et al.*, 2001, with permission of Elsevier.)

on the size of the necks between the sintered powder particles; those of the foam shown in Fig. 8.1f, made by freeze-casting, would be expected to be highly anisotropic. Lines of slope 2 and 3/2, representing the models for, respectively, the Young's modulus and compressive strength of open-cell foams – see (3.16) and (3.22) – are superimposed on Figs. 8.4 and 8.5. Most of the data for the Young's moduli of the metal foams lie well below the line of slope 2. The data for the compressive strength lie close to the line of slope 3/2 at low relative densities, but are above the line (and tending towards the solid value) at relative densities close to 1, as would be expected. Both the Young's modulus and compressive strengths of the metal foams are greater than those of trabecular bone, reflecting the fact that the solid moduli of the metals are higher than that of trabecular bone as well as the higher relative densities of the metal foams.

The transport of nutrients and biochemical agents such as growth factors to bone growing into a metal coating on an implant requires fluid flow. The permeability of tantalum and titanium foams and of trabecular bone are listed in Table 8.2. For both, the permeability decreases with increasing relative density. The permeabilities of the metal foams are in the same range as those of dense trabecular bone.

8.3 Tissue engineering scaffolds

8.3.1 Introduction

The goal of tissue engineering is to regenerate diseased or damaged tissues by providing a porous scaffold that mimics the body's own extracellular matrix (ECM), allowing cell attachment, proliferation, migration, differentiation and function. Scaffolds for the regeneration of a wide range of tissues (e.g. orthopedic, cardiovascular, nervous, gastrointestinal, urogenital) are currently being developed. Scaffolds for the regeneration of skin in burns patients have been clinically available for over ten years.

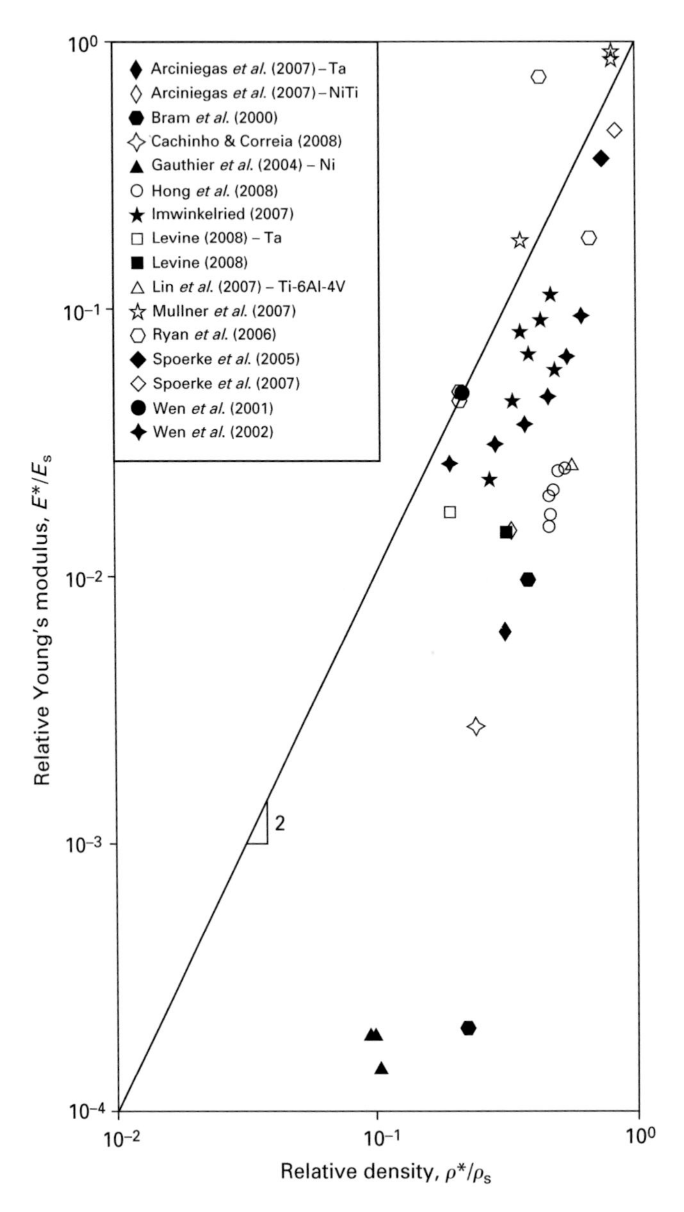

Fig. 8.4 Young's modulus of metal foam implant materials, E^*, normalized by that of the solid metal, E_s, plotted against relative density, ρ^*/ρ_s. Data for titanium foams except where noted in the legend.

Here, we first discuss the design requirements for tissue engineering scaffolds. We then review the most common materials and processes used in making scaffolds for tissue engineering studies and describe typical microstructures associated with each process. The microstructure and mechanical behavior of three scaffolds (a honeycomb-like scaffold designed for regeneration of cardiac tissue, a collagen-glycosaminoglycan (GAG) scaffold used in skin regeneration and a mineralized collagen-GAG scaffold

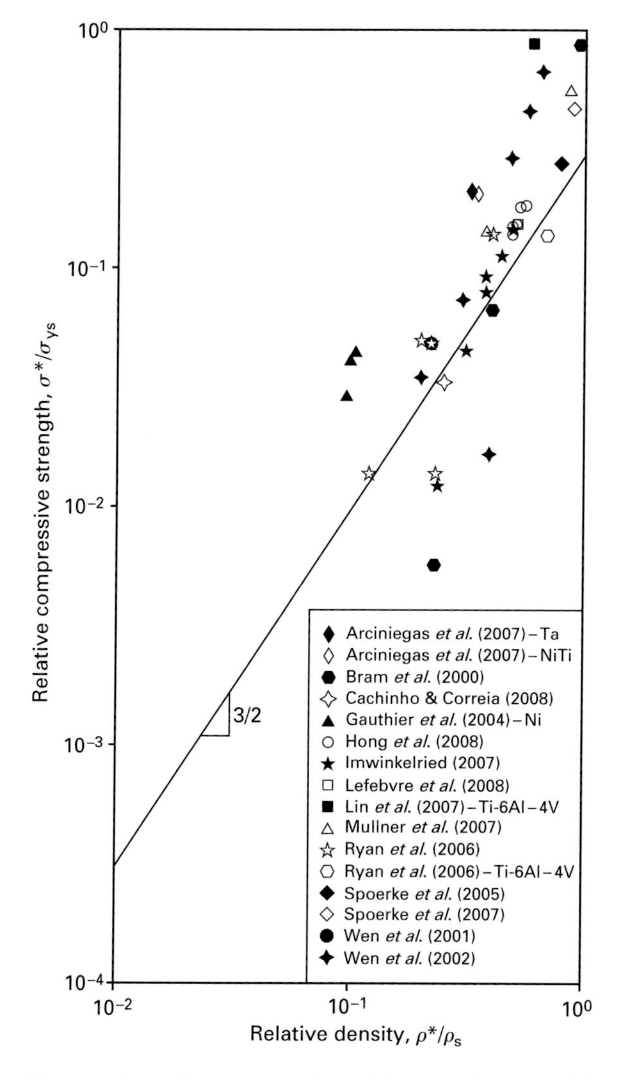

Fig. 8.5 Compressive collapse stress of metal foam implant materials, σ_{ys}^*, normalized by that of the solid metal, σ_{ys}, plotted against relative density, ρ^*/ρ_s. Data for titanium foams except where noted in the legend.

designed for bone regeneration) are reviewed along with the use of the models of Chapter 3 to describe their behavior. Methods to improve the mechanical behavior, based on the models, are also discussed.

8.3.2 Design requirements

There are a number of design requirements for tissue engineering scaffolds (Ikada, 2006; Atala, 2007; Freed and Guilak, 2007; Lanza *et al.*, 2007). The solid must be biocompatible and degrade into non-toxic components that can be eliminated from the body (or into components that can be transported out of the body at a sufficient

Table 8.2 Permeabilities of metal foams and trabecular bone

Material	Relative density ρ^*/ρ_s (-)	Permeability, k (m^2)	Reference
Tantalum foam	0.12	4.8×10^{-10}	Shimko *et al.* (2005)
Tantalum foam	0.34	2.1×10^{-10}	Shimko *et al.* (2005)
Titanium foam	0.20	10^{-10}	Imwinkelried (2007)
Titanium foam	0.45	6×10^{-12}	Imwinkelried (2007)
Trabecular bone	0.08	1.1×10^{-8}	Grimm and Williams (1997)
Trabecular bone	0.21	4.0×10^{-10}	Grimm and Williams (1997)
Cortical bone	–	2.5×10^{-13}	Mullner *et al.* (2007)

rate to maintain a tolerable level within the body). It must also promote cell attachment, proliferation, migration, differentiation and production of native ECM. The cellular structure must have a large volume fraction of interconnected pores to facilitate cell migration and transport of nutrients and regulatory factors (e.g. cell adhesion peptides, growth factors, hormones) – typical porosities are greater than 90%. The pore size must be within a critical, tissue-dependent range (e.g. skin 20–125 μm (Yannas, 1992); bone 100–500 μm (Hulbert *et al.*, 1970; Kuhne *et al.*, 1994)). The lower bound on the range is controlled by cell size (typically about 20 μm), while the upper bound is defined by the density of binding sites available for cell attachment, which, in turn, depends on the specific surface area. The pore geometry should be conducive to the cell morphology (e.g. elongated pores for nerve cells) and the overall scaffold geometry has to be able to be shaped to match irregular tissue defects. The scaffold has to have sufficient mechanical integrity as well as provide a mechanical environment appropriate for cell differentiation (Engler *et al.*, 2006). It has to degrade at a controllable rate, so that as the tissue becomes fully formed through cell proliferation and deposition of native ECM, the scaffold is completely resorbed.

8.3.3 Materials

A wide range of materials has been used for fabricating scaffolds for tissue engineering studies. Natural polymers include collagen as well as glycosaminoglycans, alginate and chitosan. Synthetic polymers include those used in resorbable sutures: polyglycolic acid (PGA), polylactic acid (PLA), the co-polymer poly(lactic-co-glycolic acid) (PLGA) and poly(ε-caprolactone). Hydrogels, produced by crosslinking water-soluble polymer chains to form insoluble networks, are attractive for scaffolds for soft tissues, which themselves, with their high water content, resemble hydrogels. Hydrogels for tissue engineering studies have been produced using both natural polymers and synthetic polymers such as polyethylene glycol (PEG), polyvinyl alcohol (PVA) and polyacrylic acid (PAA). Scaffolds for bone regeneration typically use a calcium phosphate (e.g.

hydroxyapatite or octacalcium phosphate) in a composite with collagen or a synthetic polymer. In addition, some studies have used native ECM treated to remove all cellular matter (acellular ECM). A more complete list and description are available in Ikada (2006), Boccaccini and Gough (2007) and Pachence *et al.* (2007).

Collagen is a major component of the extracellular matrix in a number of tissues (e.g. skin, bone, cartilage, tendon, ligament). As such, it has integrin-binding domains and is an excellent substrate for cell attachment and proliferation. Its low Young's modulus and rapid degradation rate can be improved by crosslinking by physical methods (e.g. UV irradiation or dehydrothermal treatment) or chemical treatments (e.g. glutaraldehyde or 1-ethyl-3-(3-dimethylaminopropyl)-carbodiimide (EDAC)). In an acetic acid solution, it can be co-precipitated with glycosaminoglycans (e.g. chondroitin sulfate); freeze-drying the solution forms a porous scaffold (Yannas, 1992). It can be used in combination with synthetic polymers to give scaffolds that combine the excellent cell adhesion and cell proliferation of collagen with the higher mechanical properties of the synthetic polymer (Pachence *et al.*, 2007). A potential risk with collagen scaffolds is the immunological response to viruses or prions in tissue obtained from animal sources; this can be minimized by appropriate testing and treatment (Ikada, 2006).

The synthetic biopolymers PLA, PGA, the co-polymer PLGA and poly(ε-caprolactone) have been widely used in resorbable sutures since the 1970s. They are among the most common synthetic polymers used in tissue engineering studies. The co-polymer PLGA is particularly attractive as the degradation rate and mechanical properties can be tailored by controlling the ratio of PLA and PGA as well as the molecular weights of each. One concern that has been raised is that they have been associated with an inflammatory reaction, thought to be a result of acidic degradation products (Pachence *et al.*, 2007).

Scaffolds for regenerating bone typically use an inorganic, typically a calcium phosphate such as hydroxyapatite (HA), tricalcium phosphate (TCP) or octacalcium phosphate (OCP), or a bioactive glass or glass-ceramic, in combination with a polymer such as collagen or PLGA (Huang and Best, 2007; Misra and Boccaccini, 2007). Hydroxyapatite most closely resembles the inorganic in bone and has a Ca/P ratio of 1.67. The rate of resorption of calcium phosphates is related to the Ca/P ratio (the lower the ratio, the higher the rate) so that the resorption rate can be controlled by combining HA with TCP or OCP, which have lower Ca/P ratios. Bioactive glass is a Na_2O-CaO-SiO_2 glass with less than 60% SiO_2 and additional P_2O_5, B_2O_3 and CaF_2. In vitro, in simulated body fluid, and in vivo, a hydroxy-carbonate apatite layer forms on the surface of bioglass, promoting bonding with bone. Apatite-wollastonite (A-W) glass ceramic offers both bioactivity and higher mechanical properties than other glasses and glass-ceramics.

Acellular scaffolds (native extracellular matrix, from which all the cellular matter has been removed) offer a number of advantages over synthetic ones (Badylak *et al.*, 2009; Fernandes *et al.*, 2009). They have a suitable porosity, pore geometry and mechanical properties. They have abundant cell adhesion sites and may retain appropriate growth factors. They inherently promote cell adhesion, proliferation and migration. Potential applications for acellular scaffolds include cardiovascular tissues, cornea, meniscus, urinary bladder, cartilage and ligaments. Most of the commercially available acellular scaffolds are made from skin, small intestine submucosa and pericardium, harvested from humans, pigs and cows (Badylak *et al.*, 2009). Decellularization is usually done by a combination of physical (e.g. freezing, pressure, agitation), chemical (e.g. alkaline and acid treatments, detergents) and enzymatic methods (e.g. trypsin) (Gilbert *et al.*, 2006). Concerns include residual cells remaining after processing, the immunological response of the host tissue to any remaining cells and degradation of the scaffold physical, chemical and biological properties by the process used to remove the cells.

8.3.4 Processing methods and microstructure

The most common methods of making porous scaffolds are freeze-drying, fiber bonding, foaming, fugitive phase leaching, rapid prototyping and electrospinning. Each of these techniques, along with micrographs showing typical scaffolds made from them, is described below. Additional methods, described in more detail in the review by Murphy and Mikos (2007), include melt molding, membrane lamination, extrusion, peptide self-assembly and in-situ polymerization.

Freeze-dried collagen-glycosaminoglycan (CG) scaffolds are used clinically for skin regeneration. Microfibrillar type I collagen is first mixed with acetic acid; the low pH (pH = 3.2) acetic acid acts to swell the collagen fibers, destroying the quarternary structure as well as the periodic banding of the collagen fibers. These structural modifications remove a large component of the immunological markers on the collagen surface, reducing host immunological response and preventing platelet aggregation to the CG scaffold surface (Yannas *et al.*, 1975; Yannas and Silver, 1975; Silver *et al.*, 1978, 1979). A glycosaminoglycan, typically chondroitin-6-sulfate, is then added to the swollen collagen-acetic acid mixture. Spontaneous crosslinks are formed between the swollen collagen fibrils and the GAG, resulting in precipitation of the collagen-GAG content out of solution and the formation of a collagen-GAG suspension in the aqueous acetic acid phase. The solution is solidified by cooling at a constant freezing rate from room temperature to the final freezing temperature; the temperature is held at this temperature until the entire solution has solidified (Yannas, 1992; O'Brien *et al.*, 2004). As the solution cools, the co-precipitate is concentrated into the regions between the growing ice crystals. The pore size is controlled by the final freezing temperature; pore sizes between 96 and 151 μm have been obtained for final freezing temperatures of between −40 and −10 °C (O'Brien *et al.*, 2004). As the pore size decreases, the surface area and number of binding sites available for cell attachment increases (O'Brien *et al.*, 2005). The microstructure of a collagen-GAG scaffold made by this process is shown in Fig. 8.6a. The relative density of collagen-GAG scaffolds made by freeze-drying is typically 0.005. Collagen-GAG scaffolds for peripheral nerve regeneration, with highly aligned, elongated pores to guide axonal growth, are made by directional solidification in the freezing process (Yannas, 1995).

Freeze-drying can also be used to produce synthetic polymer scaffolds. The polymer (e.g. PGA, PLA, PLGA) is dissolved in a solution of water that is emulsified until it is homogeneous and is then freeze-dried (Murphy and Mikos, 2007). The high water content of hydrogels allow them to be freeze-dried without the use of a solvent (Ikada, 2006).

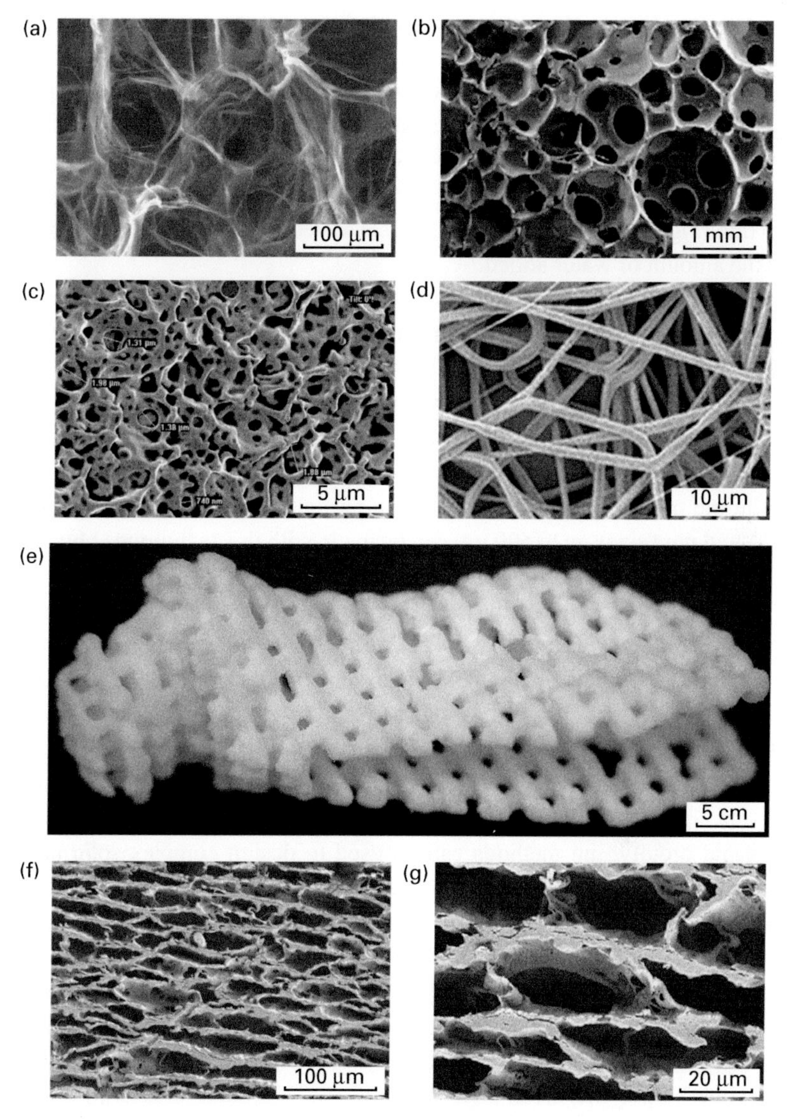

Fig. 8.6 (a) Collagen-GAG scaffold fabricated by freeze-drying; (b) synthetic polymer scaffold fabricated by foaming; (c) scaffold fabricated by salt leaching followed by emulsion coating; (d) electrospun poly(L-lactic-co-ε−caprolactone) fabrics; (e) human condyle scaffold made by selective laser sintering; (f, g) an acellular elastin scaffold, made by removing the cells and other extracellular matrix components except for elastin from porcine heart tissue. (a, Reproduced from Pek *et al.*, 2004, with permission from Elsevier; b, reproduced from Zhang *et al.*, 2003. The publisher for this copyrighted material is Mary Ann Liebert, Inc. publishers; c, reproduced from Sohier *et al.*, 2003, with permission from Elsevier; d, reproduced from Kwon *et al.*, 2005, with permission from Elsevier; e, reproduced from Partee *et al.*, 2006, with permission of the American Society of Mechanical Engineers; f,g, reproduced from Lu *et al.*, 2004, with permission of Elsevier.)

Several methods of foaming scaffolds by generating a gas have been developed. Barbucci and Leone (2004) generated carbon dioxide bubbles in formaldehyde by adding hydrochloric acid to a salt ($NaHCO_3$). The bubbles then rose to pass through a hydrogel held in a cell-culture strainer above the formaldehyde; the strainer acted as a filter to control the size of the bubbles passing through the hydrogel. Foamed polyurethane can be made by adding water to urethane prepolymers to generate carbon dioxide (Fig. 8.6b) (Zhang *et al.*, 2003). The relative density of the foam is controlled by the amount of water added.

The mechanical integrity of non-woven fiber meshes can be increased by fiber bonding. For instance, PGA fibers can be bonded by spraying them with an atomized PLGA solution, which bonds the fibers at the contact points (Mooney *et al.*, 1996). This method is limited to producing tubular structures.

Porous scaffolds can also be made by leaching of a fugitive phase such as salt or paraffin. The relative density is controlled by the volume fraction of the fugitive phase while the pore size is controlled by the particle size. Scaffolds with porosities up to 93% and pore sizes up to 500 μm can be made with this technique (Murphy and Mikos, 2007). A scaffold made by salt-leaching a co-polymer of poly(butylene terephthalate) and poly(ethylene glycol), followed by emulsion coating, is shown in Fig. 8.6c.

Electrospinning produces fibers from a polymeric solution or liquid extruded through a thin nozzle by applying a high-voltage electric field to spin the fibers (Fig. 8.6d). The technique gives an interconnected network of micron-scale diameter fibers. Porosities of 90% have been achieved with this method. Fiber diameter and pore size can be controlled through processing variables such as the polymer concentration, choice of solvent, applied voltage and nozzle diameter (Murphy and Mikos, 2007).

Rapid prototyping methods can be used to design complex solid and pore geometries in scaffolds (Hollister, 2005; Ikada, 2006; Murphy and Mikos, 2007). They are based on building up successive layers of solid, one at a time, through precise computer control of a solidification technique. For instance, in 3D printing, the material is built up from successive layers of powder, bonded by the precise application of binder. In selective laser sintering, polymer powder is selectively melted by a laser and then cooled to bind it into a solid. Stereolithography shines a light on a photosensitive liquid monomer to polymerize it. Both the solid scaffold material as well as biological cells can be integrated into the scaffold. Computer control of the desired architecture allows fabrication of complex geometries with interconnected porosity and gives improved control over the mechanical properties of the scaffold. An example of a human condyle scaffold made by selective laser sintering is shown in Fig. 8.6e.

An acellular elastin scaffold, made by removing the cells as well as the collagen and other ECM components except for elastin from a porcine heart by a chemical treatment (using cyanogen bromide) is shown in Fig. 8.6f, g. The foam-like structure of the native extracellular matrix is apparent.

Synthetic polymers used for tissue engineering scaffolds may need further surface treatment to address limitations of the surface properties of the polymer (Ikada, 2006; Murphy and Mikos, 2007). For instance, if the polymer is hydrophobic, it may need to be treated by physical or chemical coating with a hydrophilic polymer or by prewetting with ethanol or other chemicals to ensure that cell culture medium infiltrates the pores.

Synthetic polymers may also lack appropriate binding sites for cell attachment, requiring coating with ECM adhesion proteins, such as fibronectin, vitronectin, or laminin, or with short peptide sequences associated with cell binding such as RGD (consisting of the amino acids arginine, glycine and aspartate). Surface treatments may also involve incorporation of growth factors to promote cell proliferation and differentiation.

8.4 Mechanical behavior of tissue engineering scaffolds

Many tissue engineering scaffolds have a foam-like structure, allowing the models for cellular solids summarized in Chapter 3 to be used to describe their mechanical behavior. Here, we describe the mechanical behavior of a collagen-GAG scaffold used for skin regeneration, a mineralized collagen-GAG scaffold designed for bone regeneration and a novel two-dimensional honeycomb-like scaffold designed for regeneration of cardiac tissue. We compare the measured properties with the models for foams and honeycombs. Finally, we describe how we have used the models to guide the development of collagen-based scaffolds with improved mechanical properties.

8.4.1 Mechanical behavior of collagen-GAG scaffolds

A typical compressive stress–strain curve for a collagen-GAG scaffold, made by the freeze-drying technique described above, is shown in Fig. 8.7a. It exhibits the three regimes typical of a cellular solid: initial linear elasticity, dominated by strut bending, a stress plateau, corresponding to strut buckling, and, at high strains, densification, at which point the cells are nearly completely collapsed and opposing cell walls touch. The Young's modulus, E^*, the elastic collapse stress, σ_{el}^*, and the slope of the stress plateau, $\Delta\sigma/\Delta\varepsilon$, are indicated on the curve in Fig. 8.7b.

Models for open-cell foams, developed in Chapter 3, indicate that their Young's modulus and elastic collapse stress are given by

$$E^* = C_1 E_s \left(\frac{\rho^*}{\rho_s}\right)^2 \tag{8.1}$$

and

$$\sigma_{el}^* = C_2 E_s \left(\frac{\rho^*}{\rho_s}\right)^2 \tag{8.2}$$

where E_s is the Young's modulus of the solid strut material and C_1 and C_2 are constants related to the geometry of the cells, determined experimentally to be 1 and 0.05, respectively, for open-cell foams. We note that if $C_1 = 1$, then C_2 is the strain at the elastic collapse stress, $\varepsilon_{el}^* = \sigma_{el}^*/E^*$.

The compressive stress–strain curves of a range of freeze-dried collagen-GAG scaffolds, of varying strut modulus, relative density and pore size have been measured by Harley *et al.* (2007). The strut modulus was varied by increasing the crosslink density

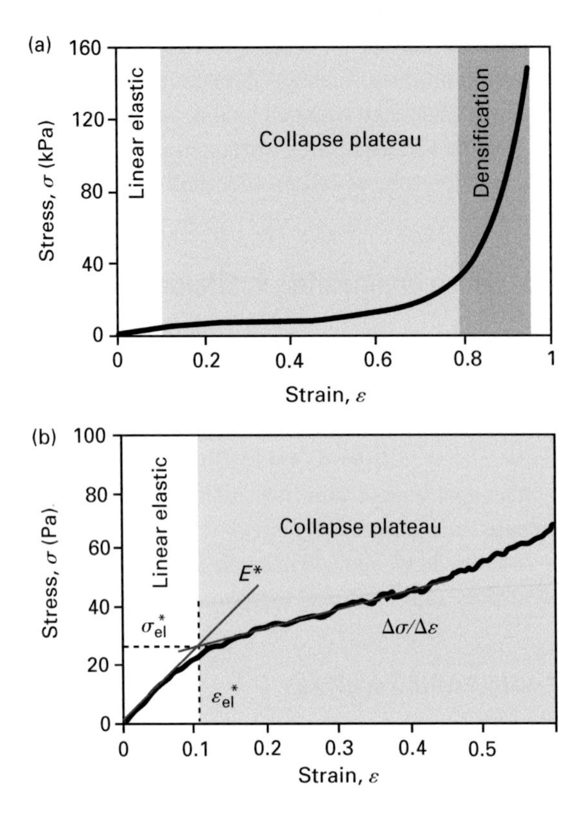

Fig. 8.7 Compressive stress–strain curves for a collagen-GAG scaffold: (a) dry; (b) wet. (Reproduced from Harley *et al.* 2007, with permission from Elsevier.)

(no crosslinking, two levels of dehydrothermal treatment (DHT) and two levels of EDAC treatment). The relative density was increased by increasing the amount of collagen and GAG in the slurry, while maintaining a constant ratio of collagen to GAG (by weight) ($\rho^*/\rho_s = 0.0058, 0.0090, 0.012, 0.018$). The pore size, d, was varied by controlling the final freezing temperature in the freezing protocol ($d = 96, 110, 121, 151$ μm).

The Young's modulus of the strut material in a collagen-GAG scaffold was measured by removing a single strut, bonding one end to a glass slide so that most of the length of the strut cantilevers out over the edge of the slide and performing a bending test using the tip of an atomic force microscope (AFM) (Fig. 8.8). The Young's modulus of the strut material, for the dry scaffold ($\rho^*/\rho_s = 0.0058$, DHT at 105 °C for 24 hours, 121 μm pore size) was measured to be $E_{s,dry} = 762$ MPa. The Young's modulus of the hydrated strut material was estimated by multiplying this value by the ratio of the hydrated scaffold modulus to the dry scaffold modulus ($E_{s,wet} = 5.28$ MPa), a ratio of less than 1/100.

The measured and calculated values of the modulus and elastic collapse stress for the 0.0058 dense scaffold are listed in Table 8.3. The model gives a good description of the compressive properties of this scaffold. The Young's modulus and elastic collapse stress of the scaffolds varied roughly linearly with relative density (to the powers 0.89 and 0.95, respectively), rather than with the square, as given by (8.1) and (8.2). The

Table 8.3 Compressive properties of a collagen-GAG scaffold ($\rho^*/\rho_s = 0.0058$)

	E^* (Pa)	σ_{el}^*(Pa)
Dry, measured	30 000	5150
Dry, calculated	25 600	5130
Wet, measured	208	21
Wet, calculated	178	18

Source: Data from Harley *et al.* (2007).

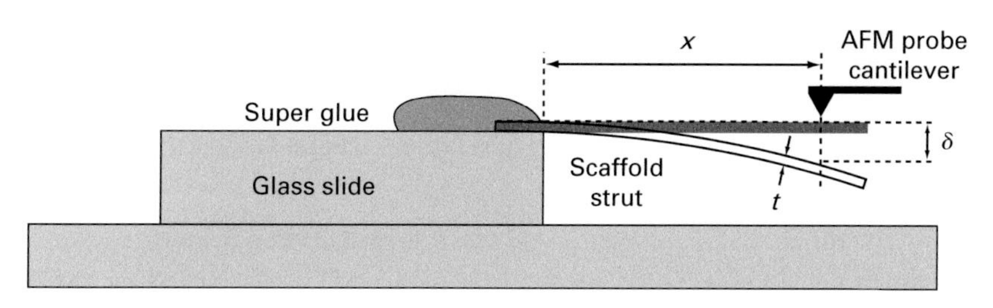

Fig. 8.8 Beam bending test on a single strut removed from a collagen-GAG scaffold using an AFM cantilever. (Reproduced from Harley *et al.*, 2007, with permission from Elsevier.)

higher density scaffolds had microstructural heterogeneities: large pores or "holes," with diameters of the order of over 500 μm as well as local regions, of the order of 100–200 μm, of higher density. Increasing the density of the solid in the CG suspension during fabrication increases the suspension viscosity, making it increasingly difficult to achieve a uniform dispersion of solid throughout the mixture. The microstructural heterogeneities observed in the higher density scaffolds are likely due to heterogeneities in the CG suspension prior to freeze drying.

The Young's modulus and elastic collapse stress of the scaffolds increased with increasing crosslink density. The solid strut modulus, E_s, was not measured for the different crosslink densities, so that a direct comparison with (8.1) and (8.2) is not possible. The compressive properties of the scaffold were found to be independent of the pore size, as expected from the cellular solids model.

8.4.2 Mechanical behavior of mineralized collagen-GAG scaffolds

Mineralized collagen-GAG scaffolds, designed for the regeneration of bone, can be produced by a modification of the freeze-drying process described above for collagen-GAG scaffolds (Harley *et al.*, 2010a). Microfibrillar, type I collagen is first blended in phosphoric acid. Chondroitin-6-sulfate, followed by calcium nitrate hydrate and calcium hydroxide, are then added and blended, to form a triple co-precipitate of collagen, GAG and calcium phosphate. The slurry is freeze-dried to form a porous scaffold with the calcium phosphate mineral brushite distributed throughout the struts (Fig. 8.9). The brushite is then converted, by hydrolysis, to octacalcium phosphate and then apatite.

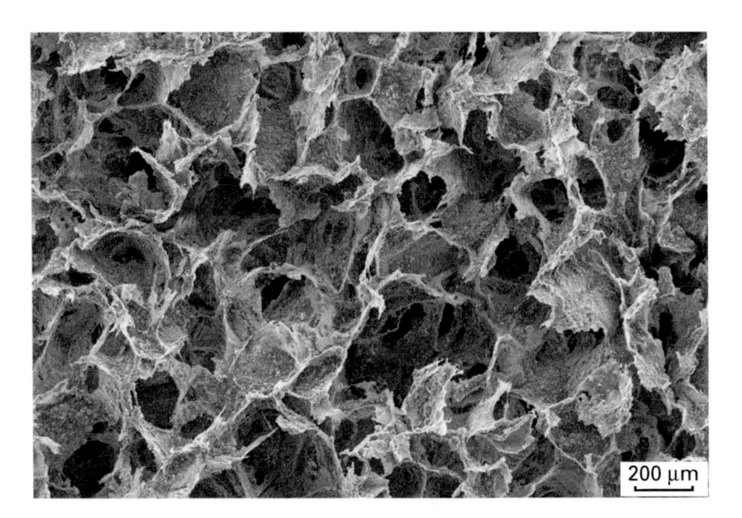

Fig. 8.9 Scanning electron micrograph of the mineralized collagen-GAG scaffold.

The mineralized collagen-GAG scaffold is also used in an osteochondral scaffold, which has two layers to regenerate bone as well as the overlying cartilage (Harley *et al.*, 2010a, 2010b; Lynn *et al.*, 2010). The scaffold is designed to mimic the composition and structure of the healthy articular joint. Cartilage is composed of type-II collagen along with proteoglycans. It is avascular and has a low cell content, making repair of damage difficult. Bone, as we saw in Chapter 2, is a composite of hydroxyapatite-like mineral in a matrix of type I collagen. The osteochondral scaffold is made by pouring the mineralized type-I collagen-GAG slurry into a mold, then adding a type-II collagen-GAG slurry on top and allowing the two to interdiffuse to form an interface with a gradient composition. The slurry is then freeze-dried to give a scaffold with two layers, one mineralized and the other not. The scaffold is designed for repair of small defects in cartilage: the surgeon removes a cylindrical plug of damaged cartilage as well as the underlying bone, replacing it with the osteochondral scaffold. Once in place, mesenchymal stem cells from the bone marrow are recruited to differentiate into either bone cells (osteoclasts and osteoblasts) or cartilage cells (chondrocytes).

A compressive stress–strain curve for the mineralized collagen-GAG scaffold is shown in Fig. 8.10. Like the unmineralized scaffold, it exhibits the three regimes characteristic of a cellular solid. In this case, cell collapse corresponding to the stress plateau is inelastic, although the plastic deformation can be recovered on hydration. Compressive tests on 3.8% dense mineralized scaffolds with 50 wt% calcium phosphate, crosslinked by treatment with EDAC, indicated that the Young's modulus and collapse stress of the dry scaffold were 762 kPa and 85.2 kPa, respectively, while those of the hydrated scaffold were 4.12 kPa and 0.29 kPa (Harley *et al.*, 2010a). The compressive strength of the scaffold is so low that it can be inelastically deformed by manual pressure between the thumb and fingers; it would be beneficial to increase the strength of the scaffolds for handling during surgery. In addition, Engler *et al.* (2006) found that mesenchymal stem cells differentiate into osteoblast-like cells when seeded

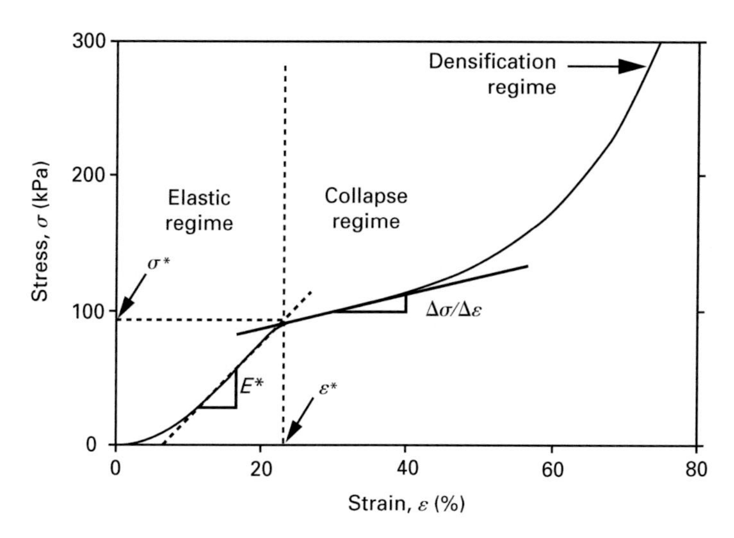

Fig. 8.10 Compressive stress–strain curve for dry mineralized collagen-GAG scaffold (50wt% mineral, non crosslinked). (Reproduced from Kanungo *et al.*, 2008, with permission from Elsevier.)

onto substrates with a Young's modulus similar to that of osteoid (25–40 kPa), the collagen matrix on which bone naturally forms in the body. For both these reasons, methods of improving the mechanical properties of the mineralized collagen-GAG scaffold have been developed.

The modulus and strength of the scaffold can be improved by increasing the solid cell-wall properties or increasing the relative density (Kanungo *et al.*, 2008). Increasing the crosslink density increases the modulus and strength of the scaffold in the hydrated state, but not in the dry state. Young's moduli and strengths of 15.2 kPa and 2.08 kPa have been achieved with EDAC crosslinking in 3.8% dense, hydrated scaffolds with 50% mineral by weight. Increasing the mineralization, from 50wt% to 75wt%, gave a decrease in modulus and strength, due to the formation of defects such as voids in the cell walls and disconnected cell walls.

The relative density of the freeze-dried mineralized collagen-based scaffolds can be increased by increasing the volume fraction of co-precipitate in the slurry. As with the collagen-GAG scaffolds, increasing the volume fraction of the constituents in the slurry increases its viscosity, making it more difficult to mix and form a homogeneous distribution of the co-precipitate. Instead, the relative density of the scaffold has been increased by using vacuum pressure to remove liquid from the low-density slurry through a filter (Kanungo and Gibson, 2009). Using this process, mineralized collagen-GAG scaffolds with relative densities of 0.045 to 0.187 have been fabricated (Fig. 8.11). The pores in the densest scaffold partially collapsed during the processing, giving a layered structure, degrading the mechanical properties. The maximum Young's modulus and collapse stress achieved for the non-crosslinked, hydrated scaffold of relative density 0.137 were 34.8 kPa and 2.12 kPa. EDAC crosslinking increased these values to 91.7 kPa and 4.15 kPa, respectively. Increasing the relative density and crosslinking gives hydrated mineralized scaffolds with the same modulus as osteoid

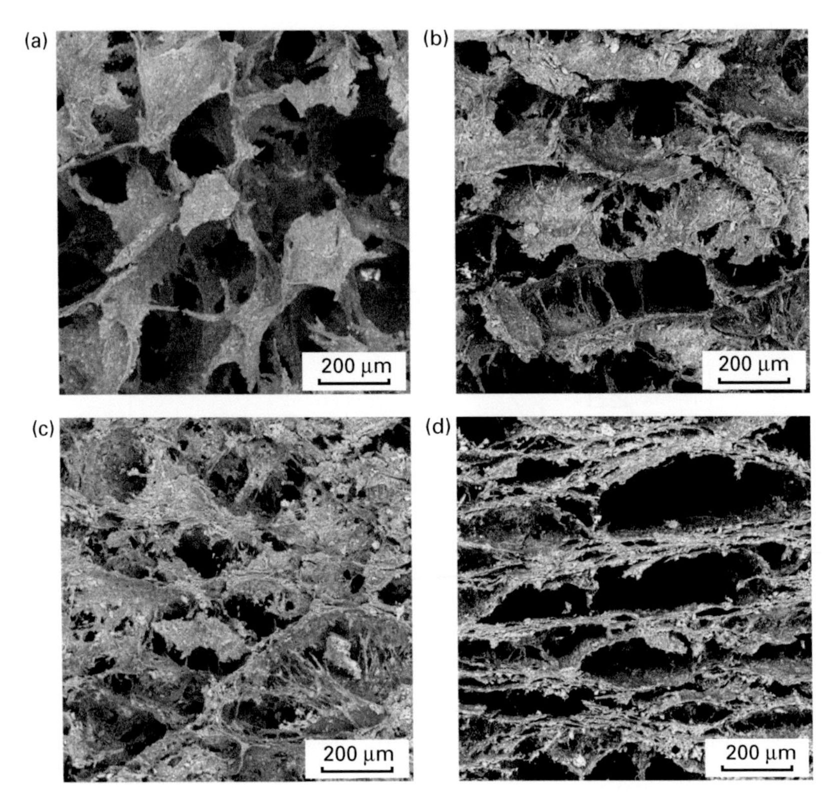

Fig. 8.11 Scanning electron micrographs of mineralized collagen-GAG scaffolds with relative densities of (a) 0.045, (b) 0.098, (c) 0.137 and (d) 0.187. (Reproduced from Kanungo and Gibson, 2009, with permission from Elsevier.)

and increases the strength by a factor of up to 14, allowing the denser scaffolds to withstand handling.

The relative Young's modulus and collapse stress of the remaining three lower density mineralized scaffolds are plotted against relative density in Fig. 8.12 (Kanungo and Gibson, 2009). The Young's modulus of the solid strut material was obtained by cantilever bending tests on individual struts using an atomic force microscope (as in Fig. 8.8): $E_s = 7.34$ GPa. This value is somewhat lower than that for human compact bone in the longitudinal direction ($E = 18$ GPa) (Currey, 2002). The strength of the solid strut material was obtained from nanoindentation tests (Kanungo *et al.*, 2008): σ_s=201 MPa, similar to the compressive strength of human compact bone in the longitudinal direction ($\sigma = 182$ MPa) (Martin *et al.*, 1998). The relative Young's modulus of the scaffolds varies with the square of relative density, as expected for the model for open-cell foams. The constant 0.047 is an order of magnitude lower than the value of 1 typically obtained for open-cell foams. The relative collapse stress of the scaffold varies with the relative density raised to the power 3/2, as expected for inelastic collapse of an open-cell foam. The constant 0.023 is again an order of magnitude lower than the values of 0.2 for brittle and 0.3 for ductile open-cell foams. The lower constants may be related to small volumes of solid aggregating at the nodes and to disconnected struts within the structure.

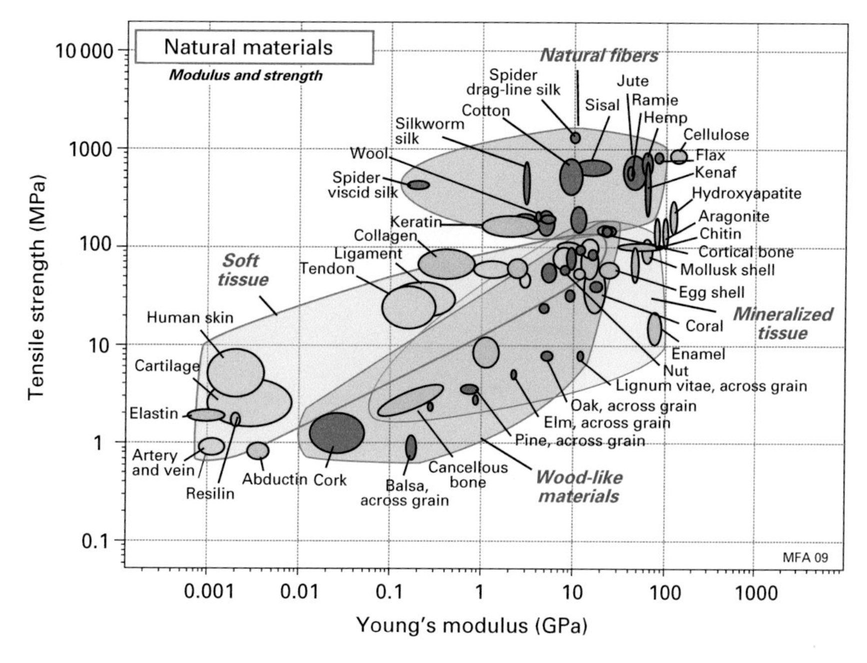

Fig. 2.9 Young's modulus and tensile strength of natural materials. For clarity the basic structural building blocks (yellow bubbles) are not enclosed in an envelope.

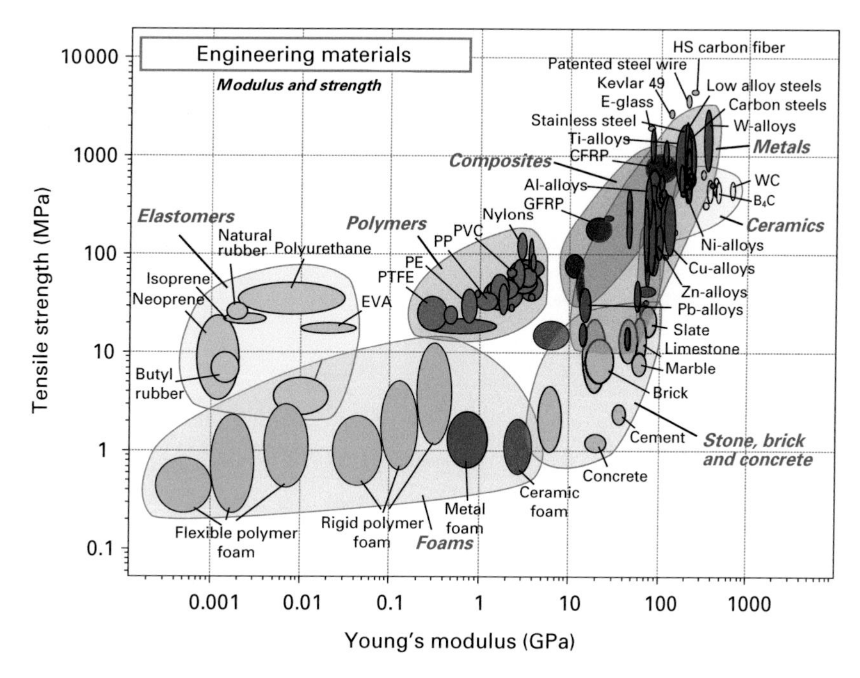

Fig. 2.10 Young's modulus and tensile strength of engineering materials plotted on exactly the same axes as Fig. 2.9.

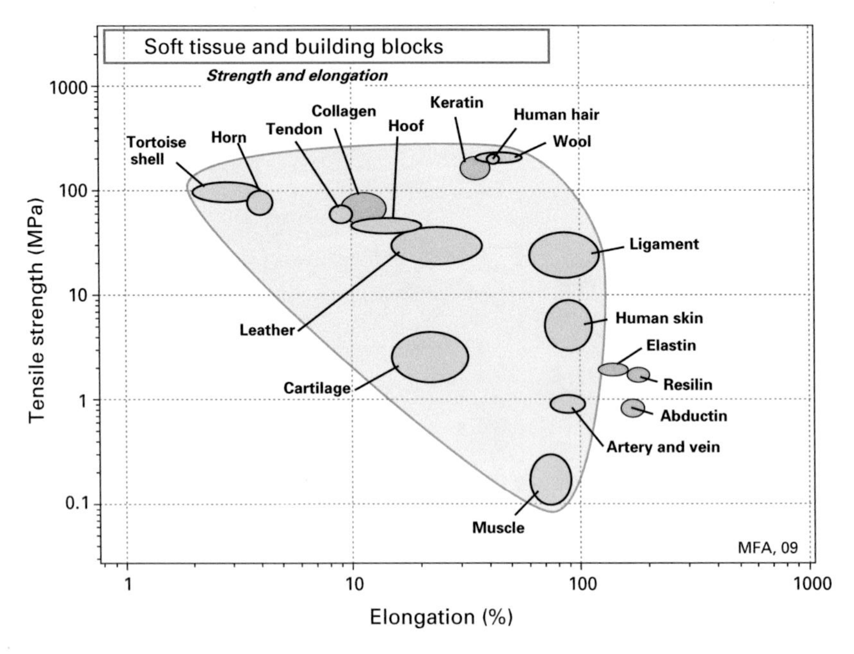

Fig. 2.11 Strength and elongation of soft tissues compared with those of the basic structural building blocks that make them up. The greater the elastin content, the greater is the elongation.

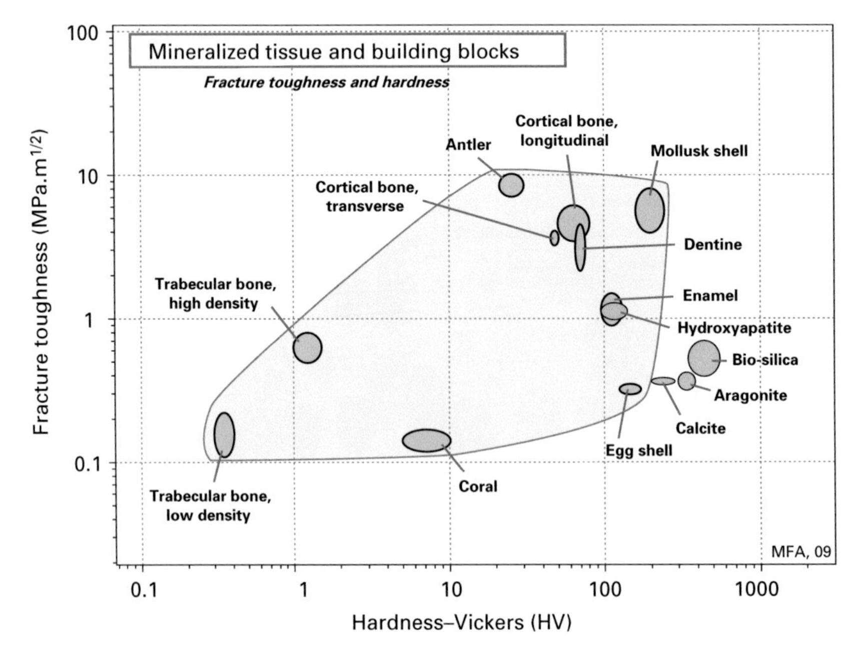

Fig. 2.12 A comparison of the fracture toughness and hardness of mineralized tissue with those of the minerals from which they are made.

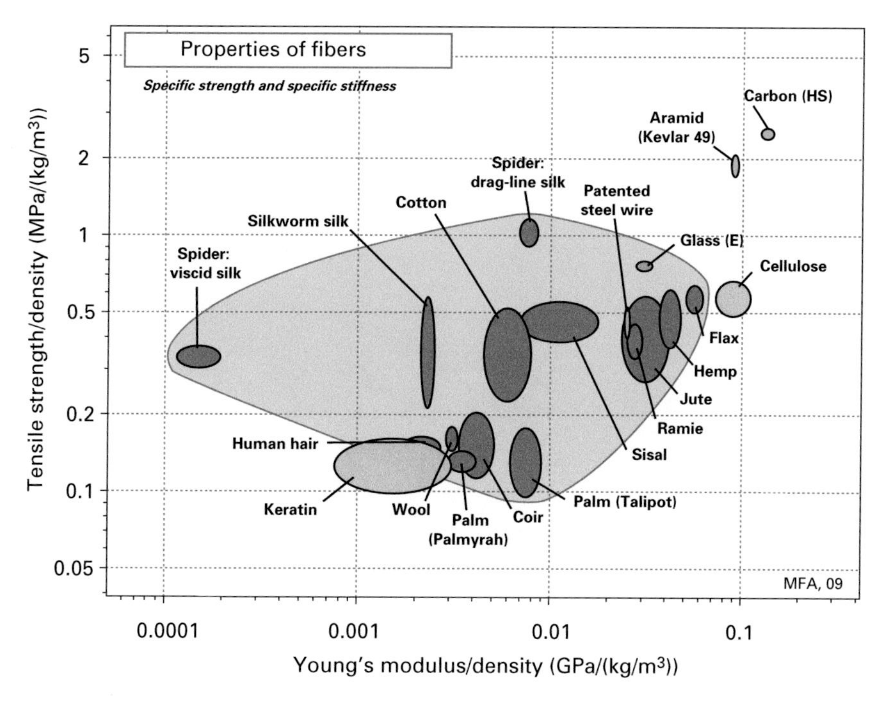

Fig. 2.13 The specific strength and specific stiffness of natural fibers compared to those of the strongest man-made fibers. Several natural fibers are as good as, or better than, steel.

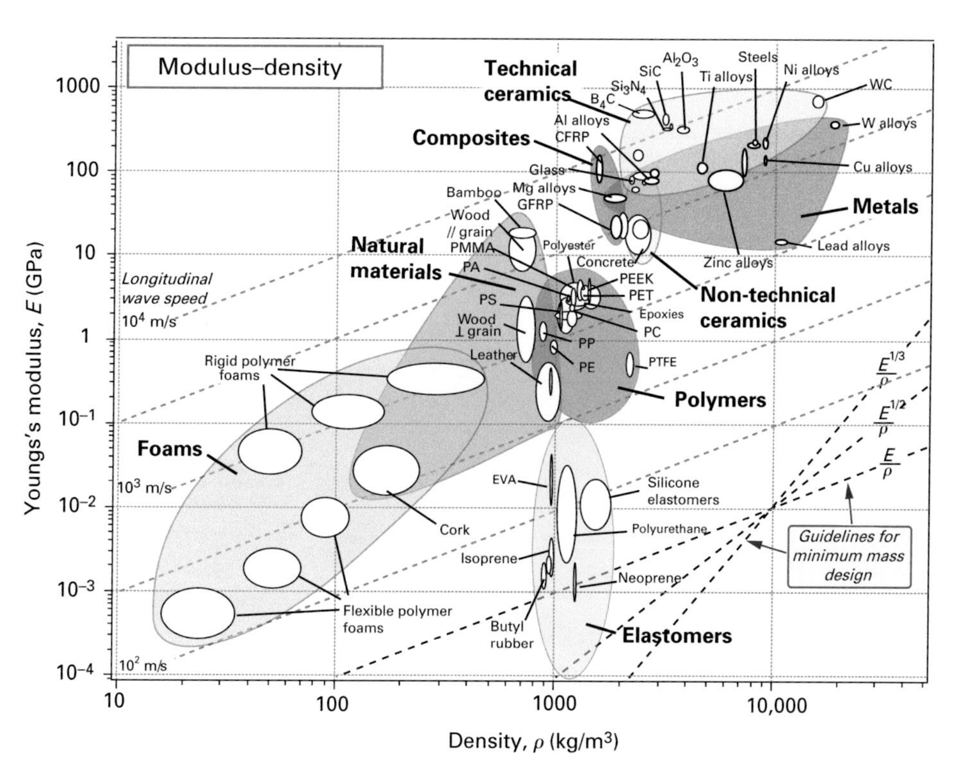

Fig. 7.1 Young's modulus, E, plotted against density, ρ. The heavy envelopes enclose data for a given class of material. The diagonal contours show the longitudinal wave velocity. The guidelines of constant E/ρ, $E^{1/2}/\rho$, and $E^{1/3}/\rho$ allow selection of materials for minimum weight, deflection-limited design.

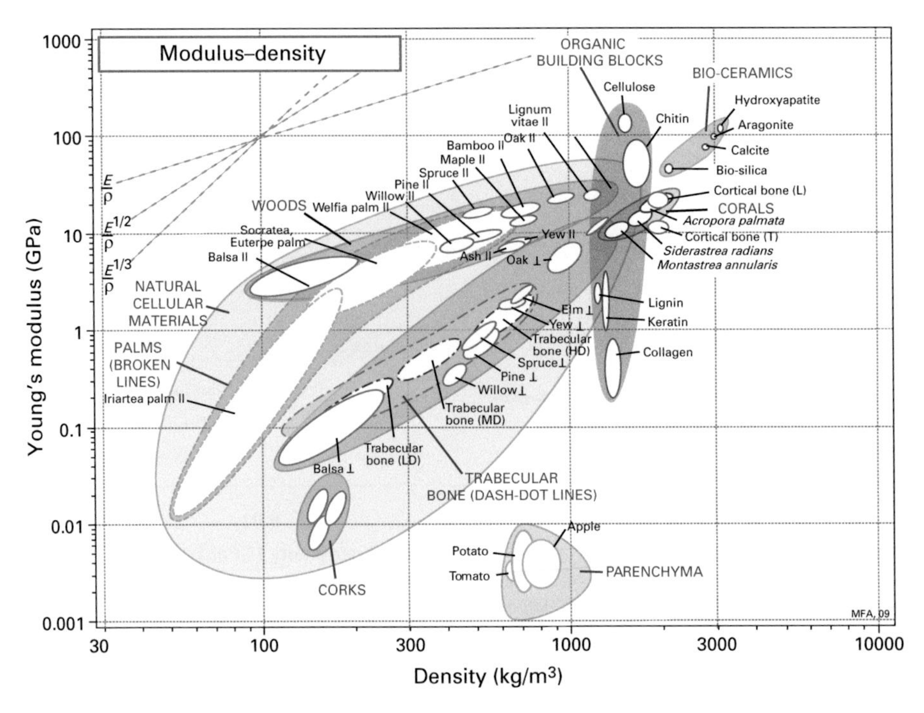

Fig. 7.3 The modulus–density chart for natural cellular materials. Materials that are stiff and light have high values of the indices indicated at the upper left.

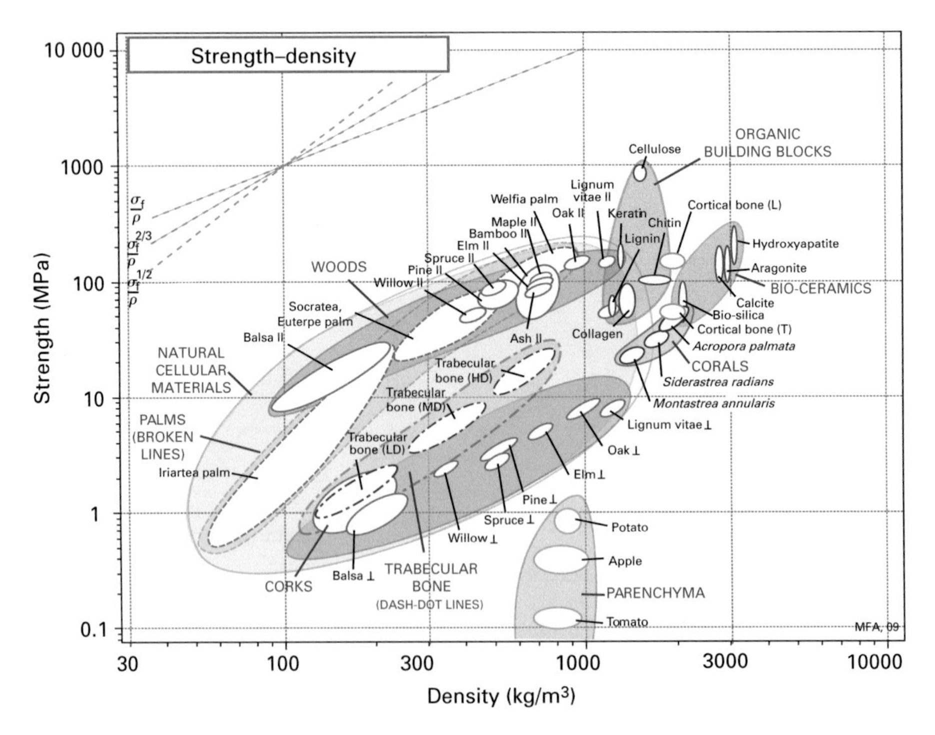

Fig. 7.4 The strength–density chart for natural cellular materials. Materials that are strong and light have high values of the indices indicated at the upper left.

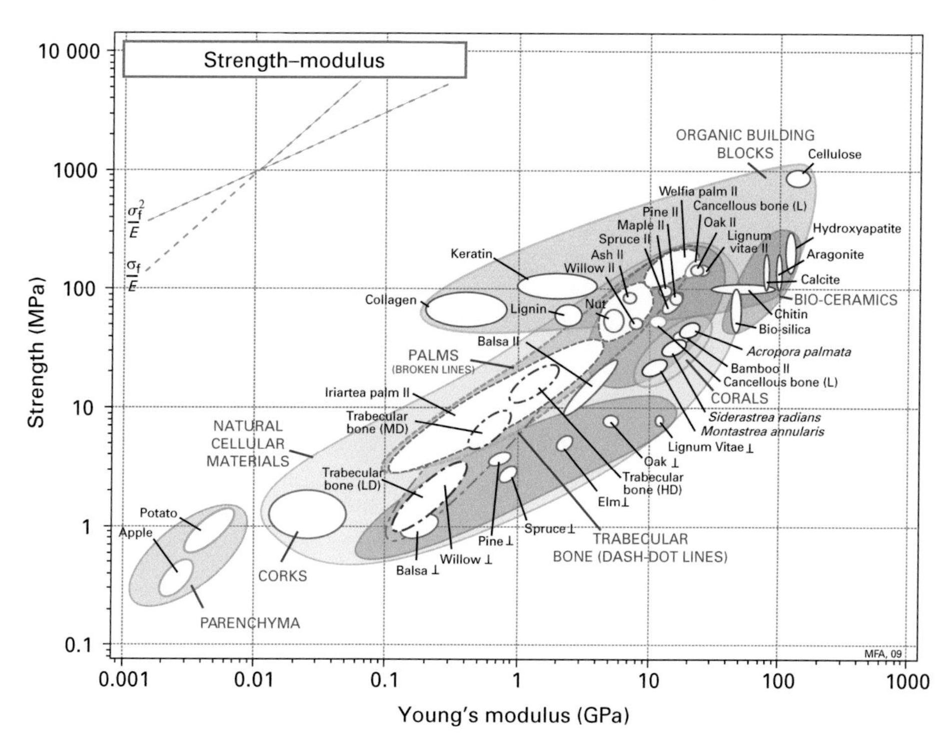

Fig. 7.5 The strength–modulus chart for natural cellular materials. Materials that excel at storing elastic strain energy per unit volume or are resilient have high values of the indices indicated at the upper left.

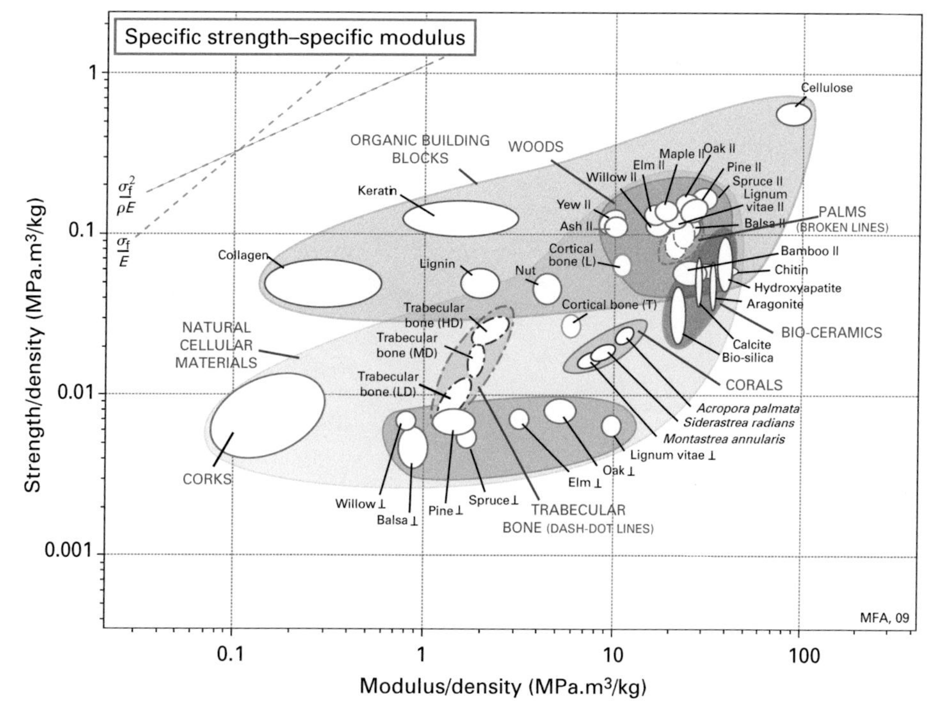

Fig. 7.6 The specific strength–specific modulus chart for natural cellular materials. Materials that excel at storing elastic strain energy per unit weight or are resilient have high values of the indices indicated at the upper left.

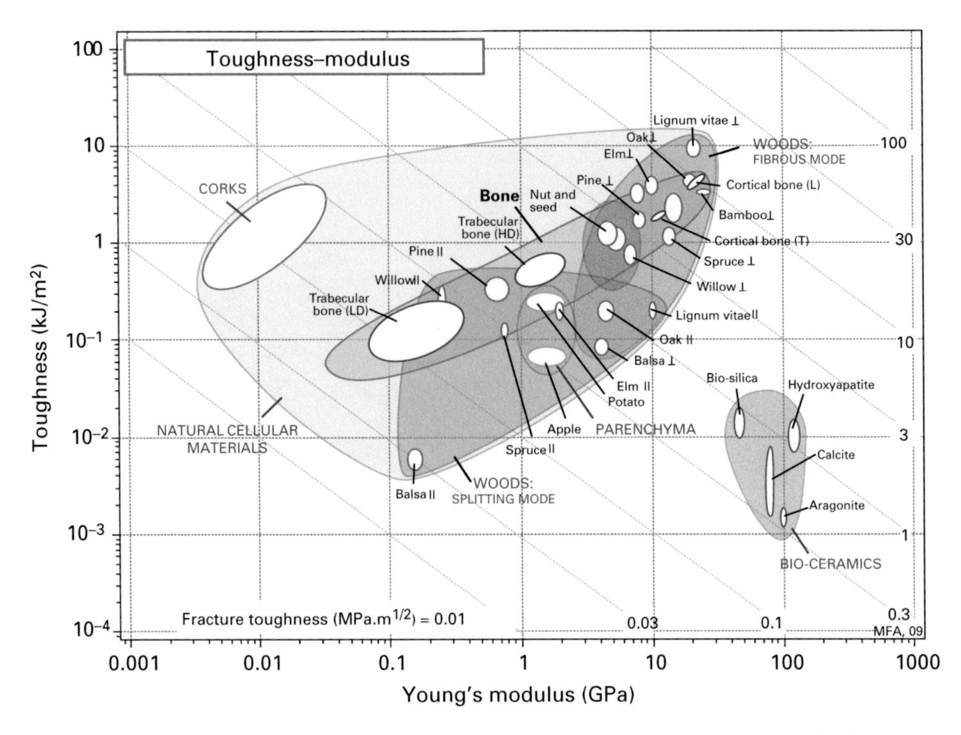

Fig. 7.7 The toughness–modulus chart for natural cellular materials. Materials that best resist fracture have high values of toughness or of fracture toughness, shown as diagonal contours.

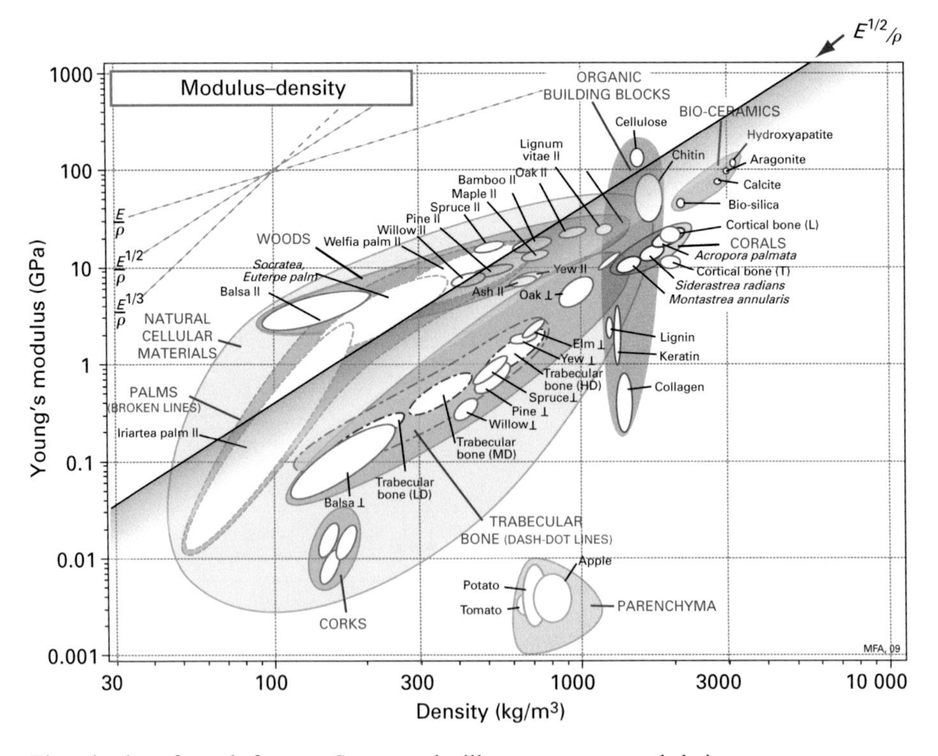

Fig. 7.9 The selection of woods for oars. Spruce and willow emerge as good choices.

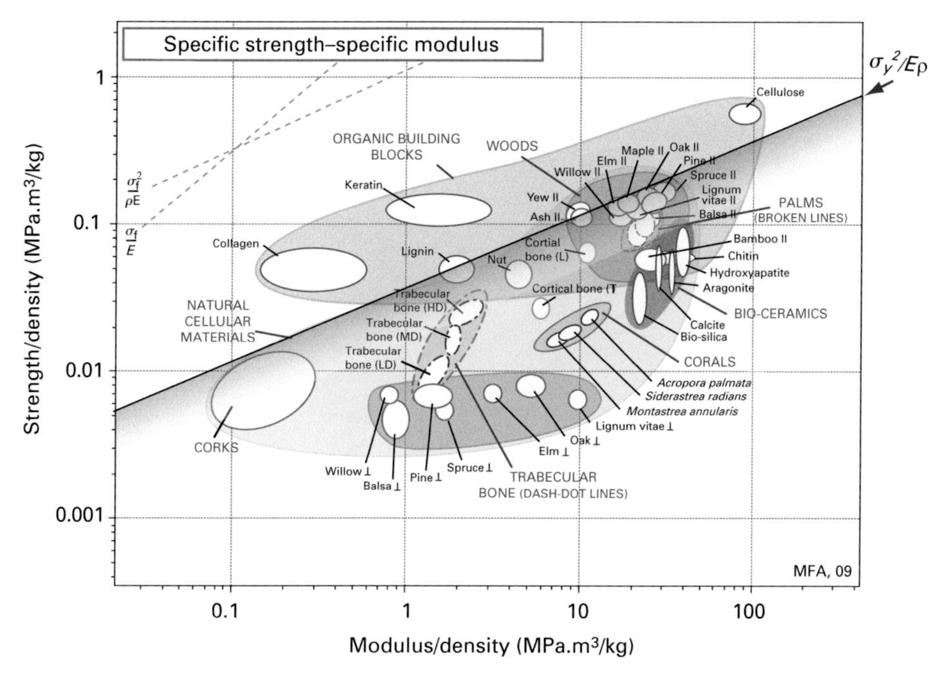

Fig. 7.11 Materials used for archery bows. As noted in the text, yew and ash stand out.

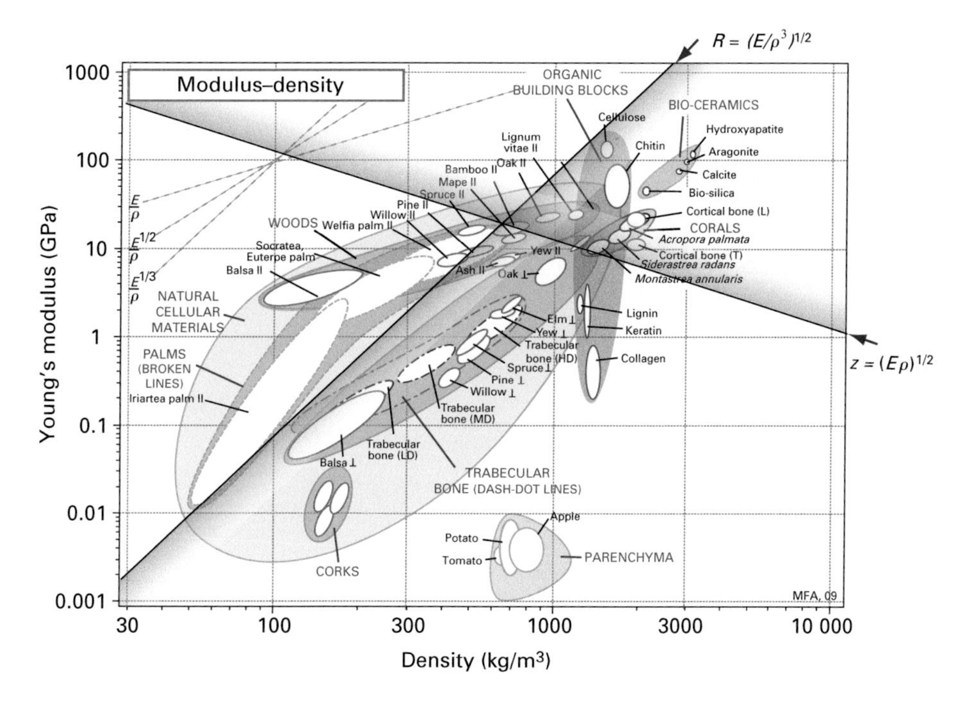

Fig. 7.13 The selection of woods for sound boards. Spruce, pine and willow stand out.

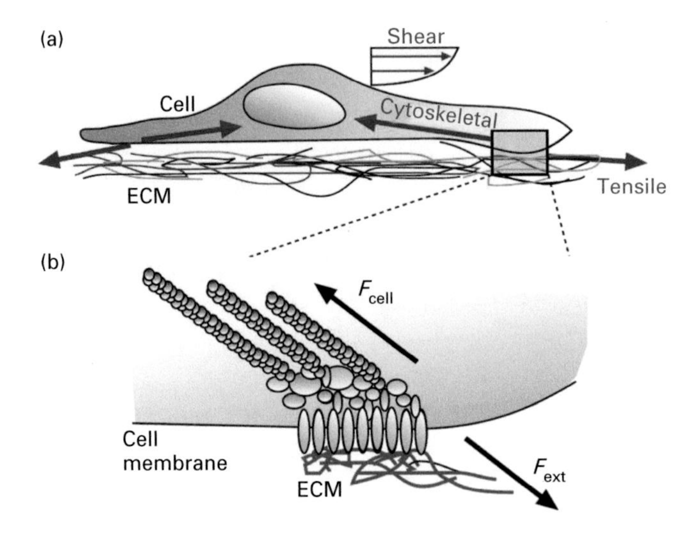

Fig. 9.1 Schematic of the types of interactions between cells and the extracellular matrix (ECM). (a) A single cell attached to a complex ECM (illustrated as a multicolored network). Shear forces arise from fluid flow adjacent to the cell. Tensile forces within the cytoskeleton arise from bonding at focal adhesion sites. (b) Close-up of an ECM-focal adhesion (FA) complex showing a balance of external and internal forces (F_{ext} and F_{cell}, respectively) mediated between actin stress fibers (orange) and FAs (multicolored array of proteins) that bind to the ECM (blue) through integrins (tan). (Reproduced with permission from Chen, 2008.)

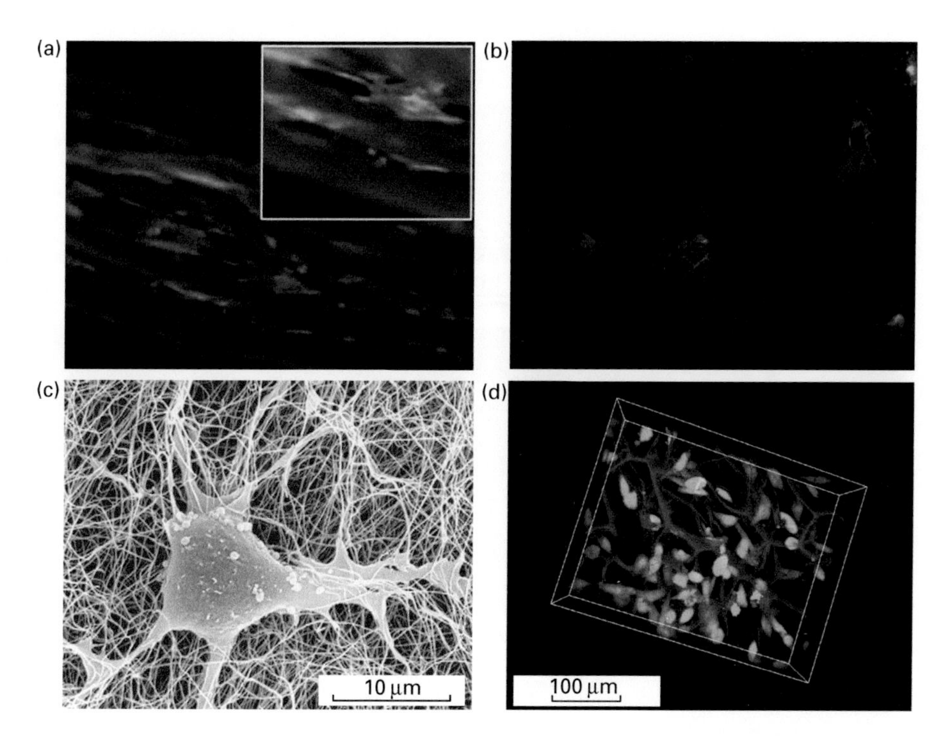

Fig. 9.2 Types of biomaterial constructs used to study cell–material interactions. (a) *One-dimensional fiber constructs*. Fibroblasts aligned along fibrin microthreads (UV crosslinked) after 1 day in culture. (b) *Impermeable, two-dimensional substrates*. MC3T3-E1 osteoblasts seeded on top of type I collagen hydrogels and fluorescently labeled for actin cytoskeleton (phalloidin, green) and counterstained for nuclear DNA (DAPI, blue). (c) *Three-dimensional, nanoporous hydrogel scaffolds*. Human fibroblasts seeded on top of 3D collagen hydrogel matrices; cell extensions penetrate into the matrix and become entangled with collagen fibrils. (d) *Three-dimensional scaffolds*. 3D confocal micrograph of the porous microstructure of a CG scaffold (red) seeded with labeled NR6 cells (green). (a, Reproduced with permission from Cornwell and Pins, 2007; b, Choi and Harley, unpublished; c, reprinted from Rhee and Grinnell, 2007, with permission from Elsevier; d, reprinted from Harley *et al*., 2008, with permission from Elsevier.)

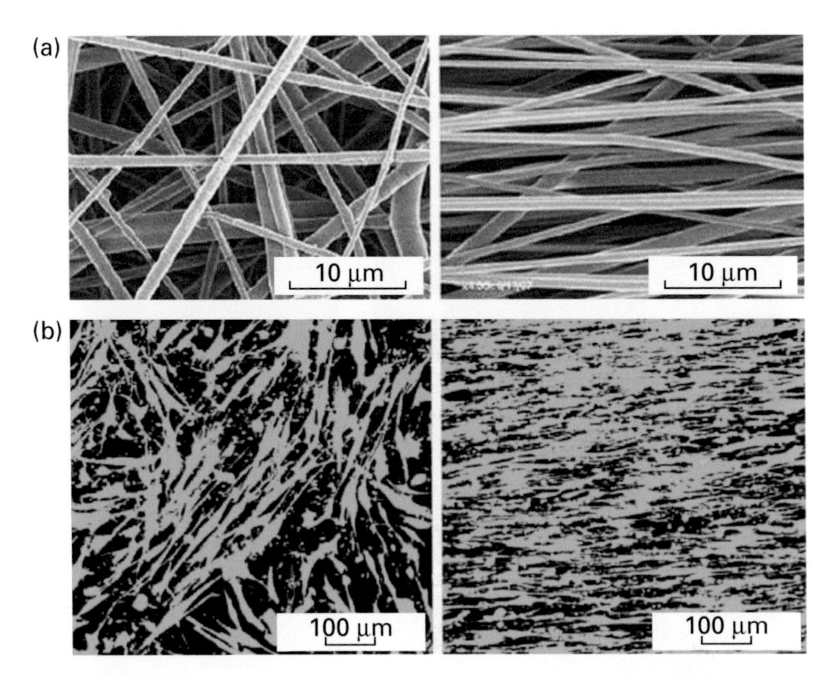

Fig. 9.6 Aligned, anisotropic cellular biomaterials designed for tissue engineering applications in anisotropic tissues. (a) Scanning electron micrographs showing unaligned versus aligned nanofiber PLGA scaffolds. (b) Rotator cuff fibroblasts cultured on unaligned versus aligned PLGA nanofiber scaffolds showed significantly different shape and alignment as early as 1 day post-seeding; cells were fluorescently labeled using a live (green) versus dead (red) stain. (Reprinted from Moffat *et al.*, 2009b, with permission from Elsevier.).

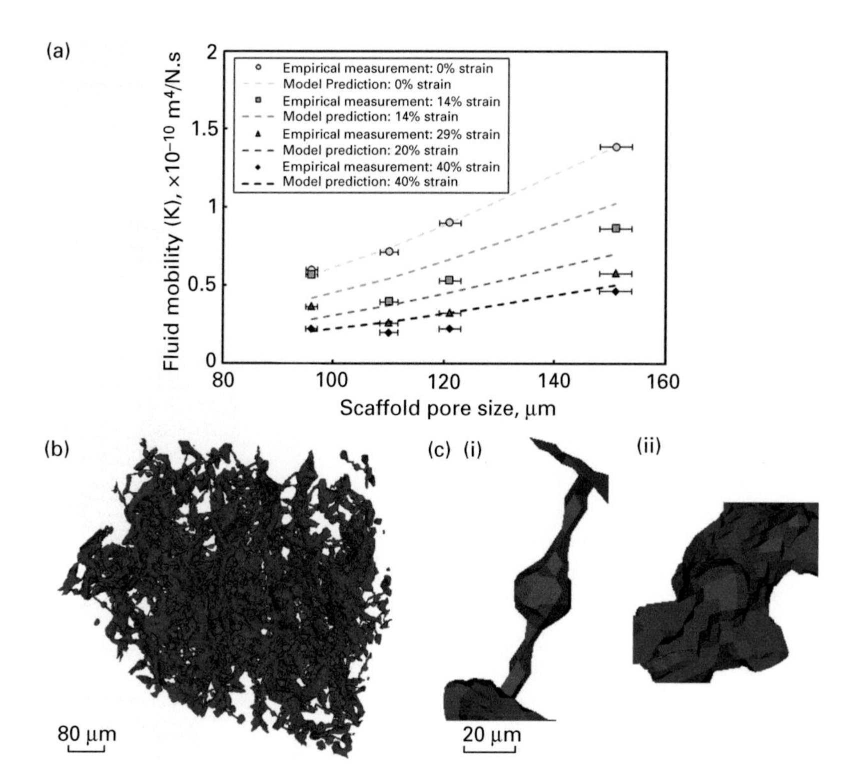

Fig. 9.7 Models of cellular biomaterials. (a) Comparison between the measured flow mobility (K_{meas}, data points) and the predicted values (curves) obtained from the cellular solids-based mathematical model for CG scaffold fluid mobility under varying compressive strains. (b) 3D mesh of the reconstructed cell-seeded CG scaffold (96 μm pore size) used in CFD simulations. (c) Typical geometry of cells (i) bridging two struts or (ii) attached to a single strut. Scaffolds are displayed in blue, simulated cells in red. (a, Adapted from data first published in O'Brien *et al.*, 2007; b,c, reprinted from Jungreuthmayer *et al.*, 2009b, with permission from Elsevier.)

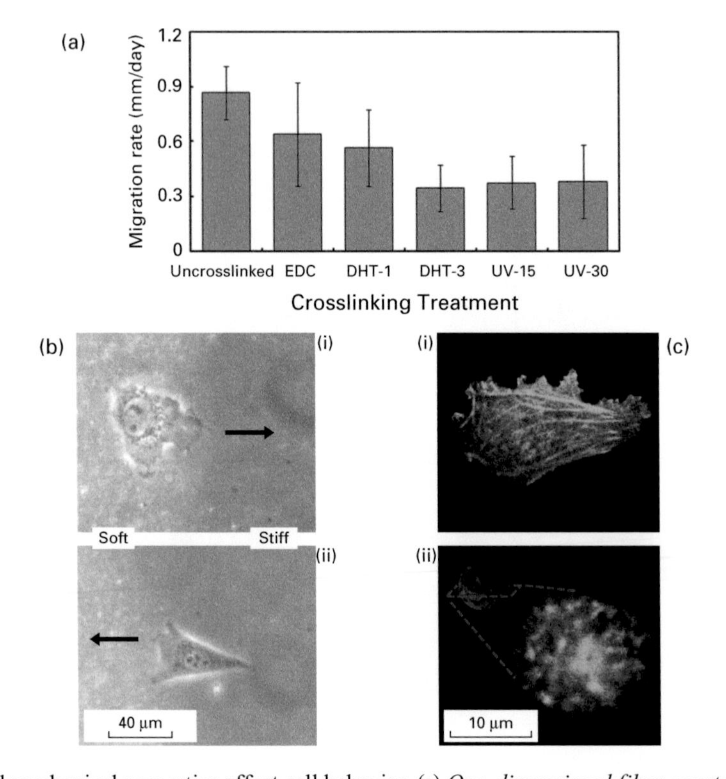

Fig. 9.8 Scaffold mechanical properties affect cell behavior. (a) *One-dimensional fiber constructs.* Fibroblast migration rate along collagen microfibers decreases with increasing crosslinking, in order of: uncrosslinked, carbodiimide crosslinking (EDC), dehydrothermal crosslinking at 110 °C under 50–100 m Torr vacuum for 1 (DHT-1) or 3 (DHT-3) days, ultraviolet light crosslinking for 15 (UV-15) or 30 (UV-30) minutes. (b) *Impermeable, two-dimensional substrates.* NIH 3T3 cells showed significantly different motility patterns on substrates with a rigidity gradient (14–30 kPa). Cells (i) move from the soft side of the substrate toward the gradient and across into the stiff side of the substrate; conversely, (ii) cells approaching the gradient from the stiff side did not cross back onto the soft substrate. A significant change in the cells spreading area of cells on the soft versus stiff substrate was also observed. (c) *Three-dimensional, nanoporous hydrogel scaffolds.* Confocal actin micrographs of metastatic prostate cancer cells (PC-3) seeded on top of (i) and embedded within (ii) 3D collagen hydrogels with identical mechanical properties ($G' = 8.73$ Pa). The actin cytoskeleton (phalloidin, green) of cells (i) seeded on top of the hydrogel (2D microenvironment) is fundamentally different from those (ii) embedded within the hydrogel (3D microenvironment); actin fibers are less elongated and defined in 3D hydrogels. (a, Reproduced with permission from Cornwell *et al.*, 2007; b, reprinted from Lo *et al.*, 2000, with permission from Elsevier; c, reprinted from Baker *et al.*, 2009, with permission from Elsevier.)

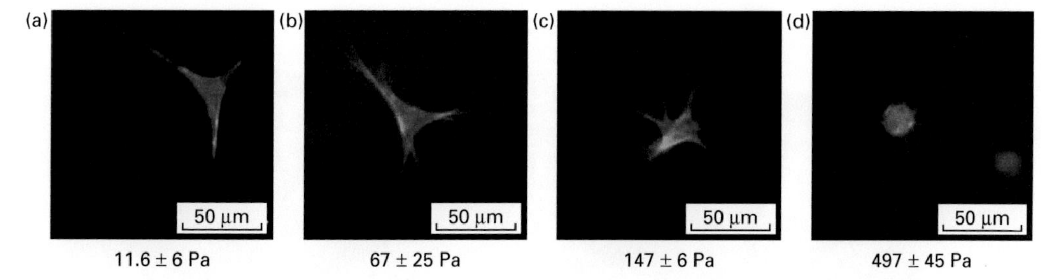

Fig. 9.9 Influence of three-dimensional hydrogel mechanical properties on cell morphology. Smooth muscle cells (SMCs) encapsulated in PEG-fibrinogen (PF) hydrogels express different morphologies depending on their local mechanical environment (cytoskeleton: green; nucleus: blue). Cell spindling decreases with increasing mechanical stiffness: (a) 11.6 ± 6 Pa (highly spindled with regular lamellipodia), (b) 67 ± 25 Pa (spindled with frayed lamellipodia), (c) 147 ± 6 Pa (nonspindled with frayed lamellipodia), (d) 497 ± 45 Pa (rounded with minor lamellipodia). (Reprinted from Dikovsky *et al.*, 2008, with permission from Elsevier.)

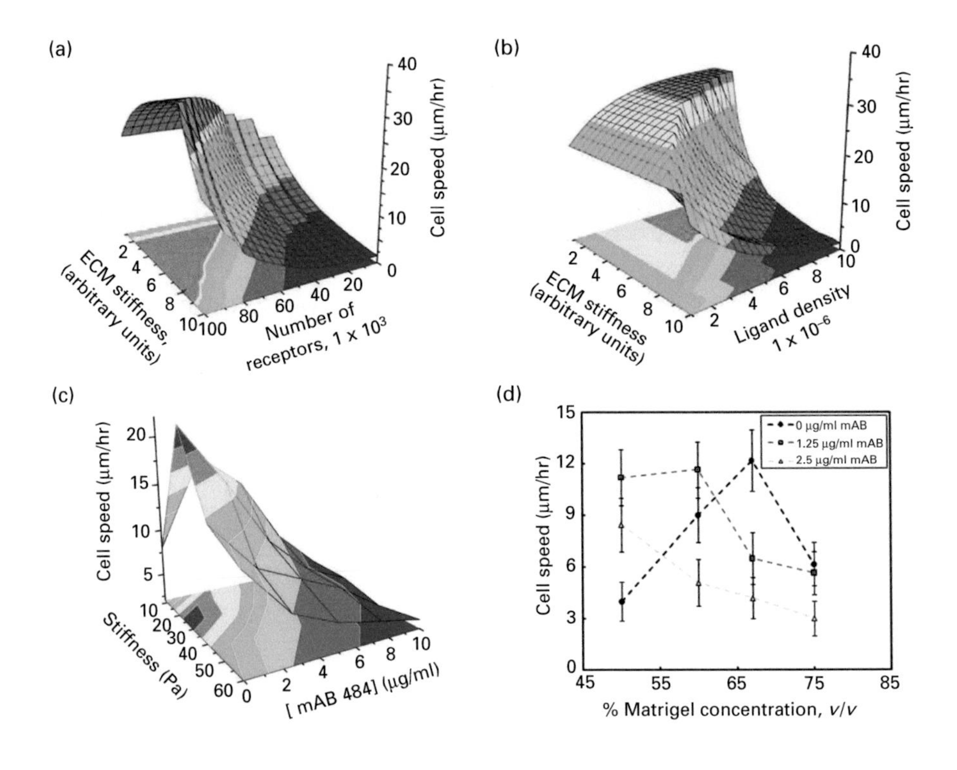

Fig. 9.13 Three-dimensional plot of cell speed as a function of matrix stiffness, ligand density and available receptors. (a) Computational model of cell migration speed as a function of matrix stiffness and number of available receptors; highest speeds occur at maximum receptors and intermediate stiffness. (b) Computation model of cell migration speed as a function of matrix stiffness and number of available ligands; highest speeds occur at intermediate stiffness and ligand concentration. (c) Experimentally measured migration speed of DU-145 human prostate carcinoma cells in a 3D Matrigel matrix as a function of receptor number and matrix stiffness. (d) Integrin inhibition shifts the maximum cell speed to lower Matrigel concentrations. The average number of motile cells decreases with increases in the concentration (0, 1.25, 2.5 μg/ml) of 4B4 anti-integrin blocking monoclonal antibody (mAB), while average speed shows somewhat of a bimodal behavior with variations in gel density, reaching a maximum with intermediate gel concentration. The presence of 4B4 antibody slowed cell speed and shifted the maximum to lower Matrigel concentrations. (a,b, Reprinted from Zaman *et al.*, 2005, with permission from Elsevier; c,d, copyright 2006 National Academy of Sciences, USA, reprinted with permission from Zaman *et al.*, 2006.)

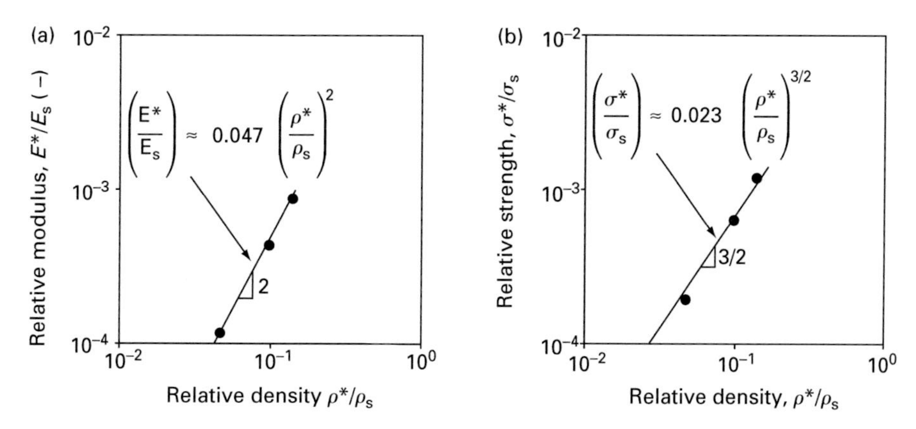

Fig. 8.12 (a) The relative Young's modulus, E^*/E_s, and (b) relative collapse stress, σ^*/σ_s, plotted against relative density for the mineralized collagen-GAG scaffolds (50wt% mineral, non crosslinked). (Reproduced from Kanungo and Gibson, 2009, with permission from Elsevier.)

8.4.3 Mechanical behavior of honeycomb-like scaffolds

Honeycomb-like scaffolds, with their more regular geometries, are attractive in some applications (Fig. 8.13). A large-scale (pore size of 3.5mm) hexagonal honeycomb, made by photopatterning, has been used to increase diffuse nutrient transport to hepatocytes for liver regeneration (Tsang *et al.*, 2007). Scaffolds with rectangular pores of varying aspect ratio and with diamond-shaped pores, made by high-resolution stereolithography, have been used to study the effect of pore geometry on fibroblast orientation (Engelmayr *et al.*, 2006). And an accordion-like honeycomb, made by excimer laser ablation, has been developed to match the anisotropy in the mechanical properties of cardiac tissue (Engelmayr *et al.*, 2008).

The triangulation of the hexagonal honeycomb in Fig. 8.13a indicates that this scaffold is a stretch-dominated rather than a bending-dominated structure, so its modulus and strength are expected to vary linearly with relative density (see the Maxwell criterion in Chapter 3). The hexagonal structure is isotropic in the plane of the hexagonal cells. The rectangular honeycomb in Fig. 8.13b is highly anisotropic: for loading along the struts, the linear-elastic behavior is controlled by stretching of the members and the Young's moduli depend linearly on the relative density. The ratio of the moduli in the two perpendicular directions corresponding to the long and short axis of the rectangle is simply the inverse of the ratio of the edge lengths in those directions. For off-axis loading (e.g. along the diagonal of the rectangular cells), bending deformations make the honeycomb much more compliant. The mechanical properties of the scaffold with the diamond-shaped pores (Fig. 8.13c) can be described by the honeycomb models of Chapter 3 (with $h = 0$). Those of the scaffold with the accordion-like pores (Fig. 8.13d) could be calculated using the same methods of structural analysis.

A less regular collagen honeycomb has been made by first preparing an acidic collagen solution. When neutralized by ammonia gas, the collagen solution turns into a

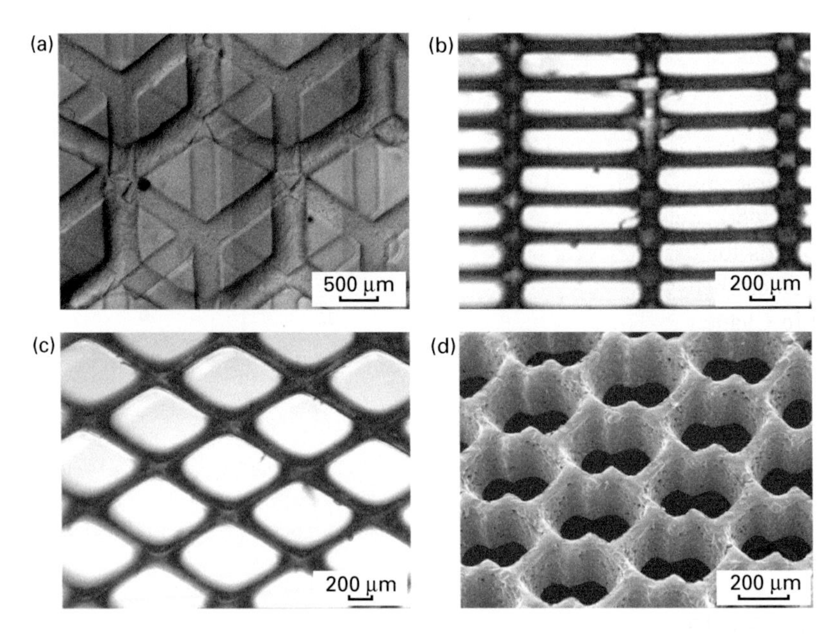

Fig. 8.13 Honeycomb-like scaffolds: (a) hexagonal honeycomb; (b) rectangular honeycomb with aspect ratio of 5:1; (c) diamond honeycomb; and (d) accordion-like honeycomb. (a, Reproduced from Tsang *et al.*, 2007, with permission from FASEB Journal; b,c, reproduced from Engelmayer *et al.*, 2006, with permission from Elsevier; d, reprinted by permission from Macmillan Publishers Ltd: Nature Materials, Engelmayer *et al.*, 2006, copyright 2008.)

white gel which is then freeze-dried to form the collagen honeycomb (Itoh *et al.*, 2001; George *et al.*, 2006, 2008). The resulting collagen honeycomb has irregular cells within the plane and more or less prismatic cells in the out-of-plane direction, so we expect the honeycomb models discussed in Chapter 3 to apply.

8.5 Summary

Metal foams offer a number of advantages over fully dense metals for bone substitute materials in some applications. The Young's moduli of metal foams more closely match that of the bone they are replacing, reducing stress shielding; their interconnected porosity allows bone ingrowth, and their high specific surface area promotes the delivery of biological agents such as cells, growth factors and proteins. The biocompatibility and low modulus of titanium make it particularly attractive; much of the work on metal foams as bone substitute materials has focused on titanium foams. The relative densities of metal foams developed as bone substitute materials are typically over 0.3. Data for the moduli are somewhat lower than expected for open-cell foam models, while the compressive strengths are somewhat higher. The scatter in the data is large, reflecting the varied microstructures obtained from the different processes used to make the materials.

Tissue engineering scaffolds are designed to provide an environment for cells to function and regenerate tissue. They are made from a wide range of materials: biological polymers, such as collagen, synthetic polymers, such as those used in resorbable sutures, or composites of a polymer and calcium phosphate or bioglass. The porous microstructure of a scaffold can resemble a honeycomb, a foam, or a fibrous network, depending on the process used to fabricate it. The relative densities of scaffolds for tissue engineering are typically low, sometimes less than 1%. The mechanical behavior of foam-like scaffolds can be described by the models developed in Chapter 3. The composition, microstructure and mechanical properties of tissue engineering scaffolds affect the attachment, morphology, migration and contractile response of biological cells, such as fibroblasts, that migrate into them. That topic is the subject of Chapter 9.

References

Metal foams

Arciniegas M, Aparicio C, Manero JM and Gil FJ (2007) Low elastic modulus metals for joint prosthesis: tantalum and nickel-titanium foams. *J. Euro. Ceram. Soc.* **27**, 3391–8.

Bobyn JD, Stackpool GJ, Hacking SA, Tanzer M and Krygier JJ (1999) Characteristics of bone ingrowth and interface mechanics of a new porous tantalum biomaterial. *J. Bone Joint Surgery Brit.* **81B**, 907–14.

Bram M, Stiller C, Buchkremer HP, Stöver D and Baur H (2000) High-porosity titanium, stainless steel, and superalloy parts. *Adv. Eng Mat.* **2**, 196–9.

Cachinho SCP and Correia RN (2008) Titanium scaffolds for osteointegration: mechanical, in vitro and corrosion behavior. *J. Mat. Sci.: Mat. Med* **19**, 451–7.

Cheung S, Gauthier M, Lefebvre L-P, Dunbar M and Filiaggi M (2007) Fibroblastic interactions with high-porosity Ti-6Al-4V metal foam. *J. Biomed. Mat. Res. Part B: Appl. Biomat.* **82B**, 440–9.

Curodeau A, Sachs E and Caldarise S (2000) Design and fabrication of cast orthopedic implants with freeform surface textures from 3-D printed ceramic shell. *J. Biomed. Mat. Res. Appl. Biomat.* **53**, 525–35.

Currey JD (2002) *Bones: Structure and Mechanics.* Princeton, NJ: Princeton University Press.

Fife JL, Li JC, Dunand DC and Voorhees PW (2009) Morphological analysis of pores in directionally freeze-cast titanium foams. *J. Mat. Res.* **24**, 117–24.

Gauthier M, Menini R, Bureau MN, So SKV, Dion M-J and Lefebvre LP (2003) Properties of novel titanium foams intended for biomedical applications. *ASM Materials and Processes for Medical Devices Conference,* Anaheim, CA, September 8–10, pp. 382–7.

Gauthier M, Lefebvre LP, Thomas Y and Bureau MN (2004) Production of metallic foams having open porosity using a powder metallurgy approach. *Mat. Manuf. Process.* **19**, 793–811.

Grimm MJ and Williams JL (1997) Measurements of permeability in human calcaneal trabecular bone. *J. Biomech.* **30**, 743–5.

Hong TF, Guo ZX and Yang R (2008) Fabrication of porous titanium scaffold materials by a fugitive filler method. *J. Mat. Sci. Mat. Med.* **19**, 3489–95.

Imwinkelried T (2007) Mechanical properties of open-pore titanium foam. *J. Biomed. Mat. Res.* **81A**, 964–70.

Kim H-M, Miyaji F, Kokubo T and Nakamura T (1996) Preparation of bioactive Ti and its alloys via simple chemical surface treatment. *J. Biomed. Mat. Res.* **32**, 409–17.
Kokubo T, Kushitani H, Sakka S, Kitsugi T and Yamamuro T (1990) Solutions able to reproduce *in vivo* surface-structure changes in bioactive glass-ceramic A-W[3]. *J. Biomed. Mat. Res.* **24**, 721–34.
Lefebvre L-P and Baril E (2008) Effect of oxygen concentration and distribution on the compression properties on titanium foams. *Adv. Eng. Mat.* **10**, 868–76.
Levine BR, Sporer S, Poggie RA, Della Valle CJ and Jacobs JJ (2006) Experimental and clinical performance of porous tantalum in orthopedic surgery. *Biomaterials* **27**, 4671–81.
Levine B (2008) A new era in porous metals: applications in orthopaedics. *Adv. Eng. Mat.* **10**, 788–92.
Li BY, Rong LJ, Li YY and Gjunter VE (2000) Synthesis of porous Ni-Ti shape-memory alloys by self-propagating high-temperature synthesis: reaction mechanism and anisotropy in pore structure. *Acta Mater.* **48**, 3895–904.
Li H, Oppenheimer SM, Stupp SI, Dunand DC and Brinson LC (2004) Effects of pore morphology and bone ingrowth on mechanical properties of microporous titanium as an orthopaedic implant material. *Mat. Trans.* **45**, 1124–31.
Lin C-Y, Wirtz T, LaMarca F and Hollister SJ (2007) Structural and mechanical evaluations of a topology optimized titanium interbody fusion cage fabricated by selective laser melting process. *J. Biomed. Mat. Res.* **83A**, 272–9.
Martin RB, Burr DB and Sharkey NA (1998) *Skeletal Tissue Mechanics*. Berlin: Springer.
Muller U, Imwinkelried T, Horst M, Sievers M and Graf-Hausner U (2006) Do human osteoblasts grow into open-porous titanium? *Euro. Cells Mat.* **11**, 8–15.
Mullner HW, Fritsch A, Kohlhauser C *et al.* (2007) Acoustical and poromechanical characterization of titanium scaffolds for biomedical applications. *Strain* **44**, 153–63.
Ryan G, Pandit A and Apatsidis DP (2006) Fabrication methods of porous metals for use in orthopaedic applications. *Biomaterials* **27**, 2651–70.
Sargeant TD, Mustafa OG, Oppenheimer SM *et al.* (2008) Hybrid bone implants: self-assembly of peptide amphiphile nanofibers within porous titanium. *Biomaterials* **29**, 161–71.
Sevilla P, Aparicio C, Planell JA and Gil FJ (2007) Comparison of the mechanical properties between tantalum and nickel-titanium foams implant materials for bone ingrowth applications. *J. Alloys Compounds* **439**, 67–73.
Shimko DA, Shimko VF, Sander EA, Dickson KF and Nauman EA (2005) Effect of porosity on the fluid flow characteristics and mechanical properties of tantalum scaffolds. *J. Biomed. Mat. Res. Part B: Appl. Biomat.* **73B**, 315–24.
Spector M (1987) Historical review of porous-coated implants. *J. Arthroplasty* **2**, 163–77.
Spoerke ED, Murray NG, Li H, Brinson LC, Dunand DC and Stupp SI (2005) A bioactive titanium foam scaffold for bone repair. *Acta Biomater.* **1**, 523–33.
Spoerke ED, Murray NGD, Li H, Brinson LC, Dunand DC and Stupp SI (2007) Titanium with aligned, elongated pores for orthopedic tissue engineering applications. *J. Biomed. Mat. Res.* **84A**, 402–12.
Wen CE, Mabuchi M, Yamada Y, Shimojima K, Chino Y and Asahina T (2001) Processing of biocompatible porous Ti and Mg. *Scripta Mater.* **45**, 1147–53.
Wen CE, Yamada Y, Shimojima K, Chino Y, Hosokawa H and Mabuchi M, (2002) Novel titanium foam for bone tissue engineering. *J. Mat. Res.* **17**, 2633–9.
Zhang X, Ayers RA, Thorne K, Moore JJ and Schowengerdt F (2001) Combustion synthesis of porous materials for bone replacement. *Biomed. Sci. Instrument.* **37**, 463–8.

Tissue engineering scaffolds

Atala A (2007) Engineering of tissues and organs. In *Tissue Engineering Using Ceramics and Polymers, ed.* Boccaccini AR and Gough JE. Cambridge, UK: Woodhead Publishing and Boca Raton, FL: CRC Press, pp. 269–93.

Badylak SF, Freytes DO and Gilbert TW (2009) Extracellular matrix as a biological scaffold material: Structure and function. *Acta Biomater.* **5**, 1–13.

Barbucci R and Leone G (2004) Formation of defined microporous 3D structures starting from cross-linked hydrogels. *J. Biomed. Mat. Res. Part B: Appl. Biomat.* **68B**, 117–26

Boccaccini AR and Gough JE (editors) (2007) *Tissue Engineering Using Ceramics and Polymers.* Cambridge, UK: Woodhead Publishing and Boca Raton, FL: CRC Press.

Currey JD (2002) *Bones: Structure and Mechanics.* Princeton, NJ: Princeton University Press.

Engelmayr GC, Papworth GD, Watkins SC, Mayer JE and Sacks MS (2006) Guidance of engineered tissue collagen orientation by large-scale scaffold microstructures *J. Biomech.* **39**, 1819–31.

Engelmayr GC, Cheng M, Bettinger CJ, Borenstein JT, Langer R and Freed LE (2008) Accordian-like honeycombs for tissue engineering of cardiac anisotropy. *Nature Mat.* **7**, 1003–10.

Engler AJ, Sen S, Sweeney HL, Discher DE (2006) Matrix elasticity directs stem cell lineage specification. *Cell* **126**, 677–89.

Fernandes H, Moroni L, van Blitterswijk C and de Boer J (2009) Extracellular matrix and tissue engineering applications. *J. Mat. Chem.* **19**, 5474–84.

Freed LE and Guilak F (2007) Engineering functional tissues. In *Principles of Tissue Engineering,* 3rd edn., ed. Lanza R, Langer R and Vacanti J. Amsterdam: Elsevier Academic Press, pp. 137–53.

George J, Kuboki Y and Miyata T (2006) Differentiation of mesenchymal stem cells into osteoblasts on honeycomb collagen scaffolds. *Biotech. Bioeng.* **95**, 404–11.

George J, Onodera J and Miyata T (2008) Biodegradable honeycomb collagen scaffold for dermal tissue engineering. *J. Biomed. Mat. Res.* **87A**, 1103–11.

Gilbert TW, Sellaro TL and Badylak SF (2006) Decellularization of tissues and organs. *Biomaterials* **27**, 3675–83.

Harley BA, Leung JH, Silva ECCM and Gibson LJ (2007) Mechanical characterization of collagen-glycoaminoglycan scaffolds. *Acta Biomater.* **3**, 463–74.

Harley BA, Lynn AK, Wissner-Gross Z, Bonfield W, Yannas IV and Gibson LJ (2010a) Design of a multiphase osteochondral scaffold II: fabrication of a mineralized collagen-GAG scaffold. *J.Biomed. Mat. Res.* **92A**, 1066–77.

Harley BA, Lynn AK, Wissner-Gross Z, Bonfield W, Yannas IV and Gibson LJ (2010b) Design of a multiphase osteochondral scaffold III: Fabrication of layered scaffolds with soft interfaces. *J.Biomed. Mat. Res.* **92A**, 1078–93.

Hollister SJ (2005) Porous scaffold design for tissue engineering. *Nature Mat.* **4**, 518–24.

Huang J and Best SM (2007) Ceramic biomaterials. In *Tissue Engineering Using Ceramics and Polymers, ed.* Boccaccini AR and Gough JE. Cambridge, UK: Woodhead Publishing and Boca Raton, FL: CRC Press, pp. 1–31.

Hulbert SF, Young FA, Mathews RS, Klawitter JJ, Talbert CD and Stelling FH (1970) Potential of ceramic materials as permanently implantable skeletal prostheses. *J. Biomed. Mat. Res.* **4**, 433–56.

Ikada Y (2006) *Tissue Engineering: Fundamentals and Applications.* Amsterdam: Elsevier Academic Press.

Itoh H, Aso Y, Furuse M, Noishiki Y and Miyata T (2001) A honeycomb collagen carrier for cell culture as a tissue engineering scaffold. *Artificial Organs* **25**, 213–17.

Kanungo B, Silva E, Van Vliet KJ and Gibson LJ (2008) Characterization of a mineralized collagen GAG scaffold for bone regeneration. *Acta Biomater.*, **4**, 490–503.

Kanungo BP and Gibson LJ (2009) Density-property relationships in mineralized collagen-GAG scaffolds. *Acta Biomater.*, **5**, 1006–18.

Kuhne JH, Bartl R, Frisch B, Hammer C, Jansson V and Zimmer M (1994) Bone formation in coralline hydroxyapatite: effects of pore size studied in rabbits. *Acta Orthop Scand.* **65**, 246–52.

Kwon IK, Kidoaki S and Matsuda T (2005) Electrospun nano- and microfiber fabrics made of biodegradable copolyesters: structural characteristics, mechanical properties and cell adhesion potential. *Biomaterials* **26**, 3929–39.

Lanza R, Langer R and Vacanti J (editors) (2007) *Principles of Tissue Engineering,* 3rd edn. Amsterdam: Elsevier Academic Press.

Lu Q, Ganesan K, Simionescu DT and Vyavahare NR (2004) Novel porous aortic elastic and collagen scaffolds for tissue engineering. *Biomaterials* **25**, 5227–37.

Lynn AK, Best SM, Cameron RE *et al.* (2010) Design of a multiphase osteochondral scaffold I: control of chemical composition. *J. Biomed. Mat. Res.* **92A**, 1057–65.

Martin RB, Burr DB and Sharkey NA (1998) *Skeletal Tissue Mechanics.* Berlin: Springer.

Misra SK and Boccaccini AR (2007) Biodegradable and bioactive polymer/ceramic composite scaffolds. In *Tissue Engineering Using Ceramics and Polymers,* ed. Boccaccini AR and Gough JE. Cambridge, UK: Woodhead Publishing and Boca Raton, FL: CRC Press, pp. 72–92.

Mooney DJ, Mazzoni CL, Breuer C *et al.* (1996) Stabilized polyglycolic acid fibre-based tubes for tissue engineering. *Biomaterials* **17**, 115–24.

Murphy MB and Mikos AG (2007) Polymer scaffold fabrication. In *Principles of Tissue Engineering*, 3rd edn., ed. Lanza R, Langer R and Vacanti J. Amsterdam: Elsevier Academic Press, pp. 309–21.

O'Brien FJ, Harley BA, Yannas IV and Gibson LJ (2004) Influence of freezing rate on pore structure in freeze-dried collagen-GAG scaffolds. *Biomaterials* **25**, 1077–86.

O'Brien FJ, Harley BA, Yannas IV and Gibson LJ (2005) The effect of pore size on cell adhesion in collagen-GAG scaffolds. *Biomaterials* **26**, 433–41.

Pachence JM, Bohrer MP and Kohn J (2007) Biodegradable polymers. In *Principles of Tissue Engineering*, 3rd edn., ed. Lanza R, Langer R and Vacanti J. Amsterdam: Elsevier Academic Press, pp. 323–39.

Partee B, Hollister SJ and Das S (2006) Selective laser sintering process optimization for layered manufacturing of CAPA® 6501 polycaprolactone bone tissue engineering scaffolds. *J. Manuf. Sci. Eng.* **128**, 531–40.

Pek YS, Spector M, Yannas IV and Gibson LJ (2004) Degradation of a collagen-chondroitin-6-sulfate matrix by collagenase and by chondroitinase. *Biomaterials* **25**, 473–82.

Silver FH, Yannas IV and Salzman EW (1978) Glycosaminoglycan inhibition of collagen-induced platelet-aggregation. *Thrombosis Research* **13**, 267–77.

Silver FH, Yannas IV and Salzman EW (1979) In vitro blood compatibility of glycosaminoglycan-precipitated collagens. *J. Biomed. Mat. Res.* **13**, 701–16.

Sohier J, Haan RE, de Groot K and Bezemer JM (2003) A novel method to obtain protein release from porous polymer scaffolds: emulsion coating. *J. Controlled Release* **87**, 57–68.

Tsang VL, Chen AA, Cho LM *et al.* (2007) Fabrication of 3D hepatic tissues by additive photopatterning of cellular hydrogels. *FASEB J.* **21**, 790–801.

Yannas IV, Burke JF, Huang C and Gordon PL (1975) Suppression of in vivo degradability and of immunogenicity of collagen by reaction with glycosaminoglycans. *Polym. Prepr. Amer. Chem. Soc.* **16**, 209–14.

Yannas IV and Silver F (1975) Thromboresistant analogs of vascular tissue. *Abstr. Papers Amer. Chem. Soc.* **170,** August 24, 122.

Yannas IV (1992) Tissue regeneration by use of collagen-GAG copolymers. *Clinical Mat.* **9**, 179–87.

Yannas IV (1995) Tissue regeneration templates based on collagen-GAG copolymers. *Adv. Polymer Sci.* **122**, 219–44.

Zhang J-Y, Doll BA, Beckman EJ and Hollinger JO (2003) Three-dimensional biocompatible ascorbic acid-containing scaffold for bone tissue engineering. *Tissue Eng.* **9**, 1143–57.

9 Interaction of biological cells with tissue engineering scaffolds

9.1 Introduction

The extracellular matrix (ECM)[1] is typically a porous, fibrillar network found within all tissues and organs in the body (Fig. 9.1). The ECM, made up of structural proteins such as collagens and a wide variety of proteoglycans, provides an insoluble structure to which cells can attach. The composition and structure of the ECM is particular to each tissue within the body. For example, the ECM of a tendon is made of type I collagen fibers banded together in an aligned structure oriented in the direction of loading; the ECM of dermis is made up an amorphous organization of type I collagen and proteoglycans; and the ECM of cartilage is made up of an organized structure of type II collagen and a high concentration of proteoglycans (Buckwalter, 1983). The ECM plays a significant role in defining the overall mechanics of a tissue and is responsible for transferring mechanical stimuli from the scale of the organ to the individual cell; the details of the mechanics of the ECM itself depends on its chemical composition, density and structural organization. The surface of the ECM is also functionalized with ligands, biomolecular targets that cells can bind to via integrins, transmembrane proteins that extend from the interior of the cell to the exterior. Integrin–ligand interactions are specific: binding, detachment and deformation of an integrin at the cell surface is translated through signaling pathways within the cell to its intracellular machinery and nucleus, providing extrinsic cues to which the cell responds.

The microstructure of the ECM varies with the type of tissue: for instance, the pore size ranges from tens to hundreds of nanometers in dense connective tissues such as tendon and ligaments to units, tens, or hundreds of microns in dermis and intestinal mucosal membranes (Even-Ram and Yamada, 2005; Tanzer, 2006; Cheng *et al.*, 2009; Wolf *et al.*, 2009). The ECM is typically inhomogeneous and anisotropic. Cortical bone, for example, is mostly a mineralized collagen extracellular matrix, containing a low volume fraction of bone cells; its mechanical properties are higher in the longitudinal direction than in the transverse direction (Currey, 2002). The size, shape and volume fraction of the pores within the ECM affect the rate of diffusion of soluble regulators (for example, cytokines, growth factors, hormones, other paracrine and endocrine signals) to cells. Our understanding of the intricate mechanisms by which the native ECM influences cell biology is by no means complete; tissue engineering scaffolds, designed to mimic the

[1] Biological abbreviations used throughout this chapter are listed in Table 9.1.

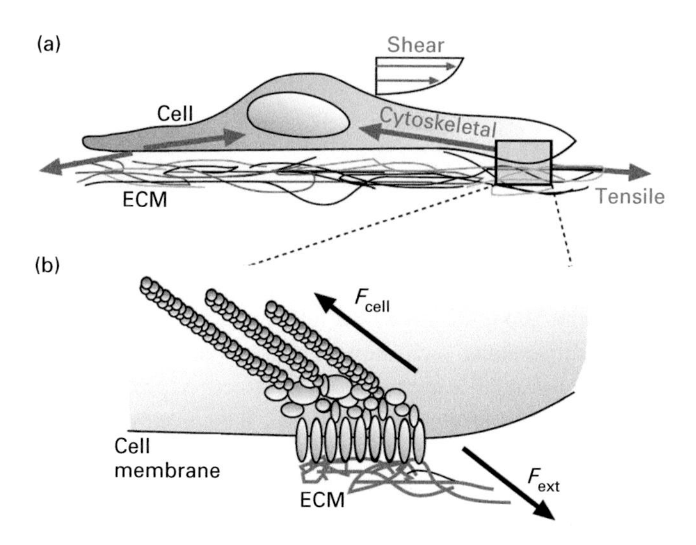

Fig. 9.1 Schematic of the types of interactions between cells and the extracellular matrix (ECM). (a) A single cell attached to a complex ECM (illustrated as a multicolored network). Shear forces arise from fluid flow adjacent to the cell. Tensile forces within the cytoskeleton arise from bonding at focal adhesion sites. (b) Close-up of an ECM-focal adhesion (FA) complex showing a balance of external and internal forces (F_{ext} and F_{cell}, respectively) mediated between actin stress fibers (orange) and FAs (multicolored array of proteins) that bind to the ECM (blue) through integrins (tan). (Reproduced with permission from Chen, 2008.) See plate section for color version.

extracellular matrix, provide an ideal model system to investigate the effect of different features of the native ECM on cell biology.

In this chapter, we describe how a feature of a scaffold may affect "cell biology" or "cell behavior." There are a multitude of potential cell behaviors at a range of scales that can be affected. Some notable behaviors that occur at the level of a whole cell include migration, proliferation (cell division), differentiation (in the case of stem cells), apoptosis (organized cell death), cytoskeletal remodeling (changes in cell shape and organization), matrix protein biosynthesis and contraction. Many changes in cell behavior occur at a subcellular level; examples include changes in gene expression (DNA transcription, RNA translation), protein synthesis and phosphorylation, as well as signal transduction (sequences of biochemical reactions within the cell that convert one kind of signal or stimulus into another). The common thread is that a cell, or a group of cells, exhibits a different level of one or more of these behaviors in response to an external cue provided by the tissue engineering scaffold under consideration. The terms "cell biology" and "cell behavior" will therefore often be used as a generic place-holder for one of many types of cell behavior; using specific examples, we will indicate the type of behavior or biological mechanisms under study. Additionally, as there is such a multitude of potential behaviors to consider and the field of tissue engineering has considered many of these behaviors via many different tests, we have attempted to compare or contrast particular types of behaviors within each section consistently, but in some cases have to compare closely related behaviors. At times bio-materials will be referred to as "bioactive" within the text to indicate that their formulation (chemical, microstructural, mechanical) has been shown to induce specific changes in cell phenotype or genotype.

With the understanding that *in situ* cells exist within a complex, three-dimensional ECM, a wide variety of tissue engineering bio-materials have been created to mimic the ECM of particular anatomical sites. The primary application of these tissue engineering scaffolds has traditionally been to provide a template to induce regeneration or improved healing of a damaged tissue in vivo by modifying the characteristic healing process following injury. Today, however, scaffolds are being increasingly used in vitro as a controlled three-dimensional environment to probe cell–scaffold interactions and the ways in which cell behavior (e.g. migration, contraction, protein biosynthesis, differentiation and division) may be governed by the local environment. Both applications will be considered here.

In this chapter, we summarize the use of a range of cellular bio-materials for in vivo and in vitro studies of how cells interact with them. We examine how scaffold properties, notably chemical composition, microstructure and mechanical properties, each independently affect cell biology. We also consider the combined effects of multiple scaffold parameters as an approach to understand how cells sense and integrate multidimensional information. The goal of this chapter is to provide the reader with a broad overview of how cellular materials can be used for a wide range of tissue engineering studies and the important role the features of such materials play in their successful utilization.

9.2 Scaffold chemical composition

Scaffolds for tissue engineering applications have been made from metals, glasses, ceramics, as well as both synthetic (human-made) and biological (or native) polymers. Details of the materials and processing methods were described in Chapter 8. In this section we discuss the way in which the chemical composition of the scaffold, in particular the degradation rate and degradation products, as well as the type and density of ligands, can significantly affect cell biology.

9.2.1 Scaffold residence time and degradation products

The native ECM is not a static structure; cells produce enzymes to degrade the ECM proteins in their local microenvironment and then synthesize new matrix proteins to replace the old ECM. This process is more marked in some tissues than others, the most notable example being bone. For instance, astronauts are known to lose significant bone mass during orbit and athletes build bone mass, sometimes asymmetrically, as in the case of tennis players (Kontulainen *et al.*, 2003). As cells can remodel their native ECM structure, it is of no surprise that they can similarly degrade (and remodel) many tissue engineering scaffolds. An intact scaffold cannot diffuse away; instead, enzymes produced endogenously by cells within the scaffold degrade it into low molecular weight fragments. The length of time that the scaffold remains insoluble (termed the "residence time") can be critical in defining its bioactivity. This is especially true in the case of in vivo applications; for new tissue to be synthesized at a wound site, the scaffold must initially support cell migration, proliferation and organization, but must then degrade in such a manner that it does not interfere with the native tissue synthesis and remodeling processes that are underway (Yannas, 2001). These considerations require a scaffold residence time with an upper and lower bound, a concept that has been formalized as the isomorphous tissue replacement model: the scaffold residence time must be approximately equal to the time required to synthesize mature tissue via regeneration at the specific tissue site under study (Williams *et al.*, 1983; Yannas *et al.*, 1989; Yannas, 2005).

The lifetime of the scaffold is typically defined by the degradation time constant (t_d) while the time it takes for the native tissue to synthesize the required volume of tissue to replace the scaffold is defined by the healing time constant (t_h). Conventionally, t_d and t_h are defined in terms of half-lives, but provided that t_d and t_h correspond to equivalent end points (for example, total scaffold degradation and complete wound healing) a quantitative comparison (t_h/t_d) can be made. In cases where $t_h/t_d < 1$ or $t_h/t_d > 1$, the scaffold degrades too slowly and interferes with the healing process or degrades too rapidly to support successful tissue repair, respectively. For most in vivo tissue engineering applications, the conventional strategy is to control scaffold degradation rates to meet the requirements of isomorphous tissue replacement ($t_h/t_d = 1$) (Yannas, 2001). Because different wound sites and even the same wound site in different species may have different time constants for healing (t_h) (Yannas, 2001), it is necessary to adjust the degradation rate of the ECM analog for each wound site and species in order to maximize scaffold bioactivity.

The chemical composition of a scaffold significantly influences its degradation kinetics. However, the degradation kinetics of a wide range of synthetic and biological polymer scaffolds can be controlled in several ways. Some common synthetic polymers used to fabricate porous scaffolds include polylactic acid (PLA, polylactide), poly-L-lactide (PLLA), polyglycolic acid (PGA, polyglycolide), its co-polymer with PLA poly(lactic-co-glycolic acid) (PLGA), polyvinyl alcohol (PVA), poly(ε-caprolactone) (ε-CPL) and polycaprolactone, polyethylene (PE), poly(ethylene glycol) (PEG) and their co-polymers (Hutmacher *et al.*, 2001; Nuttelman *et al.*, 2002; Webb *et al.*, 2004; Wang *et al.*, 2005; Alvarez-Barreto *et al.*, 2007; Tessmar and Gopferich, 2007). In many cases, degradation can be modified by changing the length of the individual elements that make up the polymer backbone of these polymers. A classic example of this strategy is the PEG system (Holland *et al.*, 2004; Kong *et al.*, 2004a; Seliktar *et al.*, 2004; Liu *et al.*, 2006a). Another strategy is to form co-polymers between degradable and non-degradable synthetic polymers, thus forming a complex polymer structure whose degradation characteristics can be controlled. For example, PLGA degrades via hydrolysis of its ester linkages, allowing its degradation kinetics to be controlled by changes in the ratio of glycolic (the hydrolyzable polymer component) to lactic (the non-degradable polymer component) acid: the GA:LA ratio. Higher GA:LA ratio co-polymers degrade more rapidly (Burdick *et al.*, 2001; Chen *et al.*, 2001; Karp *et al.*, 2003; Nakanishi *et al.*, 2003; Ryu *et al.*, 2007). Co-polymerization of PVA and PEG as well as modification of PEG backbone molecular weight has been used to control overall scaffold degradation kinetics (Bryant and Anseth, 2003; Martens *et al.*, 2003; Nuttelman *et al.*, 2004; Almany and Seliktar, 2005). Another technique applicable across these synthetic polymer platforms is to change the relative density, or amount of solid per volume of the scaffold; increasing scaffold relative density increases the scaffold degradation time (t_d).

A significant benefit to using synthetic polymers is that a host of processing techniques, including those that utilize high temperatures and pressures, can be applied to produce porous bio-materials with complex microstructures. These techniques include solid free-form fabrication, electrospinning and other fibrillar molding techniques, as well as a variety of liquid state molding techniques (Sachlos and Czernuszka, 2003), all described in Chapter 8. A particular concern for synthetic polymers is that the insoluble scaffold structure cannot diffuse away as a single unit; degradation breaks the scaffold up into much smaller sub-units, some of which may be cytotoxic (even in cases where the non-degraded polymer is non-cytotoxic) or may invoke an increased inflammatory response (Bergsma *et al.*, 1995; Agrawal and Athanasiou, 1997).

Biological polymers are perhaps the most intriguing materials used to create scaffolds for tissue engineering applications as they are native proteins found within the ECM. As materials isolated directly from the natural ECM, biological polymers contain a host of surface ligands (e.g. fibronectin, laminin and vitronectin) and peptides (e.g. RGD, a peptide with a sequence of arginine-glycine-aspartic acid amino acids, Table 2.1) that enhance cell attachment. Synthetic polymers, on the other hand, typically have to be functionalized (selectively or ubiquitously) with synthetic analogs of such native ligands (Nuttelman *et al.*, 2001; Kong *et al.*, 2007a; Hsiong *et al.*, 2008). Biological polymers typically degrade into non-toxic products at a rate that can be controlled. Like synthetic polymers, scaffolds made from biological polymers typically do not have sufficient strength for many load-bearing applications (e.g. tendons).

A wide range of natural polymers can be used to create bioactive scaffolds, notably collagen (Badylak, 2007; Al-Munajjed *et al.*, 2008; Harley and Gibson, 2008), fibrin (Dikovsky *et al.*, 2006; Willerth *et al.*, 2006; Cornwell and Pins, 2007) and hyaluronic acid (Hutmacher *et al.*, 2001; Kim and Valentini, 2002; Masters *et al.*, 2005; Ifkovits and Burdick, 2007). Alginate scaffolds, formed from a naturally occurring polymer that is not found in the ECM, have also found wide applicability (Kisiday *et al.*, 2002, 2004; Kong *et al.*, 2003, 2004b; Grimmer *et al.*, 2004; Boontheekul *et al.*, 2005; Augst *et al.*, 2006; Comisar *et al.*, 2006; Jeon *et al.*, 2009). However, alginate scaffolds do not contain native ligands appropriate for specific cell–scaffold adhesion complexes and, like synthetic polymer scaffolds, have to be functionalized with appropriate chemical moieties to induce cell adhesion and overall construct bioactivity (Rowley and Mooney, 2002; Kong *et al.*, 2003, 2005, 2006, 2007a,b, 2008; Boontheekul *et al.*, 2005; Augst *et al.*, 2006; Silva *et al.*, 2008; Jeon *et al.*, 2009).

The degradation of both synthetic and biological polymer scaffolds can be decreased by a number of strategies. Most straightforward is increasing scaffold relative density; the unit degradation rate of the scaffold is not changed, but the amount of material to degrade increases. This strategy is suitable for moderate changes in degradation times, but not order of magnitude changes due to decreased bioactivity and nutrient transport at high relative densities. However, a more elegant option is also available: resistance to

Table 9.1 Glossary of biological and specialized abbreviations used in this chapter.

Abbreviation	Term, definition
C_o	Initial nutrient concentration in a quantitative model used to describe relative rate of nutrient supply (via diffusion) versus consumption by a cell; units: moles
CaP	Calcium phosphate
CFD	Computational fluid dynamics
CFM	Cell force monitor; used to measure the macroscopic deformation of a bio-material to quantify the contractility of a population of cells
CG	Collagen-glycosaminoglycan co-precipitate
CGCaP	Collagen-glycosaminoglycan-calcium phosphate triple co-precipitate
COL*x*A*y*	Family of genes that provide instructions for synthesizing distinct collagen sub-unit chains; *x*, *y* correspond to collagen type and sub-unit chain, respectively (e.g. COL1A2, COL3A2, COL2A1, COL4A4, COL4A5)
CSK	Cytoskeleton; a dynamic protein structure in the cell cytoplasm that maintains cell shape, protects the cell, enables cellular motion, and plays important roles in both intracellular transport and cellular division
D	Diffusion coefficient (diffusivity) in a cellular material; units: $m^2 \cdot s^{-1}$
D_{jxn}	Strut junction spacing; defines the distance between points in the scaffold where multiple struts meet
d	Scaffold pore size
DHT	Dehydrothermal crosslinking
ECM	Extracellular matrix; a fibrillar network made up of structural proteins such as collagens and proteoglycans found in all tissues and organs in the body that provides an insoluble structure to which cells can attach. The composition and structure of the ECM is particular to each tissue within the body
ε-CPL	ε-Caprolactone; synthetic polymer used to fabricate scaffolds
E_s	Elastic modulus of individual scaffold strut
E^*	Elastic modulus of the entire cellular material
$E_s \cdot I$	Scaffold strut flexural rigidity
ε	Compressive strain
F_c	Average contractile force generated by an individual cell within a 3D scaffold
FA	Focal adhesions; large, multi-protein, mechanosensitive complexes that provide a mechanical link between the cell's cytoskeleton and the extracellular matrix
GAGs	Glycosaminoglycans
GTPase	A family of enzymes that can bind and hydrolyze guanosine triphosphate (GTP), a nucleotide involved in energy transfer within the cell
HGF	Hepatocyte growth factor; a paracrine signaling factor involved in cellular growth, motility, morphogenesis and angiogenesis
HMOX	Heme Oxygenase; an enzyme that catalyzes the degradation of heme and is critically involved in angiogenesis
I	Second moment of inertia
K	Fluid mobility in a cellular material (permeability normalized by the liquid viscosity); units: $m^4/N\ s$
k	Scaffold permeability; units: m^2;

Table 9.1 *(cont.)*

Abbreviation	Term, definition
L	Distance over which nutrient diffusion takes place in a quantitative model used to describe relative rate of nutrient supply (via diffusion) versus consumption by a cell; units: m
L_c	Critical cell path length; the longest distance away from the wound bed that the cell can exist within a cellular material without requiring nutrients in excess of that supplied by diffusion; units: m
MC3T3-E1	Osteoblast cell line
MMP	Matrix metalloproteinase; a family of enzymes that can degrade extracellular matrix proteins. Key MMPs include: MMP-2, MMP-12, MMP-19. Key MMP inhibitors include: TIMP1, TIMP3
ρ_{jxn}	Strut junction density; the number of strut junctions per unit cell divided by the volume of the unit cell in a cellular material
ρ^*/ρ_s	Relative density of the cellular material
PE	Polyethylene; synthetic polymer used to fabricate scaffolds
PEG	Poly(ethylene glycol) ; synthetic polymer used to fabricate scaffolds
PGA	Polyglycolic acid; synthetic polymer used to fabricate scaffolds
PLA	Polylactide; synthetic polymer used to fabricate scaffolds
PLGA	Poly(lactic-co-glycolic acid) ; synthetic polymer used to fabricate scaffolds
PLLA	Poly-L-lactide; synthetic polymer used to fabricate scaffolds
PVA	Polyvinyl alcohol; synthetic polymer used to fabricate scaffolds
R	Nutrient metabolization rate (unit: moles·m^{-3}·s^{-1}) in a quantitative model used to describe relative rate of nutrient supply (via diffusion) versus consumption by a cell
RGD	Arginine-glycine-aspartic acid amino acid sequence; the cell attachment ligand found on large number of adhesive extracellular matrix, blood and cell surface proteins. Many of the known integrins recognize this sequence.
RhoA	Ras homolog gene family, member A; a small GTPase protein that can regulate actin cytoskeleton formation
S	Cell lifeline number; the relative ratio of the rate of nutrient consumption to that of nutrient supply by the cell
SA/V	Specific surface area; total surface area per unit volume for a given unit cell
SMC	Smooth muscle cells
τ	Time constant for cell contraction
t_d	Degradation time constant; the time (half-life) it takes for an implanted biomaterial to degrade inside the body
t_h	Healing time constant; the time (half-life) it takes for the native tissue to synthesize the required volume of tissue to replace a degrading biomaterial
UC	Unit cell
UV	Ultraviolet light (crosslinking)
VEGF	Vascular endothelial growth factor; a chemical signal produced by cells that stimulates the growth of new blood vessels (angiogenesis)

degradation increases with increasing crosslink density between the fibers that make up the scaffold structure (Yannas *et al.*, 1975a, 1989; Harley *et al.*, 2004; Pek *et al.*, 2004). Such phenomena have been most thoroughly studied using collagen-based scaffolds, but similar crosslinking approaches exist for most synthetic and biological polymers. In the case of collagen-based scaffolds, decreased degradation rates have also been achieved by introducing glycosaminoglycans (GAGs) or other glycans into the collagen in order to form additional crosslinks to and within the collagenous network (Yannas *et al.*, 1975a,b, 1980; Hunter *et al.*, 2005).

Scaffold crosslink density can be increased via both physical and chemical crosslinking processes. Examples of physical crosslinking techniques are dehydrothermal (DHT) and ultraviolet light (UV) crosslinking. DHT crosslinking is performed by exposing the scaffold to a high temperature (below the scaffold melting point or glass transition temperature) under vacuum (Yannas and Tobolsky, 1967), leading to the removal of water from the scaffold. Drastic dehydration of the scaffold (to less than 1% water content) leads to the formation of inter-chain amide bonds through condensation (Yannas and Tobolsky, 1967); targets for such amide bond formation (free amines) are plentiful on collagen, as well as the majority of other biological polymer scaffolds, and can be patterned onto the surface of synthetic polymers. While not particularly robust, DHT crosslinking is highly adjustable; exposure to higher temperatures or longer lengths of time produces higher crosslink densities and slower degradation rates (Yannas *et al.*, 1989; Torres *et al.*, 2000; Yannas, 2001; Harley *et al.*, 2004; Cornwell *et al.*, 2007; Haugh *et al.*, 2008). UV crosslinking has also been widely used to form crosslinks between collagen fibers within the scaffold structure (Lee *et al.*, 2001b; Ohan *et al.*, 2002; Cornwell *et al.*, 2007). Photopolymerization schemes are also quite common in synthetic systems such as PVA (Nuttelman *et al.*, 2002).

Examples of chemical crosslinking methods are glutaraldehyde and carbodiimide based treatments; these methods induce the formation of covalent bonds between collagen fibers and other biological proteins (Olde Damink *et al.*, 1996a,b; Pieper *et al.*, 1999, 2000; Lee *et al.*, 2001a; Kikuchi *et al.*, 2004; Hunter *et al.*, 2005; Everaerts *et al.*, 2007a,b). These crosslinking techniques are considerably more powerful than physical methods, resulting in significantly higher crosslink densities and slower degradation rates; in many cases greater than an order of magnitude change can be achieved (Harley *et al.*, 2004; Kong *et al.*, 2004a; Hunter *et al.*, 2005; Cornwell *et al.*, 2007; Everaerts *et al.*, 2007a). Recently, these chemical crosslinking techniques have been significantly refined; careful control over the concentration of crosslinking agent relative to the amount of target chemical moiety on the surface of the scaffold has enabled delicate control over the final scaffold crosslink density (Vickers *et al.*, 2006; Haugh *et al.*, 2008). Many of the chemicals used to crosslink both synthetic and biological polymer scaffolds are themselves cytotoxic. In the case of cytotoxic crosslinking agents that act as catalysts of the crosslinking reaction, such as carbodiimides, the scaffolds must be extensively washed after crosslinking to remove all remaining traces of the chemicals (Olde Damink *et al.*, 1996a,b; Pieper *et al.*, 1999, 2000; Everaerts *et al.*, 2007a). However, some chemical crosslinkers, notably glutaraldehyde, integrate a portion of the chemical compound into the crosslink; degradation of these scaffolds, while slow (in vivo degradation half lives are typically greater than 1 year), resolubilizes the cytotoxic crosslinking agent and can trigger immunological responses as well as other undesirable side-effects, so these techniques must be used with care (Olde Damink *et al.*, 1996a; Lee *et al.*, 2001a). Adding some chemical moieties such as titania nanoparticles has been shown to significantly reduce this resultant cell toxicity (Liu *et al.*, 2006c), with the added benefit of further improving bio-material mechanical strength (Hung-Jen Shao *et al.*, 2009); however, integration of cytotoxic chemicals into scaffold bio-materials remains an area of active concern.

Regardless of polymer type, scaffold degradation rate has been shown to be tightly linked to its bioactivity, both in vitro and in vivo (Freed *et al.*, 1994; Lu *et al.*, 2001; Yannas, 2001; Harley *et al.*, 2004; Almany and Seliktar, 2005; Mahoney and Anseth, 2006; Cornwell *et al.*, 2007), necessitating careful control over scaffold degradation kinetics, particularly for in vivo applications where the isomorphous tissue replacement requirement needs to be met.

9.2.2 Integrin–ligand interactions

Cells attach and interact with ECM proteins through the formation of multi-molecular complexes referred to as focal adhesions (FAs). The FAs are large, multi-protein, mechanosensitive complexes that provide a mechanical link between the cell's cytoskeleton (CSK) and the ECM (Chen, 2008). The FA has trans-membrane integrin molecules that interact directly with the ECM and a multi-protein sub-membrane plaque that connects the integrin to the cellular CSK. Integrin specificity for particular ECM proteins had been shown to be a key regulator of cell biology. Cell behaviors such as attachment, migration, proliferation and contraction are all mediated by interactions between the focal adhesions and integrins expressed on the cell surface and the ligands available on the scaffold surface (Guan, 1997a,b; Schlaepfer and Hunter, 1998; Friedl and Brocker, 2000; Geiger *et al.*, 2001). However, misregulation of integrin expression is often associated with pathological conditions such as cancer metastasis and leukemogenesis (Logsdon *et al.*, 2003; Tabe *et al.*, 2007). A bioactive scaffold must take into account the complexity of integrin–ligand interaction and be fabricated in a manner, and from specific materials, to produce a chemical environment that enables desired cell–scaffold interactions.

The biological activity of scaffolds used in tissue engineering applications therefore hypothetically depends on the available density of ligands particular to the application of interest. Ligand density is established by the composition of the scaffold, which defines the type and surface density of the ligand(s) of interest, and by the specific surface area of the scaffold, which defines the total surface of the bio-material exposed to the cells. Specific surface area will be discussed in detail in Section 9.3. Biological polymers, like collagen, hyaluronic acid, proteoglycans and fibrin/fibronectin, provide an ideal substrate for tissue engineering applications because they are significant constituents of the native ECM and therefore express a wide range of native ligand binding sites. Studies of biological polymer scaffolds have shown the specific importance of scaffold surface chemistry and chemical modification on cell behavior. Collagen and collagen-glycosaminoglycan (CG) scaffolds were among the first biological polymer scaffolds developed, and have been used successfully for both in vivo and in vitro tissue engineering (Yannas *et al.*, 1989; Pieper *et al.*, 2000; Yannas, 2001; van Tienen *et al.*, 2002; O'Brien *et al.*, 2005). Collagen fiber structure has also been implicated as a significant regulator of the kinetics of cell integrin–collagen bonding (Zaman, 2007). Collagen type, notably type I versus II, has been shown to significantly affect collagen scaffold in vivo regenerative potential (Yannas, 1992; Louie *et al.*, 1997; Mueller *et al.*, 1999; Saad and Spector, 2004), as well as in vitro cell attachment (Grzesiak *et al.*, 1997; Sethi *et al.*, 2002), proliferation and biosynthesis (Chen *et al.*, 2003; Cool and Nurcombe, 2005; Vickers *et al.*, 2006; Farrell *et al.*, 2007) and gene expression (Kim *et al.*, 1999; Veilleux *et al.*, 2004). Adding glycosaminoglycans to collagen scaffolds not only influences scaffold degradation rate (Yannas *et al.*, 1975b), but also affects scaffold regenerative potential (Silver *et al.*, 1974; Shafritz *et al.*, 1994). However, for successful use in vivo to induce tissue regeneration, the periodic banding (~67 nm) of the collagen fiber structure must be selectively abolished to prevent platelet aggregation (Yannas, 2001).

Fibronectin, vitronectin and hyaluronic acid have long been identified as critical ligand targets for cell–matrix interactions (Allingham *et al.*, 2006; Shah *et al.*, 2008, Weber *et al.*, 2008). Biomaterials created from these constituents have allowed investigation and identification of the optimal composition(s) and spatial distribution(s) of these materials in 3D scaffolds and hydrogels, with distinct geometries and concentrations required for different cell types (Kim *et al.*, 1999; Schmidt *et al.*, 2006; Takai *et al.*, 2006; Willerth *et al.*, 2006; Weber *et al.*, 2008). Collagen scaffolds were used as the basis for identifying the sequential utilization of fibronectin, vitronectin and collagen via distinct integrin subcomponents in the generation of cell-mediated contractile force (Sethi *et al.*, 2002).

Cell interactions with synthetic polymer scaffolds such as PLGA, PCL and PE, which do not express native ligands, are typically mediated by the introduction of native ECM proteins and/or biological polymers to the scaffold to provide targets for specific cells (Huttenlocher *et al.*, 1996; Bryant and Anseth, 2003; Bryant *et al.*, 2003, 2005; Kong *et al.*, 2006, 2007a; Petrie *et al.*, 2006; Rydholm *et al.*, 2008). Modifying the surface chemistry of synthetic polymers, or blocking specific ligand targets on biological polymers has been shown to significantly affect cell attachment to the scaffold as well as related cell behaviors such as migration, contraction and gene uptake (Castner and Ratner, 2002; Sethi *et al.*, 2002; Bryant *et al.*, 2003; Hwang *et al.*, 2006; Kong *et al.*, 2007a; Vasita *et al.*, 2008; Hsiong *et al.*, 2008).

9.3 Scaffold microstructure

A wide range of biomaterial constructs have been used to study cell–material interactions. Grossly, they can be divided into four types of microstructures (Fig. 9.2), each with associated advantages and disadvantages:

(1) *One-dimensional (1D) fiber constructs.* Here cells are cultured on individual fibers of variable length and diameter. Such substrates have been used for ease of analysis to investigate the role substrate mechanical properties play on motility.

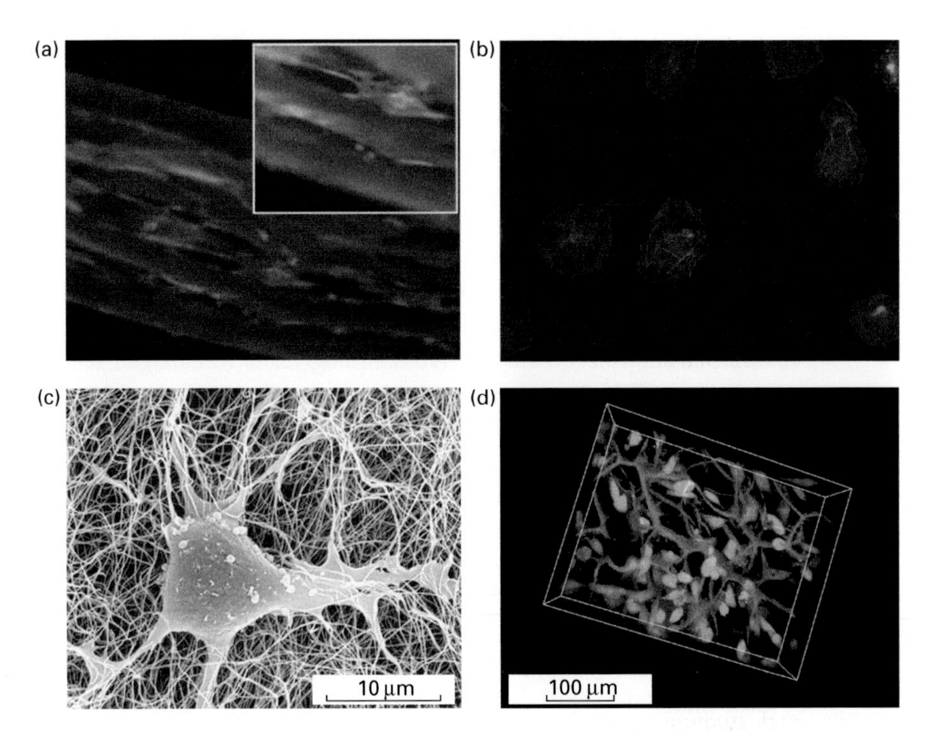

Fig. 9.2 Types of biomaterial constructs used to study cell–material interactions. (a) *One-dimensional fiber constructs.* Fibroblasts aligned along fibrin microthreads (UV crosslinked) after 1 day in culture. (b) *Impermeable, two-dimensional substrates.* MC3T3-E1 osteoblasts seeded on top of type I collagen hydrogels and fluorescently labeled for actin cytoskeleton (phalloidin, green) and counterstained for nuclear DNA (DAPI, blue). (c) *Three-dimensional, nanoporous hydrogel scaffolds.* Human fibroblasts seeded on top of 3D collagen hydrogel matrices; cell extensions penetrate into the matrix and become entangled with collagen fibrils. (d) *Three-dimensional scaffolds.* 3D confocal micrograph of the porous microstructure of a CG scaffold (red) seeded with labeled NR6 cells (green). (a, Reproduced with permission from Cornwell and Pins, 2007; b, Choi and Harley, unpublished; c, reprinted from Rhee and Grinnell, 2007, with permission from Elsevier; d, reprinted from Harley *et al.*, 2008, with permission from Elsevier.) See plate section for color version.

(2) *Impermeable, two-dimensional (2D) substrates.* Cells are exposed to a truly 2D environment; such substrates have been used to assess the role of material chemistry, mechanics and micro-scale patterning – both topological patterns and regional patterning of specific ligands. However, because they fail to capture the 3D structure of native ECM, 2D substrates are limited for some investigations.
(3) *Three-dimensional (3D), nanoporous hydrogel scaffolds.* Cells can either be placed on the surface of the hydrogel, in essence interacting with a 2D substrate with a nano-scale surface topology, or can be encapsulated within the 3D structure. In the latter case, there is a high degree of steric hindrance, so that cells must degrade the hydrogel enzymatically to move or extend processes. These substrates allow investigation of material chemistry and mechanics in a fully 3D material that mimics aspects of the native ECM. However, the nanoporosity and high density introduce some experimental difficulties in terms of material permeability. In addition,

hydrogel mechanics, density and microstructure are often innately linked, making it difficult to tease apart their independent effects on cell behavior.

(4) *Three-dimensional (micro-porous) scaffolds.* The larger pore sizes of these materials avoids the steric hindrance seen in nanoporous hydrogels. Depending on scaffold pore size, cells can either be aligned along essentially one-dimensional scaffold struts or attached to multiple struts simultaneously and be spread in three dimensions. While these materials are often more porous than the native ECM, their large porosity (typically >90%) enables sufficient diffusion to metabolically support cells within the structure. Fabrication techniques have been developed to allow independent control of their material chemistry, mechanics and microstructure.

All of these microstructures have played a valuable role in understanding how cells respond to their local environment. In this section, we primarily focus on 3D scaffolds as they most closely approximate the cellular nature of the materials described in the rest of this book, but at times we also discuss 2D substrates and hydrogel constructs.

For scaffolds with roughly equiaxed pores, the scaffold surface area and permeability depend on the relative density and pore size, while the mechanical properties depend on the relative density, the properties of the solid making up the cell wall and a constant that is related to the pore geometry, as we saw in Chapter 3. In this section we examine the effect of the scaffold microstructure on cell behavior. We summarize selected early work studying cell–material interactions using 2D materials that are the basis for later investigations using 3D scaffolds. We then show how scaffold surface area and pore shape affect cell attachment via the spatial presentation of available ligands. Finally, we explore how the cellular nature of scaffold-based biomaterials influences construct permeability, hence the ability to deliver nutrients, metabolites and paracrine signaling factors to cells. The permeability also affects the magnitude of applied shear stresses on cells due to fluid movement within the scaffold. In the next section, we describe the role of scaffold mechanical properties, such as the Young's modulus and compressive strength, in affecting cell behaviors such as motility and contractility.

9.3.1 Two-dimensional vs. three-dimensional structures

Cells are mechanically coupled to their extracellular environment; changing the dimensionality of the local microenvironment has been shown to significantly affect in vitro cell behavior and mechanobiology; for a review see Pedersen and Swartz (2005). Cells behaviors affected by construct dimensionality include cell morphology (Elsdale and Bard, 1972; Cukierman *et al.*, 2001; Tamariz and Grinnell, 2002), adhesion (Cukierman *et al.*, 2001) and motility (Brown, 1982; Maaser *et al.*, 1999). Similar effects have been seen in stem cell populations.

In two separate studies, Liu and co-workers showed that the hematopoietic differentiation efficiency (Liu and Roy, 2005) as well as the global gene expression profile during differentiation (Liu *et al.*, 2006b) of mouse embryonic stem cells were significantly different when they were cultured on 2D substrates versus 3D tantalum foams. Notably, higher differentiation efficiencies as well as higher expression levels of genes related to ECM production, cell growth, proliferation and differentiation were observed for embryonic stem cells cultured in 3D tantalum foams compared with those cultured on traditional two-dimensional tissue culture plates. Further, these effects were increased in dynamic culture, with fluid flow through the scaffold network. The additional effect of flow conditions through the scaffold on embryonic stem cell gene profiles will be discussed in more detail later in this section. Taken together, these studies suggest that the 3D scaffold structure is a critical feature in regulating cell behavior and stimulating differential gene expression profiles.

Unfortunately, due to limitations in biomaterials fabrication, the majority of experiments investigating the role of construct dimensionality have compared a 3D scaffold of one chemical composition to a 2D substrate with separate chemical and mechanical properties, or have used hydrogel-based 3D materials, not scaffolds. In a recent study by Jaworski and Klapperich (2006), collagen-glycosaminoglycan (CG) scaffolds have been used to probe the role ECM dimensionality, independent of material properties, plays in influencing gene expression profiles. Gene expression profiles for genes involved in angiogenesis (i.e. VEGF, HMOX and HGF) and ECM remodeling (i.e. matrix metalloproteinases, ECM components) were compared for fibroblasts cultured within CG scaffolds fabricated via a phase separation method versus fibroblasts cultured on deposited 2D CG surfaces. These target genes were chosen because cell-mediated remodeling of the local extracellular environment and induced angiogenesis are two critical types of cell behavior to study (and eventually control) for many tissue engineering applications. Differential expression of these genes was observed for cells in 3D constructs compared with those on 2D surfaces. Genes for some matrix metalloproteinases (e.g. MMP-2, MMP-12, MMP-19), ECM component synthesis (e.g. COL1A2, COL3A2, COL2A1, COL4A4, COL4A5) and pro-angiogenic factors were up-regulated while some MMP inhibitors (e.g. TIMP1, TIMP3) were down-regulated for cells cultured in the 3D CG constructs versus on the 2D CG surfaces (Jaworski and Klapperich, 2006) (Fig. 9.3). These results confirm those of previous studies comparing the effect of material dimensionality across materials systems: the dimensional parameters of a biomaterial alone can significantly affect a wide range of cell behaviors.

9.3.2 Pore size and specific surface area

The biological activity of scaffolds used in tissue engineering hypothetically depends on the density of available ligands, scaffold sites at which specific cell binding occurs. Ligand density is determined by the composition of the scaffold, which defines the surface density of ligands, and by the surface area of the scaffold available to cells, which is determined by scaffold relative density and pore size. The specific surface area, SA/V, the total surface area per unit volume, can be readily calculated for tissue engineering scaffolds with a structure similar to an open-cell foam. Using dimensional arguments, the specific surface area of an open-cell foam of pore size, d, and relative density, ρ^*/ρ_s, is given by (Gibson and Ashby, 1997):

$$\frac{SA}{V} \propto \frac{1}{d}\left(\frac{\rho^*}{\rho_s}\right)^{1/2} \tag{9.1}$$

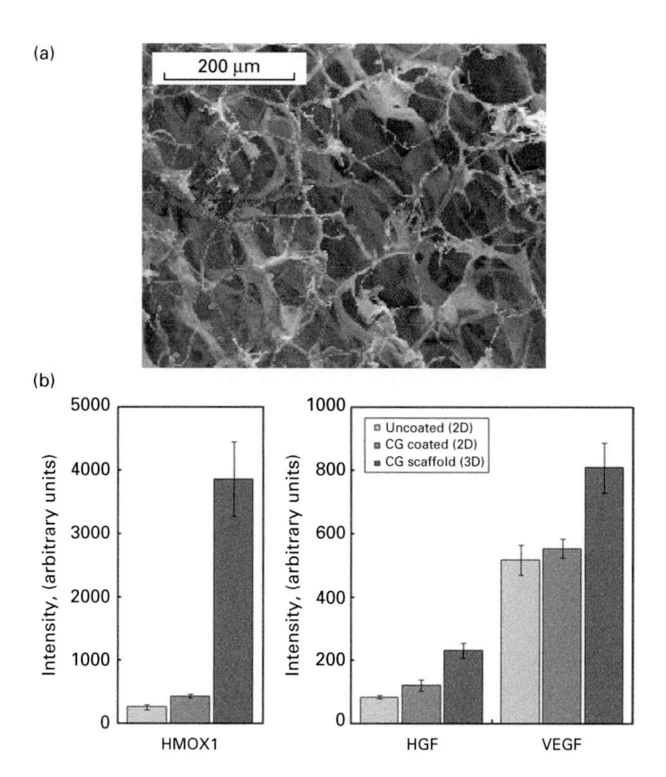

Fig. 9.3 Biomaterial dimensionality (2D vs. 3D) affects cell behavior. (a) Scanning electron microscope image of a representative collagen-GAG scaffold used to assess scaffold influence on lung fibroblast gene expression. (b) Increased gene expression (HMOX, HGF, VEGF shown here) was observed for lung fibroblasts cultured within 3D collagen-GAG scaffolds, compared with those cultured on 2D uncoated glass substrates or on 2D collagen-GAG coated glass substrates. (Reprinted from Jaworski and Klapperich, 2006, with permission from Elsevier.)

For an open-cell foam with a tetrakaidecahedral unit cell with edges of circular cross-section, the constant of proportionality is 10.2. Specific surface area has long been hypothesized to influence cell behavior in artificial ECM analogs in vitro (Pizzo *et al.*, 2005; Alvarez-Barreto *et al.*, 2007) as well as in scaffolds developed for in vivo regenerative medicine applications such as skin (Yannas *et al.*, 1989) and peripheral nerves (Chamberlain *et al.*, 1998).

In scaffolds implanted into a wound site, pore size plays a significant role in defining scaffold bioactivity (regenerative potential). For instance, Yannas and co-workers found that there is an optimum pore size to inhibit wound contraction in skin (Fig. 9.4a); increasing the wound defect contraction half life to 30 days allows normal dermal tissue to form instead of scar tissue (Yannas *et al.*, 1989). The optimum pore size varies for different tissues, and is typically in the range 20–500 μm (Hulbert *et al.*, 1970; Yannas *et al.*, 1989; Kuhne *et al.*, 1994; Chamberlain *et al.*, 1998; Yannas, 2001, 2005; Harley and Yannas, 2006).

Pore size is also a significant regulator of overall scaffold bioactivity for in vitro applications. Byrne and co-workers recently examined differential expression of RNA markers of osteogenic differentiation by mesenchymal stem cells cultured in

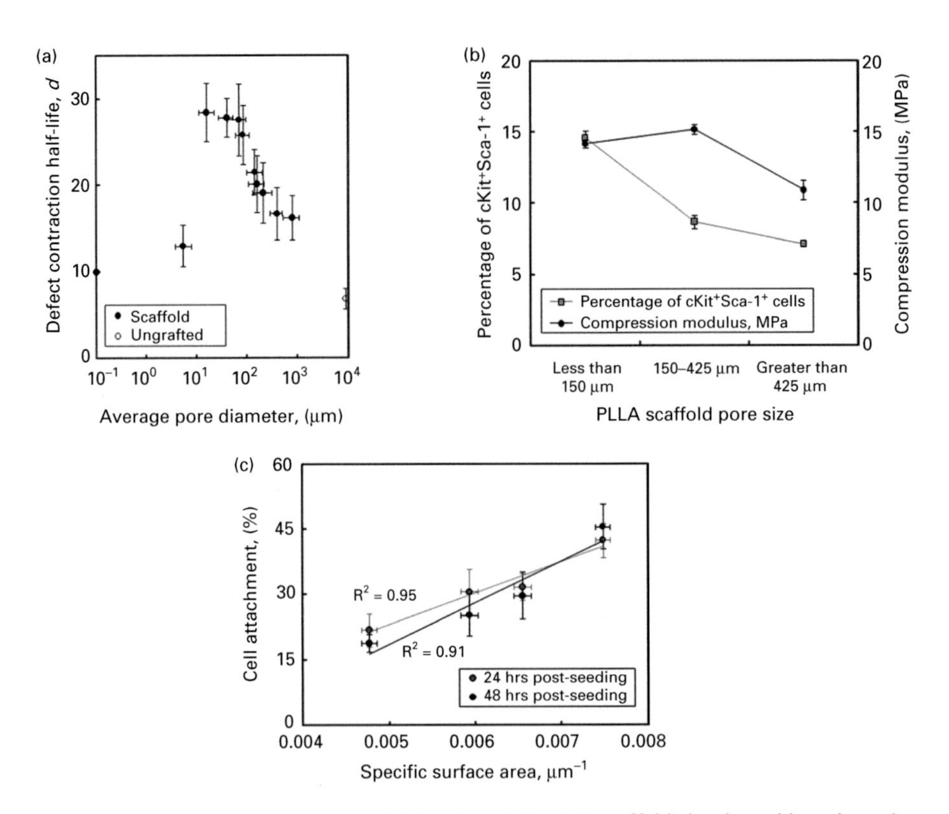

Fig. 9.4 Influence of scaffold pore size on cell behavior. (a) For CG scaffolds implanted into dermal wounds, the time required for a wound to contract to 50% of the original area, varies significantly in response to changes in scaffold pore size. (b) Embryonic stem cells (ESCs) cultured in PLLA scaffolds showed significantly different differentiation rates into hematopoietic progenitor cells (HPC – $cKit^+Sca\text{-}1^+$ cells), depending on pore size. HPC generation decreases with increasing scaffold pore size. Percentage of $cKit^+Sca\text{-}1^+$ cells at various pore sizes (gray squares) is simultaneously plotted with the corresponding scaffold compression moduli (filled circles). (c) MC3T3-E1 pre-osteoblasts seeded into CG scaffolds show a strong linear relationship between cell attachment and scaffold specific surface area (SA/V) at 24 hours (gray line) and 48 hours post-seeding (black line). (a, Reproduced with permission from Yannas *et al.*, 1989; b, reprinted from Taqvi and Roy, 2006, with permission from Elsevier; c, reprinted from O'Brien *et al.*, 2005, with permission from Elsevier.)

a series of CG scaffolds with variable pore size (Byrne *et al.*, 2008). In this study, they noted a significant effect of scaffold pore size on osteogenic differentiation of MC3T3-E1 pre-osteoblasts; increased RNA transcripts for osteocalcin, osteopontin and collagen I were observed for increasing CG scaffold pore size (96–151 μm). In another study, examining the role of scaffold pore size on differentiation potential of a stem cell population, Taqvi and Roy (2006) recently showed the significance of scaffold pore size on the hematopoietic differentiation capacity of mouse embryonic stem cells. While not showing precise control over scaffold pore size, variants with pore sizes of <150 μm, 150–425 μm, or >425 μm were used. Here, decreasing scaffold pore size increased the hematopoietic differentiation of ESCs seeded within the structure (Fig. 9.4b).

Table 9.2 Pore size and relative density of the CG scaffold variants (O'Brien *et al.*, 2005).

Freezing temperature	Pore size (μm)	Relative density
T_f (°C)	Mean ± standard deviation	Mean ± standard deviation
–10	151 ± 32	0.0062 ± 0.0005
–20	121 ± 23	0.0061 ± 0.0003
–30	110 ± 18	0.0059 ± 0.0003
–40	96 ± 12	0.0058 ± 0.0003

To empirically study the relationship between cell attachment, cell viability and scaffold microstructure, collagen-glycosaminoglycan (CG) scaffolds with a constant chemical composition and relative density ($\rho^*/\rho_s = 0.006$), but with four different pore sizes (96, 110, 121, 151 μm) corresponding to four levels of specific surface area were seeded with MC3T3-E1 mouse clonal osteogenic cells (Table 9.2) (O'Brien *et al.*, 2005). The cells and scaffold were maintained in culture and the number of viable cells that remained attached to the scaffold was counted after 24 or 48 hours. The fraction of viable cells attached to the CG scaffold decreased with increasing mean pore size, increasing linearly with the scaffold specific surface area, SA/V (Fig. 9.4c). The strong correlation between CG scaffold specific surface area and cell attachment indicates that initial cell attachment and viability are primarily influenced by scaffold specific surface area over this range (96–151 μm) of pore sizes for MC3T3 cells. A parallel study, using the same CG scaffolds, but varying both the relative density and pore size, seeded with the same MC3T3 cells, also found that the cell attachment increased linearly with specific surface area (Kanungo and Gibson, 2009). Longer culture times show more complex relationships between cell attachment and viability in scaffolds with pore sizes from 96 to 325 μm (Murphy *et al.*, 2009); these results indicate that while specific surface area may be important for initial cell adhesion, the improved cell migration and reduced cell aggregation capacity provided by scaffolds with pores larger than 300 μm overcomes this effect.

Cell motility is critical in many physiological and pathological processes as well as for tissue engineering applications. Cell migration is influenced by a complex, spatiotemporally integrated set of biophysical mechanisms that are influenced not only by the biochemistry of extracellular and intracellular signaling, but also by the biophysics of the surrounding extracellular environment. Using a CG scaffold system with pore sizes ranging from 95 to 150 μm, Harley and co-workers recently showed that fibroblast migration, characterized by motile fraction (fraction of all fibroblasts migrating at any one time) as well as speed and overall directional displacement, is strongly influenced by scaffold pore size: motile fraction, speed, displacement all decrease as scaffold pore size increases (Harley *et al.*, 2008). In-depth analysis of cell locomotion paths further revealed that the distribution (density, spacing) of junction points between scaffold struts played a significant role in this modulation. Further, strut junction interactions strongly influenced fibroblast directional persistence as well as cell speed at and away

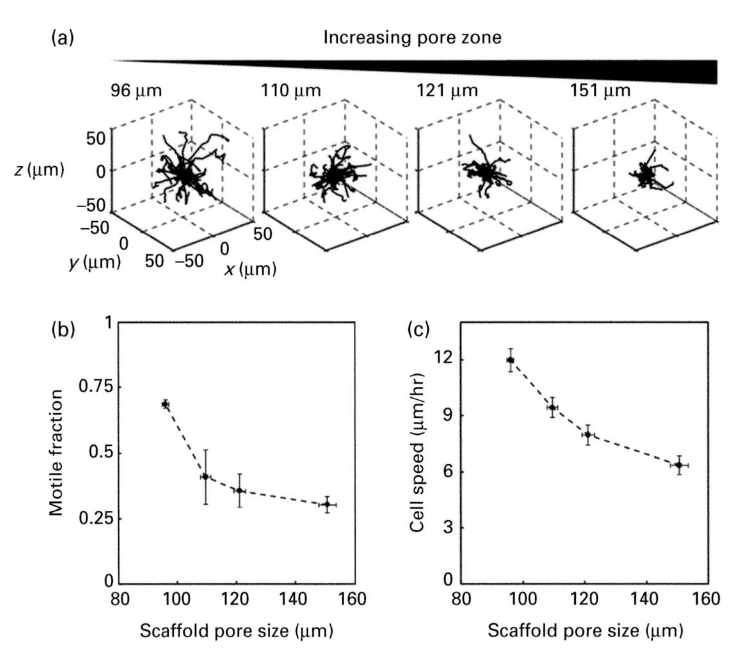

Fig. 9.5 Migration of NR6 fibroblasts in CG scaffolds with four distinct mean pore sizes (96–151 μm) determined via 3D time-lapse confocal microscopy. (a) 3D Wind–Rose plots of randomly chosen cell tracks (40 per pore size) graphically represent the average cell dispersion from its starting point. Decreased dispersion is seen as scaffold mean pore size increases, from left to right. (b) Fraction of motile (versus non-moving) cells, and (c) average migration speed plotted against scaffold pore size. (Reprinted from Harley *et al.*, 2008, with permission from Elsevier.)

from the junction points, providing a new biophysical mechanism for the governance of cell motility by the extracellular microstructure (Fig. 9.5) (Harley *et al.*, 2008). Taken together, these results, along with many others in the literature, suggest that scaffold pore size must be uniquely optimized for each application.

The pore size of the polymer network in hydrogel biomaterials has also been shown to significantly influence cell behavior. Here the cellular network is typically of a much smaller scale (less than 5 μm) than most tissue engineering scaffolds (with pore sizes of the order of 100 μm). The hydrogel scaffold network encapsulates the cell, providing a level of steric hindrance: cell motility or changes in cell shape therefore require active enzymatic degradation of the hydrogel. While the mechanisms might be different due to the distinction between scaffolds and hydrogels, the influence of pore microstructure on cell behavior is important nonetheless. Dikovsky *et al.* (2006) have shown the significance of network structure (porosity, pore size) on cell invasiveness into PEG hydrogel scaffolds, with reduced cellular invasiveness with increasing hydrogel density due to steric hindrances placed on the cells. Collagen hydrogel scaffold fibril density (which scales with *SA*/*V* for fibrils of constant size) has also been shown to affect fibroblast morphology and behavior. Fibroblasts seeded in collagen hydrogels with higher relative fibril densities appear more rounded with an increased number of cellular protrusions and overall cell surface area; those same fibroblasts seeded into collagen hydrogel scaffolds with lower fibril densities, but identical chemical composition, showed increased proliferation and local matrix remodeling with associated upregulation of the β_1 integrin expression and localization (Pizzo *et al.*, 2005).

For scaffolds and hydrogels, over a critical range of ligand densities (or specific surface areas), cell motility (speed, directional persistence) is observed to increase, reach a maximum, and then decrease (Kuntz and Saltzman, 1997; Friedl *et al.*, 1998; Burgess *et al.*, 2000; Friedl and Brocker, 2000; Gunzer *et al.*, 2000; Cukierman *et al.*, 2002; Lutolf *et al.*, 2003; Even-Ram and Yamada, 2005; Raeber *et al.*, 2005; Zaman *et al.*, 2006). The exact magnitude and quantitative transitions of this behavior depend critically on the type of cells and scaffolds used in each study. This biphasic relationship has been observed with or without native matrix metalloproteinase (MMP), the enzymes a cell produces to degrade the surrounding matrix environment and hence create room for migration. The conclusions drawn by the Harley *et al.* study (2008) showing an inverse relationship between collagen scaffold pore size and cell motility are consistent with this relationship, suggesting future experiments to identify decreased pore sizes or increased ligand densities to further influence cell motility. Theoretical modeling approaches have also found similar results (Zaman *et al.*, 2007), making it possible to begin to quantitatively probe cell motility parameters within well-defined systems.

9.3.3 Pore shape

Nerve, ligament, tendon and muscle are examples of tissues that are highly oriented. Scaffolds with pores aligned in the direction of the desired tissue orientation have been shown to increase the bioactivity and regenerative capacity, compared to scaffolds with equiaxed pores, for injuries in the peripheral nerve (Chang *et al.*, 1990; Chamberlain *et al.*, 1998; Kim *et al.*, 2008), myocardium (Engelmayr *et al.*, 2008), tendon and ligament (Spalazzi *et al.*, 2006; Butler *et al.*, 2008; Liu *et al.*, 2008; Moffat *et al.*, 2009a,b) (Fig. 9.6). It is thought that pore shape provides contact guidance cues to cells that are important for processes such as cell division and migration. For instance, the spatial distribution of the ECM plays a significant role in determining the orientation of the division axis of cells during cell division by influencing the localization of actin bundles at the membrane, and thus the segregation of cortical components in interphase (Thery *et al.*, 2005). Empirical studies using PLGA polymer scaffolds with differential levels of pore alignment have shown that mesenchymal stem cells adopt levels of alignment that match scaffold pore alignment (Wang *et al.*, 2009). Computational modeling has also suggested that heterogeneities in the local ECM structure play a role in the formation and evolution of stress fibers within the cell (Deshpande *et al.*, 2006).

9.3.4 Permeability

There are two mechanisms available for transport of metabolites to and waste products from the cells: diffusion through the scaffold microstructure or, in the case of in vivo applications, transport along capillaries that have sprouted into the scaffold as a result of angiogenesis. While angiogenesis becomes the limiting factor for long-term cell survival and growth, significant angiogenesis is not observed for the first few days after

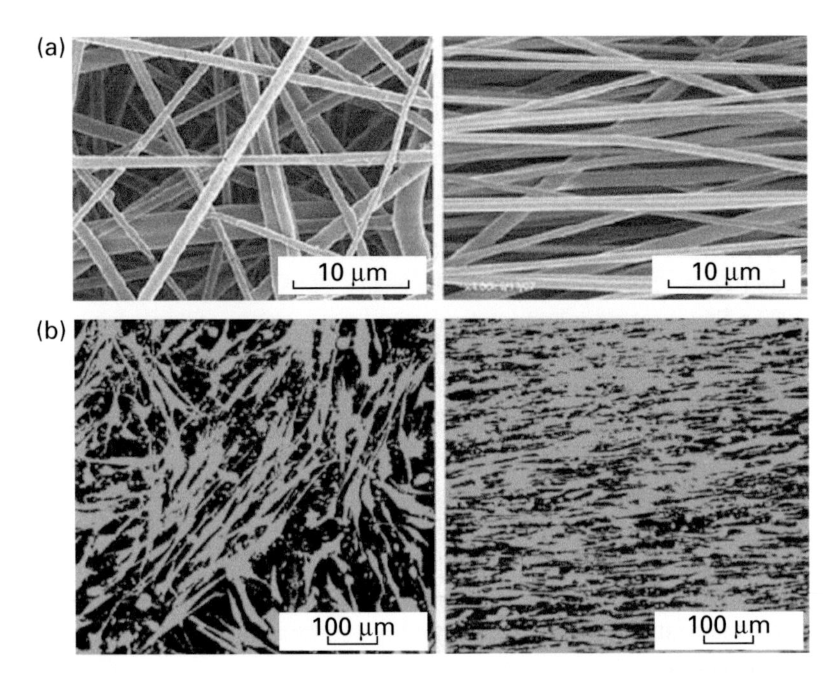

Fig. 9.6 Aligned, anisotropic cellular biomaterials designed for tissue engineering applications in anisotropic tissues. (a) Scanning electron micrographs showing unaligned versus aligned nanofiber PLGA scaffolds. (b) Rotator cuff fibroblasts cultured on unaligned versus aligned PLGA nanofiber scaffolds showed significantly different shape and alignment as early as 1 day post-seeding; cells were fluorescently labeled using a live (green) versus dead (red) stain. (Reprinted from Moffat *et al.*, 2009b, with permission from Elsevier.) See plate section for color version.

implantation and is not present at all in in vitro experiments. Therefore, early cell survival inside the scaffold is defined solely by diffusion. The critical scaffold thickness, the maximum scaffold thickness that can be supported by metabolite diffusion, has been observed empirically to be of the order of a few millimeters (Yannas, 2001).

A quantitative model of the rate of nutrient supply to that of nutrient consumption by a cell in a scaffold is described by Yannas (2000). Here, the complexity of the nutritional requirements of the cell is simplified by considering a generic critical nutrient required for normal cell function that is assumed to be metabolized by the cell at a rate R (in moles m^{-3} s^{-1}). The nutrient is pictured as being transported from the wound bed, where the concentration of nutrient is assumed to be constant, at a value of C_o, over a distance L via the wound exudate until it reaches the cell within the scaffold. Immediately following implantation of the scaffold, nutrient transport is performed exclusively via diffusion that can be modeled using the scaffold diffusivity D (in m^2/s). The cell lifeline number, S, characterizes the relative ratio of the rate of nutrient *consumption* to that of nutrient *supply* by the cell. Dimensional analysis readily yields

$$S = RL^2 / DC_o \tag{9.2}$$

If the rate of consumption of the critical nutrient greatly exceeds the rate of supply, $S >> 1$, and the cell will soon die. At steady state ($S = 1$) the rate of consumption of nutrient by the cell equals the rate of transport via diffusion over a distance L; at steady state, the value of L is the critical cell path length, L_c, the longest distance away from the wound bed that the cell can exist within a scaffold without requiring nutrient in excess of that supplied by diffusion. For many cell nutrients of low molecular weight, L_c is of the order of a few hundred micrometers to a few millimeters (Yannas, 2001).

The flow of fluid through porous media such as tissue engineering scaffolds can be characterized by three parameters: k, scaffold permeability (in m^2); K, fluid mobility in the scaffold (in $m^4/N{\cdot}s$), which is the permeability normalized by the liquid viscosity; or D, the diffusion coefficient (diffusivity) (in m^2/s). Fluid flow through the scaffold affects the ability of metabolites and nutrients to reach cells within the scaffold. Fluid flow through the scaffold network also significantly modifies the induced shear stresses applied to cells attached to the scaffold, another significant regulator of cell biology (Davies, 1995; Li *et al.*, 1997; Brown, 2000; Imberti *et al.*, 2002; Engelmayr *et al.*, 2006).

Models of the permeability, k, of open-cell foams indicate that it depends on the pore size, d, and relative density, ρ^*/ρ_s, of the cellular material (Dawson *et al.*, 2007):

$$k = Ad^2\left(1-\frac{\rho^*}{\rho_s}\right)^3 \tag{9.3}$$

where A is a dimensionless constant. Using a somewhat different model, O'Brien and co-workers derived an expression for the permeability of CG scaffolds as (O'Brien *et al.*, 2007):

$$k = A'd^2\left(1-\varepsilon\right)^2\left(1-\frac{\rho^*}{\rho_s}\right)^{3/2} \tag{9.4}$$

where ε is the compressive strain (taking compression as positive) and A' is a constant. The permeabilities of CG scaffolds of varying pore size (96–151 μm) at constant relative density (0.006) and at different levels of compressive strain (0, 14, 29 and 40%) were measured empirically by O'Brien and co-workers using saline solution as the fluid. A single constant, A' in (9.4), could be used to describe scaffold permeability. Fluid mobility, K, was calculated for the entire tested range of pore sizes and applied uniaxial compressive strains by dividing the permeability by the constant fluid viscosity. Excellent correlation between model predictions and experimental results was observed (Fig. 9.7).

The permeability of the scaffold also affects shear stresses acting on cells within the scaffold structure. This shear stress arises from fluid flow: the shear stresses in the fluid are linearly proportional to the velocity gradient (by Newton's law), and the fluid velocity increases linearly with permeability under a constant pressure gradient (by Darcy's law) . Fluid flow induced shear stresses have been shown to be a major biophysical stimulus in activating matrix biosynthesis and mineralization processes

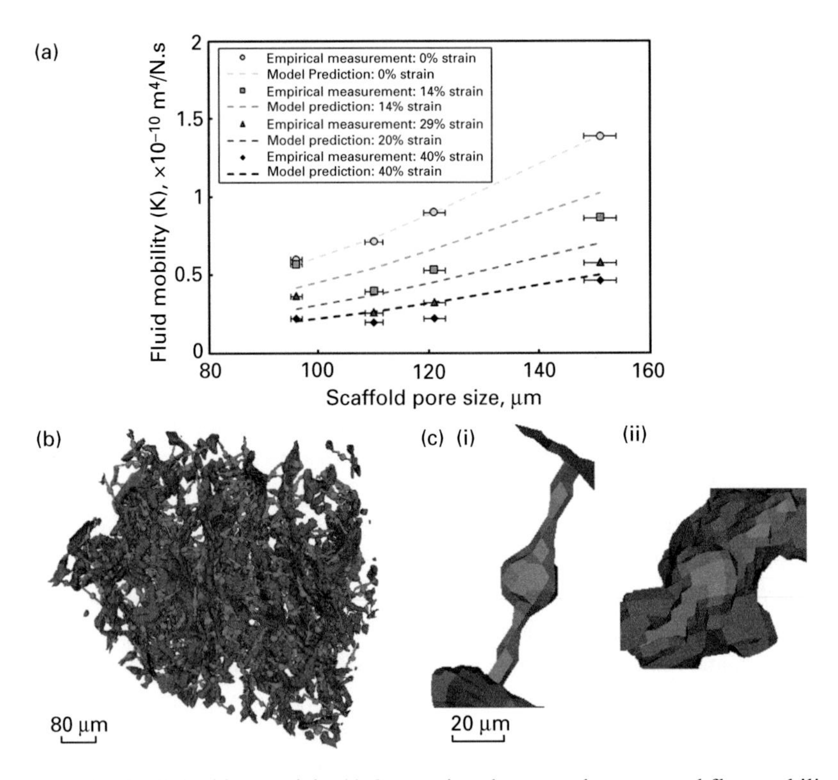

Fig. 9.7 Models of cellular biomaterials. (a) Comparison between the measured flow mobility (K_{meas}, data points) and the predicted values (curves) obtained from the cellular solids-based mathematical model for CG scaffold fluid mobility under varying compressive strains. (b) 3D mesh of the reconstructed cell-seeded CG scaffold (96 μm pore size) used in CFD simulations. (c) Typical geometry of cells (i) bridging two struts or (ii) attached to a single strut. Scaffolds are displayed in blue, simulated cells in red. (a, Adapted from data first published in O'Brien *et al.*, 2007; b,c, reprinted from Jungreuthmayer *et al.*, 2009b, with permission from Elsevier.) See plate section for color version.

in bone regeneration studies (Prendergast *et al.*, 1997; McGarry *et al.*, 2005; Wang *et al.*, 2007). Using the collagen-glycosaminoglycan (CG) scaffold system, oscillatory, pulsatile and steady flow (as low as 0.05 ml/min) have all been shown to significantly increase the osteogenic potential, measured via ECM protein biosynthesis gene expression and prostaglandin expression, of osteoblasts versus static culture (Jaasma and O'Brien, 2008). In another study, gene expression levels were quantified for mouse embryonic stem cells cultured in three-dimensional tantalum foams with or without active fluid flow. Cells in the tantalum foams placed in spinner-flasks, which induce steady-state flow profiles throughout the foam, expressed significantly higher expression levels for genes associated with ECM protein biosynthesis and cell growth compared to cells cultured in the same foams in static culture (no fluid flow) (Liu *et al.*, 2006b). Matrix porosity and pore structure have also been found to significantly affect fluid shear stresses along matrix fibers in generic ECM matrix structures (Pedersen *et al.*, 2007). These results reinforce long-standing studies which identified fluid-flow induced shear stresses as significant regulators of cell behavior in 2D culture systems (Satcher *et al.*, 1997 Brown, 2000; Imberti *et al.*, 2002).

With empirical evidence that fluid-flow induced shear stresses affect cell behavior in cellular materials, homogeneous scaffold structures offer an ideal platform to develop modeling approaches to quantify the local shear stresses applied to individual cells within the scaffold. Here, shear stresses arising from fluid flow through a scaffold can be calculated via computational fluid dynamics methods based on micro-computed tomography (micro-CT) images of the scaffold microstructure. Jungreuthmayer *et al.* (2009a) applied this method to consider the magnitude of the applied shear stress due to identical macroscopic flow rates (normalized by scaffold cross-sectional area) through two distinct scaffold microstructures, a conventional CG scaffold (1% relative density, 96 μm pore size) and a calcium phosphate (CaP) scaffold (40% relative density, 350 μm pore size). This work showed that an almost threefold increase in input fluid flow rate (normalized by the cross-sectional area of each scaffold) is required in the CaP scaffold in order to obtain a comparable level of wall shear stress as that seen in the CG scaffold. This analysis also showed that wall shear stresses are approximately 40-fold higher in CaP scaffolds than in CG scaffolds when using flow rates experimentally determined in each scaffold system to stimulate osteoblast bioactivity for bone tissue engineering applications (Vance *et al.*, 2005; Jaasma and O'Brien, 2008).

Scaffold microstructure can also affect cell mechano-regulation due to fluid flow via a mechanism separate from the flow-induced wall shear stress. The local scaffold microstructure also influences cell attachment morphology. There are two main modes of cell attachment: cells either align along a single scaffold strut or bridge multiple struts (Harley *et al.*, 2008). With upwards of 75% of cells able to bridge multiple struts in the case of some CG scaffold variants (Stops *et al.*, 2008), it is important to consider the deformation profiles of cells spread along or across struts due to fluid flow. Results of CFD analysis based upon micro-computed tomography scans of CG scaffolds have been used in conjunction with an elastostatics model of cell deformation to suggest that under identical flow parameters, cells bridging struts can experience up to a 500-fold higher level of deformation relative to those cells attached along a single strut (Fig. 9.7) (Jungreuthmayer *et al.*, 2009b). Well-characterized cellular biomaterials combined with rigorous computational methodologies have laid the foundation for quantitative exploration of the different mechanisms by which fluid flow can influence cell behavior, providing a basis for improving the design of culture systems that more accurately model physiologically relevant conditions.

9.4 Substrate mechanical properties

The rapidly increasing use of scaffolds requires better understanding of the role their mechanical properties play in influencing cell behavior and overall scaffold bioactivity. Scaffold mechanical properties influence cell behavior in two distinct ways: direct mechanical effects (for instance, the effect of the substrate stiffness) and indirect mechanical effects (for instance, the effect of mechanical stimuli transmitted through the scaffold structure).

The effect of substrate mechanical properties on cell behavior (attachment, morphology, proliferation, contractility and migration speed) has been studied using one-dimensional fibers, two-dimensional flat substrates, three-dimensional hydrogels where cells are encapsulated in a nanoporous fibrillar structure, and three-dimensional tissue engineering scaffolds where cells are attached to struts of lengths of the order of tens to hundreds of microns. For all systems, substrate mechanical properties have been shown to significantly influence cell behaviors, such as adhesion, growth and differentiation, in vitro as well as affect in vivo scaffold bioactivity. Here, we first describe selected earlier results for 1D and 2D substrates as well as hydrogels, before focusing on the tissue engineering scaffolds.

9.4.1 One-dimensional fibers

One-dimensional collagen fibers, used alone or as a composite with a collagen hydrogel, have been used to explore the influence of fiber mechanical properties on cell bioactivity. In one series of experiments, Cornwell and co-workers created collagen fibers with diameters ranging between 55 and 125 μm; the fiber elastic modulus was controlled via crosslinking techniques to between 4 and 350 MPa (Fig. 9.8a) (Cornwell *et al.*, 2004; Cornwell *et al.*, 2007; Cornwell and Pins, 2007). Fibroblast viability, proliferation, migration speed along the fiber and overall contractile capacity were all found to be regulated by local mechanical properties. Fibroblast migration speed as well as viability and proliferation were found to increase with decreasing collagen fiber elastic modulus (Cornwell *et al.*, 2004; Cornwell *et al.*, 2007; Cornwell and Pins, 2007). Adding increasing amounts of 1D collagen fibers to 3D collagen hydrogels to form fiber-hydrogel composites was shown to reduce overall fibroblast contractility without affecting cell viability and proliferation (Gentleman *et al.*, 2003, 2007).

9.4.2 Two-dimensional substrates

Two dimensional substrates have long formed the basis for our understanding of the critical role substrate mechanics plays in regulating cell behavior. Substrate mechanical properties have previously been shown to be a critical regulator of cell motility (speed and directional persistence), DNA biosynthesis, cell-mediated contraction and applied traction forces, cell-cell interactions, cell proliferation and gene expression, to name just a few (Wang *et al.*, 1993, 2001; Pelham and Wang, 1997; Chen and Ingber, 1999; Beningo and Wang, 2002; Engler *et al.*, 2004a; Discher *et al.*, 2005; Rehfeldt *et al.*, 2007). Key works utilizing 2D substrates to study the influence of mechanical properties on cell behavior are summarized in Table 9.3. A classic example was shown by Lo and co-workers, where they observed that NIH 3T3 fibroblasts showed significantly different motility patterns on collagen-coated polyacrylamide membrane with a rigidity gradient (14–30 kPa) (Fig. 9.8b). Cells could be seeded onto either the stiff or soft region, though a significant increase in cell spreading was observed on the softer region of the membrane; however, while cells readily migrated from the soft side of the substrate toward the gradient and across into the stiff side of the substrate, cells approaching the gradient from the stiff side did not cross back onto the soft substrate.

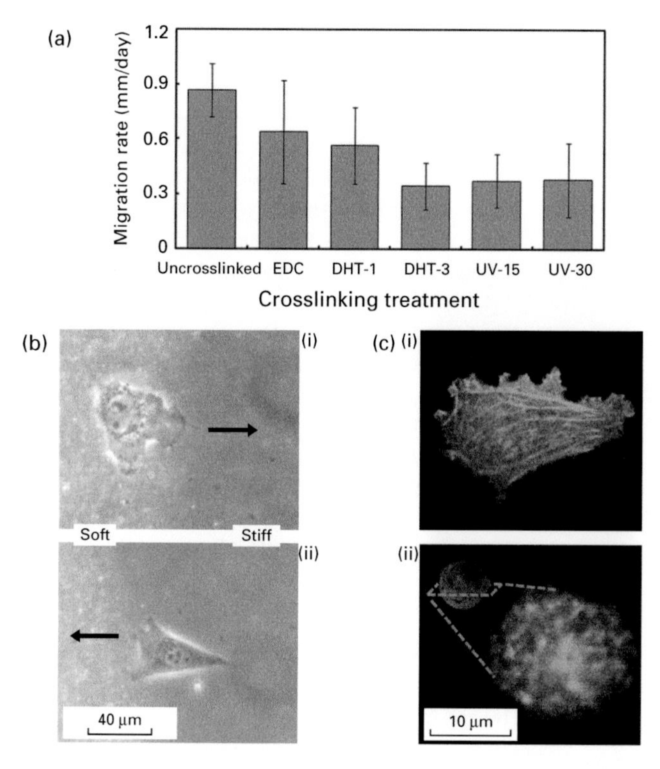

Fig. 9.8 Scaffold mechanical properties affect cell behavior. (a) *One-dimensional fiber constructs.* Fibroblast migration rate along collagen microfibers decreases with increasing crosslinking, in order of: uncrosslinked, carbodiimide crosslinking (EDC), dehydrothermal crosslinking at 110 °C under 50–100 m Torr vacuum for 1 (DHT-1) or 3 (DHT-3) days, ultraviolet light crosslinking for 15 (UV-15) or 30 (UV-30) minutes. (b) *Impermeable, two-dimensional substrates.* NIH 3T3 cells showed significantly different motility patterns on substrates with a rigidity gradient (14–30 kPa). Cells (i) move from the soft side of the substrate toward the gradient and across into the stiff side of the substrate; conversely, (ii) cells approaching the gradient from the stiff side did not cross back onto the soft substrate. A significant change in the cells spreading area of cells on the soft versus stiff substrate was also observed. (c) *Three-dimensional, nanoporous hydrogel scaffolds.* Confocal actin micrographs of metastatic prostate cancer cells (PC-3) seeded on top of (i) and embedded within (ii) 3D collagen hydrogels with identical mechanical properties (G' = 8.73 Pa). The actin cytoskeleton (phalloidin, green) of cells (i) seeded on top of the hydrogel (2D microenvironment) is fundamentally different from those (ii) embedded within the hydrogel (3D microenvironment); actin fibers are less elongated and defined in 3D hydrogels. (a, Reproduced with permission from Cornwell *et al.*, 2007; b, reprinted from Lo *et al.*, 2000, with permission from Elsevier; c, reprinted from Baker *et al.*, 2009, with permission from Elsevier.) See plate section for color version.

While such experiments have provided valuable information regarding the mechanobiology of cell–matrix interactions, these techniques have only limited applicability in rigorously understanding cellular processes in the three-dimensional structure found in native tissues and organs. Exposing cells to three-dimensional structures induces significant differences in cell morphology, cytoskeletal

Table 9.3 Influence of substrate mechanical properties in regulating cell behavior

Cell Behavior	Reference
Attachment	Engler *et al.* (2004a), (2006); Discher *et al.* (2005); Thompson *et al.* (2005); Takai *et al.* (2006)
Viability, proliferation	Wang *et al.* (1993), (2000), (2002); Engler *et al.* (2004a); Discher *et al.* (2005); Pirone *et al.* 2006) ; Takai *et al.* (2006); Rehfeldt *et al.* (2007); Shapira-Schweitzer and Seliktar (2007); Wei *et al.* (2008)
Motility: speed, directional persistence	Harris (1980); Pelham and Wang (1997), (1999); Dembo and Wang (1999) ; Roy *et al.* (1999); Lo *et al.* (2000); Beningo *et al.* (2001); Munevar *et al.* (2001b); Lo *et al.* (2004); Peyton and Putnam (2005)
Applied traction forces	Harris *et al.* (1981); Lee *et al.* (1994); Oliver *et al.* (1995); Burton and Taylor (1997); Galbraith and Sheetz (1997); Roy *et al.* (1997); Dembo and Wang (1999); Pelham and Wang (1999); Beningo *et al.* (2001); Munevar *et al.* (2001a); Beningo and Wang (2002)
Contractility	Harris (1980); Harris *et al.* (1981); Wrobel *et al.* (2002)
DNA biosynthesis, gene expression	Mochitate *et al.* (1991) ; Cool and Nurcombe (2005); Kong *et al.* (2005)
Cell-cell interactions	Hui and Bhatia (2007); Winer *et al.* (2009)
Stem cell differentiation	Engler *et al.* (2004b), (2006), (2007), (2008); Discher *et al.* (2009)

organization and integrin–ligand complexes versus two-dimensional substrates. For instance, the amorphous, rounded shape of cells in suspension is quite different than the polygonal cell shape typically observed on two-dimensional substrates, and different still from the spindle-shaped cells often observed for contractile cells in in vivo wound sites (Yannas, 2001) and within three-dimensional constructs (Freyman *et al.*, 2001b,c). Further, the cytoskeletal stress fibers typically seen in cells seeded on two-dimensional substrates are often lacking in cells seeded in three-dimensional hydrogels (Fig. 9.8) (Panorchan *et al.*, 2006; Zaman *et al.*, 2006; Peyton *et al.*, 2008; Baker *et al.*, 2009).

9.4.3 Three-dimensional hydrogels

Hydrogels are networks of water-insoluble polymer chains dispersed in an aqueous media; these stable, three-dimensional structures exhibit "pore sizes" typically of the order of tens to hundreds of nanometers. In comparison, cells typically have dimensions of the order of tens of microns, so that cells within a hydrogel are sterically hindered: they are surrounded or encapsulated by the hydrogel with a high density of available adhesion sites on many fibers. Recently, the effect of

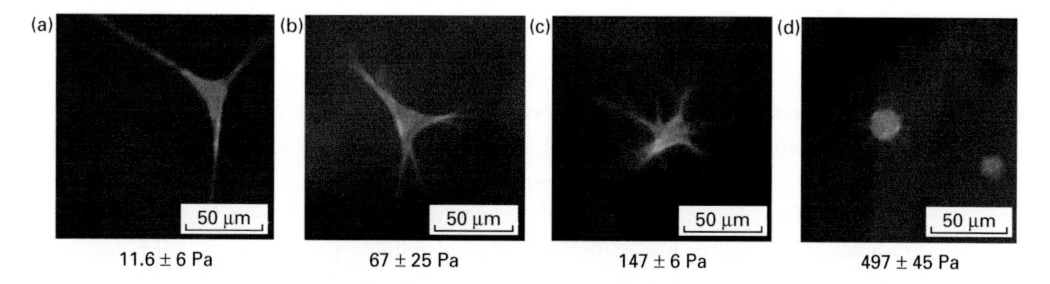

Fig. 9.9 Influence of three-dimensional hydrogel mechanical properties on cell morphology. Smooth muscle cells (SMCs) encapsulated in PEG-fibrinogen (PF) hydrogels express different morphologies depending on their local mechanical environment (cytoskeleton: green; nucleus: blue). Cell spindling decreases with increasing mechanical stiffness: (a) 11.6 ± 6 Pa (highly spindled with regular lamellipodia), (b) 67 ± 25 Pa (spindled with frayed lamellipodia), (c) 147 ± 6 Pa (nonspindled with frayed lamellipodia), (d) 497 ± 45 Pa (rounded with minor lamellipodia). (Reprinted from Dikovsky *et al.*, 2008, with permission from Elsevier.) See plate section for color version.

hydrogel modulus on cell behavior has become an area of intense study; while a multitude of material systems have been employed, here we restrict our discussion to a single material system: PEG-fibrinogen hydrogels. Dikovsky and co-workers have shown that changing the elastic modulus of PEG-fibrinogen hydrogels from 10 to 700 Pa induces a significant change in the morphology of encapsulated smooth muscle cells (SMCs) from spindle-shaped (10 Pa) to amorphous-rounded (700 Pa) (Fig. 9.9). Changing the hydrogel modulus over the range of 10 to 700 Pa also affected the cell-mediated matrix remodeling processes, with decreased hydrogel compaction by cells in hydrogels with increased elastic modulus (Dikovsky *et al.*, 2008). Similar results have been seen for fibroblasts in low- and high-density collagen hydrogels (Tamariz and Grinnell, 2002). Peyton and co-workers developed a series of PEG-fibrinogen conjugate hydrogels with tunable elastic moduli (from 450 to 5800 Pa). Here, varying the hydrogel modulus was shown to have little effect on the proliferation or cytoskeletal organization of encapsulated SMCs; however, this range of materials was utilized to identify RhoA, a small GTPase, as a critical element of the SMC mechanosensing process (Peyton *et al.*, 2008). These results, and those by many other investigators, have shown that the mechanical properties of three-dimensional materials can significantly affect the behavior of cells within them, but that these effects can only be seen over discrete – and often narrow – ranges of mechanical properties. Cells on the two-dimensional versions of the PEG-fibrinogen hydrogels described here showed significantly higher vinculin expression levels, less-organized cytoskeletal structures, and reduced motility than when encapsulated within the hydrogel matrix (Dikovsky *et al.*, 2008; Peyton *et al.*, 2008). These experiments further reinforce the concept that the same cells seeded within (3D) versus on (2D) hydrogel constructs show significantly different behaviors.

9.4.4 Three-dimensional scaffolds

Engineered cellular materials present an advantageous platform to study the role of material mechanical properties on cell behavior. As described in Chapter 8, a wide range of elastic properties are attainable; further, some materials allow mechanical properties to be varied independently of other microstructural features. Scaffolds with homogenous microstructures can be modeled using the results of Chapter 3. To begin to unravel the significance of the scaffold mechanical properties on cell behavior, a series of experiments have been carried out using one of the original scaffolds developed in the field of tissue engineering: collagen-glycosaminoglycan (CG) scaffolds fabricated via a freeze-drying (lyophilization) process.

A primary use of CG scaffolds has been to induce regeneration of tissues following severe injury; notably, these scaffolds have been used to regenerate skin (Yannas *et al.*, 1989; Yannas, 2001), peripheral nerves (Chang *et al.*, 1990; Yannas, 2001; Harley *et al.*, 2004), the conjunctiva (Hsu *et al.*, 2000), cartilage (Samuel *et al.*, 2002; Capito and Spector, 2003; Lee *et al.*, 2003; Kinner *et al.*, 2005; Vickers *et al.*, 2006; Capito and Spector, 2007), meniscus (Steinert *et al.*, 2007) and intervertebral disk (Saad and Spector, 2004). The typical mammalian physiological reaction to both chronic and acute severe injuries is characterized by a combination of a complex inflammatory response, cell-mediated wound contraction and scar tissue synthesis termed *repair*. Introduction of a suitable analog of the ECM into the wound site has been observed to block cell-mediated contraction of the wound site, prevent scar formation and induce *regeneration* of physiological tissue. The microstructural, chemical compositional and biodegradation properties of these CG scaffolds require separate optimization for each tissue of interest.

CG scaffolds have recently been used as experimental model systems to investigate the effects of scaffold chemical composition and microstructure, as discussed earlier in Sections 9.2 and 9.3, as well as scaffold mechanical properties, to be discussed here, on cell behavior, notably cell viability, proliferation, motility and contraction. A key advantage of the CG scaffold system is the ability to independently vary the scaffold chemical composition (Yannas *et al.*, 1980; Yannas, 1992; Nehrer *et al.*, 1997; Harley *et al.*, 2009a,b), pore size (Yannas *et al.*, 1989; Chamberlain and Yannas, 1998; O'Brien *et al.*, 2005), pore shape (Chamberlain *et al.*, 2000; Sannino *et al.*, 2005) and mechanical properties (Chen *et al.*, 1995; Harley *et al.*, 2007b) (Tables 9.2, 9.4 and 9.5). Cellular solids models, described in Chapter 3, have been applied to understand their morphology and mechanics (O'Brien *et al.*, 2005, 2007; Harley *et al.*, 2007b). Grossly, mechanical properties of CG scaffolds have been shown to affect the biological activity of primary chondrocytes (Lee *et al.*, 2001a; Vickers *et al.*, 2006; Capito and Spector, 2007), tenocytes (Schulz-Torres *et al.*, 2000) and (dermal) fibroblasts (Freyman *et al.*, 2002; Harley *et al.*, 2008).

CG scaffolds have been used to quantify the macroscopic contractile behavior of fibroblasts (Freyman *et al.*, 2001a,b,c, 2002), tenocytes (Schulz-Torres *et al.*, 2000) and chondrocytes (Zaleskas *et al.*, 2004), as well as the contractile behavior of individual fibroblasts within the CG scaffold network (Harley *et al.*, 2007a). These scaffolds have also helped elucidate the relationship between cell contractility and cell morphology

Table 9.4 Mechanical properties of the homogeneous CG scaffold variants (96 μm; 0.0058 relative density; DHT crosslinking at 105°C for 24 hours; hydrated)

	Hydrated CG scaffold
Property	Mean ± standard deviation
E^*	208 ± 41 Pa
σ_{el}^*	21 ± 8 Pa
ε_{el}^*	0.10 ± 0.04
$\Delta\sigma/\Delta\varepsilon$	92 ± 14 Pa
E_s	5.28 ± 0.25 MPa

Source: Harley *et al.* (2007b)

Table 9.5 Elastic moduli of individual scaffold struts within hydrated CG scaffolds crosslinked via DHT and EDAC/NHS techniques

Crosslinking treatment	Scaffold strut elastic moduli, E_s (MPa) Mean ± standard deviation	Elastic modulus relative to standard
Uncrosslinked	3.9 ± 0.20	0.74
DHT105/24 (Standard)	5.28 ± 0.25	1.0
DHT120/48	5.7 ± 0.30	1.08
EDAC1:1:5	10.6 ± 0.50	2.0
EDAC5:2:5	11.8 ± 0.56	2.24
EDAC5:2:1	38.0 ± 1.8	7.2

Source: Harley *et al.* (2007b)

as well as cytoskeletal organization (Freyman *et al.*, 2001b). Dermal fibroblasts are observed to undergo morphological reorganization while contracting the CG scaffold to which they are attached: initially rounded fibroblasts (diameter 20 μm) elongate over time into spindle-shaped cells, with the average cell aspect ratio increasing to approximately 3:1 during the first 15 hours in the scaffold (Fig. 9.10) (Freyman *et al.*, 2001b). A cell force monitor (CFM), which measures the macroscopic deformation of a CG scaffold seeded with dermal fibroblasts, was used to quantify the contractility of the fibroblast population. Dermal fibroblasts were observed to significantly contract the CG scaffold, generating a force that varies with time, t, as $(1 - \exp[-t/\tau])$, with the force reaching an asymptotic value of 1.0 ± 0.2 nN/cell after about 12 hours independent of the number of cells seeded (Freyman *et al.*, 2001c). Further, the time constant for cell contraction, τ, closely matched the time constant observed for cell elongation, suggesting that scaffold deformation occurs simultaneously with this cell elongation and presumably cytoskeletal reorganization (Freyman *et al.*, 2001b,c). Interestingly, when the stiffness of the flexible beam in the CFM used to measure scaffold deformation was

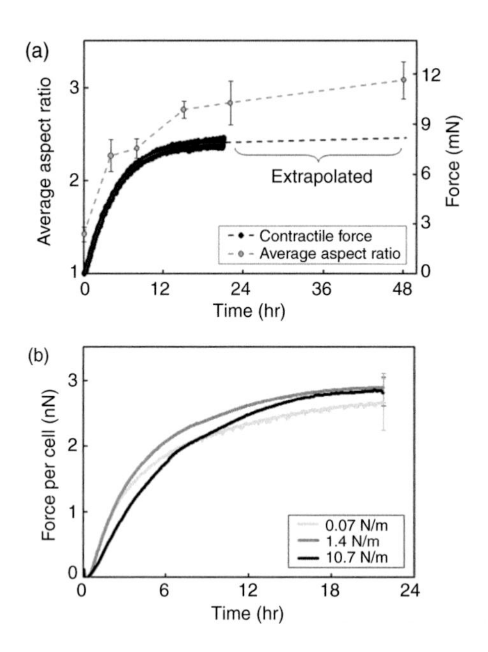

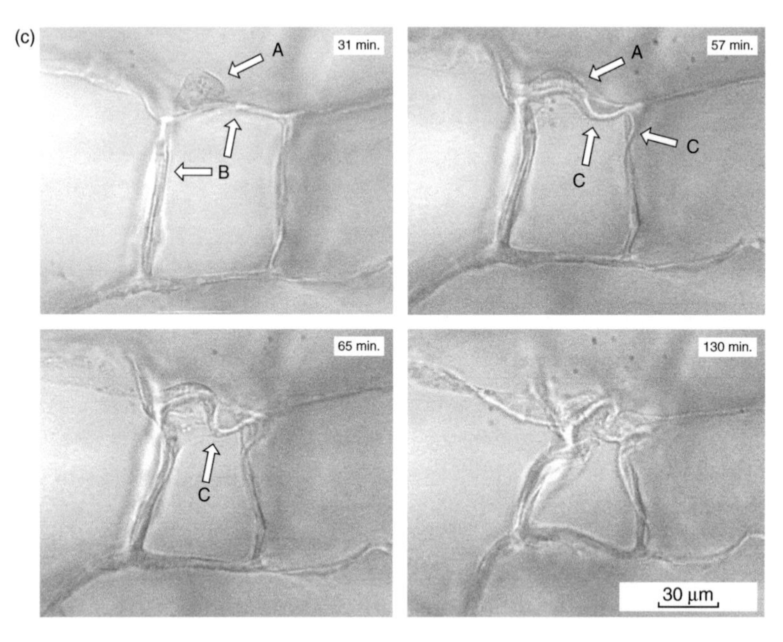

Fig. 9.10 (a) Plot of average aspect ratio of dermal fibroblasts seeded in CG scaffolds with time; the increase in aspect ratio after 15 hours is not statistically significant. A typical plot of force generated by these fibroblasts in the scaffold with time curve is also shown for comparison. (b) Plot of contractile force per cell generated by dermal fibroblasts in CG scaffolds for varying system stiffnesses of the scaffold. The force developed per cell was independent of the system stiffness. (c) Time lapse images of an individual dermal fibroblast within the CG scaffold. The sequence of images shows a dermal fibroblast (arrow A) elongating and deforming the scaffold surrounding struts (arrows B). Several struts are deformed over time (arrows C). The number in the top right corner of each image indicates the time, in hours and minutes, after cell seeding. (a, Reprinted from Freyman *et al.*, 2001b, with permission from Elsevier; b, reprinted from Freyman *et al.*, 2002, with permission from Elsevier; c, reprinted from Harley *et al.*, 2007a, with permission from Elsevier.)

increased, effectively increasing the system stiffness of the CG scaffold-cell construct, the macroscopic deformation of the scaffold by the fibroblasts decreased. However, it was observed that dermal fibroblasts apply a constant average force regardless of system stiffness. These results suggest that dermal fibroblasts apply contractile forces that are independent of the local mechanical microenvironment (Fig. 9.10) (Freyman *et al.*, 2002).

CG scaffolds have also been used to quantify the contractile forces generated by individual cells within the scaffold network. Time-lapse images of individual dermal fibroblasts within the CG scaffold show the fibroblasts generate sufficient contractile forces to buckle the individual struts to which they are attached (Fig. 9.10c) (Freyman *et al.*, 2001b,c, 2002). The force required to buckle a strut within the scaffold can be calculated using Euler's equation for buckling of a column, knowing the strut radius and length, the Young's modulus of the strut and the end constraint factor. Scaffold strut dimensions have been measured by stereology (Harley *et al.*, 2007a), while the moduli of individual struts excised from the CG scaffold have been measured by performing beam bending tests using an atomic force microscope (Harley *et al.*, 2007b). The end constraint factor can be estimated by considering the overall hydrostatic contractile loading by the cells on the scaffold (Gibson and Ashby, 1997). The contractile force generated by individual dermal fibroblasts to buckle CG scaffold struts was calculated to range between 11 and 41 nN, with an average contractile force, F_c, of 26 ± 13 nN (Harley *et al.*, 2007a). In one instance where a fibroblast was unable to buckle the strut it was attached to, the buckling load of that strut was calculated to be approximately 450 nN (Fig. 9.11) (Harley *et al.*, 2007a). These results suggest that while dermal fibroblasts can easily develop the ~25 nN force required to buckle conventional CG scaffold struts, they are unable to develop contractile forces at the level of 450 nN.

Cell migration is a complex process governed by many factors including extracellular ligands and intracellular signaling (Lauffenburger and Horwitz, 1996; Friedl *et al.*, 1998). Quantitative study of individual cell behavior within a three-dimensional scaffold construct requires understanding the local extracellular environment of the individual cells through accurate compositional, microstructural and mechanical characterization. CG scaffolds with uniform, well-characterized and independently controllable mechanical and microstructural properties are an ideal platform for in vitro studies of the effect of the extracellular matrix environment on cell behavior. CG scaffold pore sizes are significantly larger than the characteristic dimension of the fibroblasts; hence, cells are not exposed to steric hindrance as in a dense network of thin ECM fibers, as in a hydrogel. Rather, cells are forced to migrate along scaffold struts, a phenomenon known as contact guidance. Migration of NR6 fibroblasts in a series of scaffolds of constant pore size (96 μm) but variable strut modulus (E_s = 5.3–38 MPa, made by different crosslinking treatments) was tracked and the average cell speed was plotted against scaffold strut modulus (Fig. 9.12). The migration speed, increased (significantly) from 11 to 15 μm/hour for strut moduli (E_s) between 5 and 12 MPa and then decreased (significantly) back to 12 μm/hour for strut moduli of 38 MPa. This dependence of scaffold strut modulus (E_s) on cell migration speed correlates well with previous experimental and computational studies of cell motility in dense, three-dimensional hydrogel materials with a high degree of steric hindrance (Zaman *et al.*, 2005, 2006).

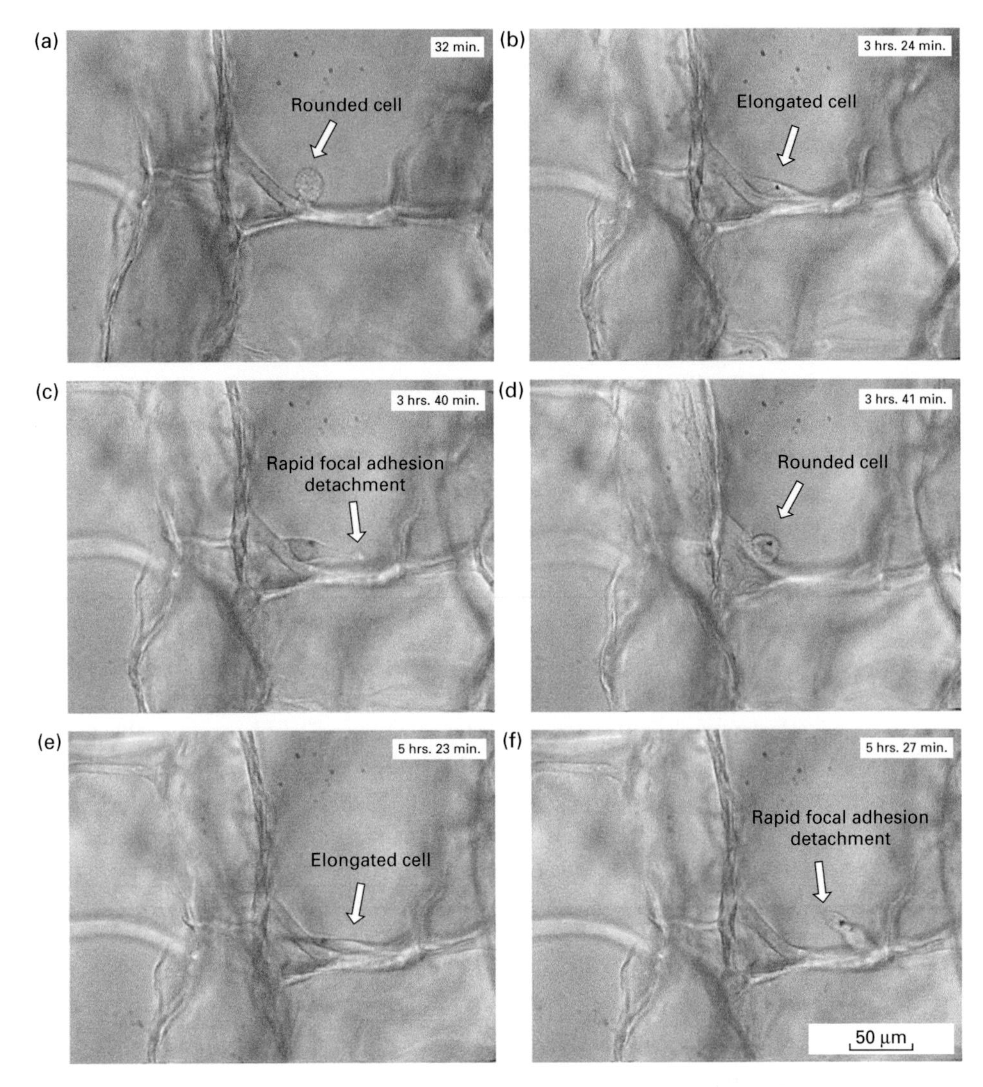

Fig. 9.11 Time-lapse light microscopy images of an individual dermal fibroblast that was unable to buckle a CG scaffold strut. Note the cell (b) elongating, (c) detaching, (d) rounding, (e) elongating and (f) detaching again. The number in the top left corner of each image indicates the time, in hours and minutes, after cell seeding. (Reprinted from Harley *et al.*, 2007a, with permission from Elsevier.)

9.4.5 Effect of force transmission through material structure: three-dimensional scaffolds

Mechanical stimuli can be applied directly to cells via macroscopic deformation of the underlying substrate or indirectly through the effects of changes in pressure and/or fluid shear stresses within a porous material (i.e. scaffold) due to deformation of the porous network. Both effects have been previously shown to be significant regulators of cell behavior. The duration, orientation and level of mechanical stimulation have previously been shown to significantly affect cell distribution within both three-dimensional biomaterials and tissues. The mean size of chondrocytes increases threefold while chondrocyte

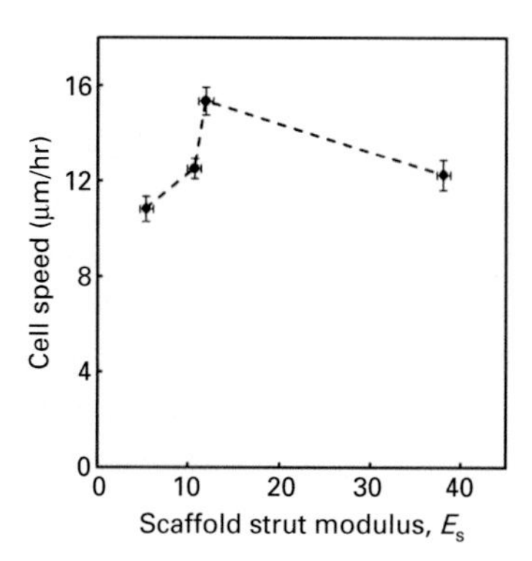

Fig. 9.12 A subtle biphasic relationship is observed between NR6 fibroblast migration speed and CG scaffold strut modulus. (Reprinted from Harley *et al.*, 2008, with permission from Elsevier.)

density decreases more than twofold in regions of native cartilage with increased mechanical loading (Eggli *et al.*, 1988); application of static or cyclic strains as low as 0.2% to fibroblast-populated collagen matrices has been shown to be sufficient to induce fibroblast alignment along the axis of applied strain (Eastwood *et al.*, 1998). Dynamic compressive loading of chondrocytes in peptide hydrogel biomaterials has been shown to induce a significant increase in chondrocyte-mediated proteoglycan synthesis, as compared to static culture, without any significant differences in cell viability or proliferation (Kisiday *et al.*, 2004); this result, in a biomaterial system, compares well with those from studies of chondrocyte metabolic activity in loaded cartilage tissue specimens (Larsson *et al.*, 1991; Vanwanseele *et al.*, 2002). Mechanical loading of fibroblasts in collagen hydrogels and scaffolds has also been shown to significantly influence cell protease production; matrix metalloproteinase synthesis increased with static loading (versus unloaded culture), and increased even more with dynamic (cyclic) loading; greatest protease production was seen for high strain, high frequency stimulation. Additionally, while still significant, the effect of static and dynamic loading was reduced with increasing initial elastic modulus of the material (Brown *et al.*, 1998; Prajapati *et al.*, 2000a,b). The interplay between scaffold loading and the resulting cell behavior has also been probed using collagen hydrogel matrices, where fibroblasts were encapsulated within prestressed or floating (non-prestressed) collagen matrix variants. The presence of different levels of matrix pre-stress was shown to significantly affect the nature and kinetics of fibroblast-mediated contractile behavior (Grinnell *et al.*, 1999). Those in prestressed matrices showed increased actin cytoskeleton organization, lamellipodia extensions and the required microtubules for polarization, while fibroblasts in non-prestressed matrices showed increased quiescence and lacked stress fibers (Fringer and Grinnell, 2001, 2003; Rhee *et al.*, 2007).

In the case of scaffolds implanted into the body for in vivo tissue engineering applications, a number of additional effects have also been noted. Of particular note is that the relationship between surrounding tissue and scaffold remodeling and mechanical properties is governed by a feedback loop where deformation of the ECM stimulates changes in cell metabolic activity and matrix expression (Carterson and Lowther, 1978; Swann and Seedhom, 1993).

9.5 Multi-parameter effects

While the previous sections have addressed the influence of scaffold chemical composition, microstructure and mechanical properties on cell–matrix interactions, in many of these investigations the interplay between such scaffold features has been ignored. Previous studies of cell mechanobiology in two-dimensional systems have suggested that there is a complex feedback mechanism by which a cell samples information from a wide variety of extrinsic factors before interpolating this information and making a genotypic or phenotypic fate decision. Studies of cell–scaffold interactions have recently begun to control multiple extrinsic factors to investigate the additive, contradictory, or synergistic effects of distinct scaffold features on cell biology.

The influence of multiple biomaterial parameters, notably combinations of chemical composition, microstructure and mechanical properties, has been shown to differentially influence cell mechanotransduction and behavior as compared to individual parameters alone. In a series of papers, Zaman *et al.* (2005) explored the extrinsic regulation of cell motility by matrix microstructural and mechanical features using both empirical and modeling methods. Using a force-based dynamics approach, a computation model was developed to describe three-dimensional cell motility in response to surrounding matrix stiffness and matrix ligand density or the number of receptors present on the cell. Maximum speed was modeled to occur for intermediate levels of stiffness and ligand concentration or at maximum number of receptors; higher ligand concentrations induce steric hindrances while low concentrations correspond to negligible traction. These results were confirmed via in vitro experiments, where tumor cell migration was observed to be influenced by local matrix stiffness, ligand concentration and cell integrin receptors in a manner consistent with computational models (Fig. 9.13) (Zaman *et al.*, 2006). Local matrix stiffness and microstructure also strongly influence the intracellular mechanical state of cancer cells. The relative intracellular stiffness of cancer cells decreases as the local collagen matrix stiffness increases and also as local ligand density (a proxy for matrix microstructure) increases; partially blocking cancer cell β1 integrins also has the effect of decreasing cancer cell intracellular stiffness (Baker *et al.*, 2009).

A series of publications by Grinnell and co-workers further established the multidimensional influence of matrix mechanics and microstructure on fibroblast motility, cytoskeletal structure and the mechanisms by which cells sense and respond to their local microenvironment (Grinnell *et al.*, 2003; Jiang and Grinnell, 2005; Rhee and Grinnell, 2007). Chondrocyte response to mechanical stimulation in PEG-based hydrogels has also been show to be differentially affected by scaffold composition and architecture (Appelman *et al.*, 2009): a significant influence of scaffold chemical composition was observed along with differential chondrocyte mechanotransduction to static versus dynamic loading for each distinct matrix composition tested. These investigations, using hydrogel-based biomaterials, suggest that there are differential and interlinked roles played by proteolysis, ligand density and matrix stiffness in influencing cell behavior.

Differential regulation of cell behavior by the chemistry, microstructure and mechanical properties of scaffolds, as opposed to sterically hindered hydrogels, has also

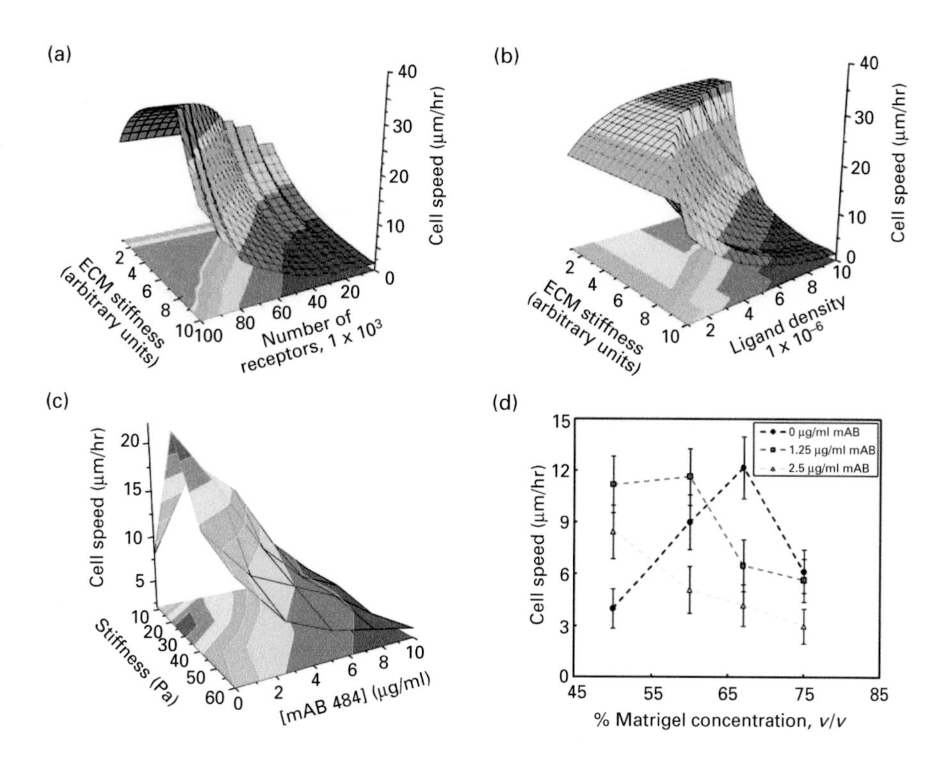

Fig. 9.13 Three-dimensional plot of cell speed as a function of matrix stiffness, ligand density and available receptors. (a) Computational model of cell migration speed as a function of matrix stiffness and number of available receptors; highest speeds occur at maximum receptors and intermediate stiffness. (b) Computation model of cell migration speed as a function of matrix stiffness and number of available ligands; highest speeds occur at intermediate stiffness and ligand concentration. (c) Experimentally measured migration speed of DU-145 human prostate carcinoma cells in a 3D Matrigel matrix as a function of receptor number and matrix stiffness. (d) Integrin inhibition shifts the maximum cell speed to lower Matrigel concentrations. The average number of motile cells decreases with increases in the concentration (0, 1.25, 2.5 μg/ml) of 4B4 anti-integrin blocking monoclonal antibody (mAB), while average speed shows somewhat of a bimodal behavior with variations in gel density, reaching a maximum with intermediate gel concentration. The presence of 4B4 antibody slowed cell speed and shifted the maximum to lower Matrigel concentrations. (a,b, Reprinted from Zaman *et al.*, 2005, with permission from Elsevier; c,d, copyright 2006 National Academy of Sciences, USA, reprinted with permission from Zaman *et al.*, 2006.) See plate section for color version.

been observed. Harley *et al.* (2008) examined how independently modulating CG scaffold mechanical properties and pore size differentially regulate fibroblast motility patterns. A significant influence of scaffold pore size (Section 9.3) and scaffold Young's modulus (Section 9.4) on the motility (motile fraction, migration speed and directional persistence) has already been described in this chapter, but probing the mechanism by which scaffold pore size (96–151 μm, significantly larger than the cells) affects the motility of cells (20 μm in size), presented the opportunity to fully leverage the cellular nature of the scaffold.

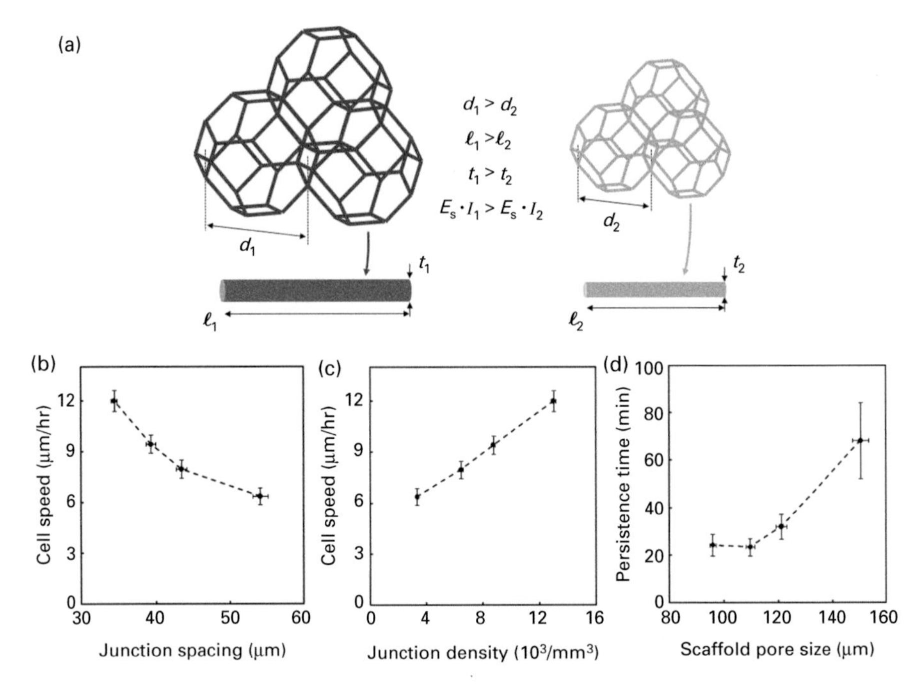

Fig. 9.14 (a) Tetrakaidecahedral unit cell model for CG scaffold. For a series of scaffolds with a constant relative density, those with larger pore sizes, d_1 (left), exhibit struts that are longer and thicker than a scaffold with smaller pore sizes, d_2 (right). The longer/thicker struts have a greater flexural rigidity ($E_s \cdot I$) than the shorter/thinner struts, and would deform less under a constant cell-applied traction force. NR6 fibroblast migration speed shows good correlation with (b) strut junction spacing (D_{jxn}) and (c) strut junction density (ρ_{jxn}), pore-size-dependent parameters calculated for the CG scaffold from the tetrakaidecahedral unit cell model. (d) For CG scaffolds with decreasing strut junction density (increasing pore size), NR6 fibroblasts were observed to have increasing persistence times. (Reprinted from Harley *et al.*, 2008, with permission from Elsevier.)

While the influence of scaffold strut elastic modulus (E_s) on cell motility was expected, since cells were not exposed to steric hindrance in these porous, CG scaffolds, the strong dependence of cell motility on pore size was not expected. To this end, potential local variations in the micromechanical and microstructural environment of individual cells within the scaffold network were considered using experimental and cellular solids modeling tools in order to better explain the experimental results. Since the scaffolds all had the same relative density and strut modulus, E_s, they all had the same scaffold modulus, E^*. However, scaffolds with a larger pore size had somewhat longer and thicker struts (Fig. 9.14). Though the moduli (E_s) of these struts was constant, the second moment of inertia (I) increased with increasing pore size due to the increase (taken then to the fourth power) in strut thickness, translating to an increasing strut flexural rigidity ($E_s I$) with pore size. Fibroblasts had previously been observed to apply a constant contractile force to the CG scaffold regardless of the system's stiffness (Freyman *et al.*, 2002), suggesting that cells probe their local mechanical environment by applying a constant traction force and measuring the resultant substrate deformation (Vogel and Sheetz, 2006). Therefore, strut flexural rigidity was hypothesized to likely be a more relevant mechanical cue rather than strut modulus alone: even though the struts have a constant Young's modulus, they may "feel" stiffer in scaffolds with larger pore sizes because of an increased resistance to deformation. The flexural rigidity (E_sI) of the CG scaffold variants was calculated to increase sixfold from the smallest (96 μm) to the largest (151 μm) pore size (Harley *et al.*, 2008). Thus, if a mechanosensitive hypothesis that posits the influence of scaffold pore size on motility as being due to changes in the strut flexural rigidity were true, it would further predict that cell motility would be reduced as scaffold elastic modulus increases independent of microstructure (modifying E_s rather than I to change the overall flexural rigidity, E_sI). However, the results of the independent influence of strut E_s on cell motility indicate that this hypothesis should be rejected. Strut modulus (E_s) in scaffolds with a pore size of 96 μm pore size was increased approximately sevenfold, closely approximating the sixfold change in strut flexural rigidity with mean pore size. However, cell speed did not decrease with increasing scaffold strut stiffness; instead cell speed exhibited the biphasic dependence on strut stiffness (Fig. 9.14). Therefore, the pore size-dependence in cell speed did not arise due to changes in apparent scaffold strut stiffness (Harley *et al.*, 2008).

After exploring the predictions made by the cellular solids models regarding local scaffold mechanics, geometric insights were then used to suggest that potential local contact guidance cues may explain the influence of pore size on cell motility (Harley *et al.*, 2008). Strut junctions, points in the scaffold microstructure where two or more struts meet, are discrete areas of significantly different extracellular morphology compared to an individual strut. With the average strut length for the different scaffolds used in the cell motility investigation of the order 30–60 μm, motile cells as well as sessile cells extending processes are expected to regularly encounter strut junctions (likely multiple junctions during the 10 hour imaging period used in this experiment). Two particular measurements of strut junctions were explored: strut junction spacing (D_{jxn}) and density (ρ_{jxn}, the number of strut junctions per unit cell divided by the volume of a unit cell). D_{jxn} and ρ_{jxn} of the different CG scaffold microstructures (mean pore size = 96–151 μm) can be described using the scaffold mean pore size (d), assuming a tetrakaidecahedral unit cell (Gibson and Ashby, 1997):

$$D_{jxn} = l = \frac{d}{2.785} \tag{9.5}$$

$$\rho_{jxn} = \frac{\left(\text{junctions}/\text{unit cell}\right)}{\left(\text{volume}/\text{unit cell}\right)} = \frac{6}{11.31 \times \left(\dfrac{d}{2.785}\right)^3} = \frac{11.459}{d^3} \tag{9.6}$$

Replotting the pore-size dependent cell speed data against strut junction density, an exceptionally strong correlation between cell speed and strut junction density was observed (Fig. 9.14); further exploration of the potential influence of this local microstructural cue was pursued. Cells migrating in scaffolds with larger pore sizes, and therefore larger D_{jxn} and lower ρ_{jxn}, were observed to exhibit greater persistence times, indicating more directional motion along a scaffold strut (Fig. 9.14). In contrast, persistence times of cells migrating in scaffolds with smaller pore sizes and greater ρ_{jxn} are significantly lower, representative of erratic movement that likely occurs more often at junctions when cells can probe their local environment along multiple struts. This behavior is likely not solely due to the influence of ligand availability, as in the case of sterically hindered cells within hydrogels; for cells attached to the scaffold, the spacing between struts was significantly larger than the cell itself, precluding availability of additional ligand sites on struts other than that to which the cell was attached except at the junction points. Further, the density of ligands available to cells within the scaffold structure, even at the junctions, is far smaller than that seen in sterically hindering structures; the density of available ligands never approaches the concentration seen in hydrogels where adhesive interactions between cells and the surrounding ligands can retard cell motility. These results provided a mechanistic explanation for the initially counter-intuitive observation that cell motility decreases as scaffold pore size increases and provide a link between junction density and cell migration behavior (Harley *et al.*, 2008).

9.6 Summary

Porous, three-dimensional biomaterials have been used extensively for a variety of tissue engineering applications. The primary application is their use as an analog of the ECM capable of inducing regeneration of damaged tissues and organs. An important evolving application is their use as constructs to quantitatively study cell behavior and cell–scaffold interactions. Scaffold material, microstructural and mechanical properties have all been observed to significantly affect individual cell behavior as well as overall scaffold bioactivity and regenerative capacity. Models for cellular solids contribute to our understanding of cell–scaffold interactions and hold promise for future studies of in vitro cell mechanobiology and in vivo tissue engineering.

References

Agrawal CM and Athanasiou KA (1997) Technique to control pH in vicinity of biodegrading PLA-PGA implants. *J. Biomed. Mat. Res.* **38**, 105–14.

Al-Munajjed AA, Gleeson JP and O'Brien FJ (2008) Development of a collagen calcium-phosphate scaffold as a novel bone graft substitute. *Stud. Health Technol. Inform* **133**, 11–20.

Allingham PG, Brownlee GR, Harper GS, Pho M, Nilsson SK and Brown TJ (2006) Gene expression, synthesis and degradation of hyaluronan during differentiation of 3T3-L1 adipocytes. *Arch Biochem. Biophys.* **452**, 83–91.

Almany L and Seliktar D (2005) Biosynthetic hydrogel scaffolds made from fibrinogen and polyethylene glycol for 3D cell cultures. *Biomaterials* **26**, 2467–77.

Alvarez-Barreto JF, Shreve MC, Deangelis PL and Sikavitsas VI (2007) Preparation of a functionally flexible, three-dimensional, biomimetic poly(l-lactic acid) scaffold with improved cell adhesion. *Tissue Eng* **13**, 1205–17.

Appelman TP, Mizrahi J, Elisseeff JH and Seliktar D (2009) The differential effect of scaffold composition and architecture on chondrocyte response to mechanical stimulation. *Biomaterials* **30**, 518–25.

Augst AD, Kong HJ and Mooney DJ (2006) Alginate hydrogels as biomaterials. *Macromol. Biosci.* **6**, 623–33.

Badylak SF (2007) The extracellular matrix as a biologic scaffold material. *Biomaterials* **28**, 3587–93.

Baker EL, Bonnecaze RT and Zaman MH (2009) Extracellular matrix stiffness and architecture govern intracellular rheology in cancer. *Biophys. J.* **97**, 1013–21.

Beningo KA, Dembo M, Kaverina I, Small JV and Wang Y-L (2001) Nascent focal adhesions are responsible for the generation of strong propulsive forces in migrating fibroblasts. *J. Cell Biol.* **153**, 881–7.

Beningo KA and Wang Y-L (2002) Flexible substrata for the detection of cellular traction forces. *Trends Cell Biol.* **12**, 79–84.

Bergsma JE, de Bruijn WC, Rozema FR, Bos RR and Boering G (1995) Late degradation tissue response to poly(L-lactide) bone plates and screws. *Biomaterials* **16**, 25–31.

Boontheekul T, Kong HJ and Mooney DJ (2005) Controlling alginate gel degradation utilizing partial oxidation and bimodal molecular weight distribution. *Biomaterials* **26**, 2455–65.

Brown AF (1982) Neutrophil granulocytes: adhesion and locomotion on collagen substrata and in collagen matrices. *J. Cell Sci.* **58**, 455–67.

Brown RA, Prajapati R, McGrouther DA, Yannas IV and Eastwood M (1998) Tensional homeostasis in dermal fibroblasts: mechanical responses to mechanical loading in three-dimensional substrates. *J. Cell Physiol.* **175**, 323–32.

Brown TD (2000) Techniques for mechanical stimulation of cells in vitro: a review. *J. Biomech.* **33**, 3–14.

Bryant SJ and Anseth KS (2003) Controlling the spatial distribution of ECM components in degradable PEG hydrogels for tissue engineering cartilage. *J. Biomed. Mat. Res. A* **64**, 70–9.

Bryant SJ, Durand KL and Anseth KS (2003) Manipulations in hydrogel chemistry control photoencapsulated chondrocyte behavior and their extracellular matrix production. *J. Biomed. Mat. Res. A* **67**, 1430–6.

Bryant SJ, Arthur JA and Anseth KS (2005) Incorporation of tissue-specific molecules alters chondrocyte metabolism and gene expression in photocrosslinked hydrogels. *Acta Biomater.* **1**, 243–52.

Buckwalter JA (1983) Articular Cartilage. *AAOS Instructional Course Lectures* **32**, 349–70.

Burdick JA, Mason MN and Anseth KS (2001) In situ forming lactic acid based orthopaedic biomaterials: influence of oligomer chemistry on osteoblast attachment and function. *J. Biomat. Sci. Polymer Ed.* **12**, 1253–65.

Burgess BT, Myles JL and Dickinson RB (2000) Quantitative analysis of adhesion-mediated cell migration in three-dimensional gels of RGD-grafted collagen. *Ann. Biomed. Eng.* **28**, 110–8.

Burton K and Taylor DL (1997) Traction forces of cytokinesis measured with optically modified elastic substrata. *Nature* **385**, 450–4.

Butler DL, Juncosa-Melvin N, Boivin GP *et al.* (2008) Functional tissue engineering for tendon repair: A multidisciplinary strategy using mesenchymal stem cells, bioscaffolds, and mechanical stimulation. *J. Orthop. Res.* **26**, 1–9.

Byrne EM, Farrell E, McMahon LA *et al.* (2008) Gene expression by marrow stromal cells in a porous collagen-glycosaminoglycan scaffold is affected by pore size and mechanical stimulation. *J. Mat. Sci. Mat. Med.* **19**, 3455–63.

Capito RM and Spector M (2003) Scaffold-based articular cartilage repair. *IEEE Eng. Med. Biol. Mag.* **22**, 42–50.

Capito RM and Spector M (2007) Collagen scaffolds for nonviral IGF-1 gene delivery in articular cartilage tissue engineering. *Gene Therapy* **14**, 721–32.

Carterson B and Lowther DA (1978) Changes in the metabolism of the proteoglycans from sheep articular cartilage in response to mechanical stress. *Biochim. Biophys. Acta* **540**, 412–22.

Castner DG and Ratner BD (2002) Biomedical surface science: foundations to frontiers. *Surface Sci.* **500**, 28–60.

Chamberlain LJ and Yannas IV (1998) Preparation of collagen–glycosaminoglycan copolymers for tissue regeneration. In *Methods of Molecular Medicine*, ed.Morgan JR and Yarmush ML. Tolowa, NJ: Humana Press.

Chamberlain LJ, Yannas IV, Hsu H-P, Strichartz G and Spector M (1998) Collagen-GAG substrate enhances the quality of nerve regeneration through collagen tubes up to level of autograft. *Exper. Neurol.* **154**, 315–29.

Chamberlain LJ, Yannas IV, Hsu HP, Strichartz GR and Spector M (2000) Near-terminus axonal structure and function following rat sciatic nerve regeneration through a collagen-GAG matrix in a ten-millimeter gap. *J. Neurosci. Res.* **60**, 666–77.

Chang AS, Yannas IV, Perutz S *et al.* (1990) Electrophysiological study of recovery of peripheral nerves regenerated by a collagen-glycosaminoglycan copolymer matrix. In *Progress in Biomedical Polymers*, ed. Gebelin CG and Dunn RL. New York: Plenum Press.

Chen CS, Yannas IV and Spector M (1995) Pore strain behaviour of collagen-glycosaminoglycan analogues of extracellular matrix. *Biomaterials* **16**, 777–83.

Chen CS and Ingber DE (1999) Tensegrity and mechanoregulation: from skeleton to cytoskeleton. *Osteoarth. Cartil.* **7**, 81–94.

Chen G, Ushida T and Tateishi T (2001) Preparation of poly(L-lactic acid) and poly(DL-lactic-co-glycolic acid) foams by use of ice microparticulates. *Biomaterials* **22**, 2563–7.

Chen CS (2008) Mechanotransduction – a field pulling together? *J J. Cell Sci Sci.* **121**, 3285–92.

Chen G, Sato T, Ushida T, Hirochika R and Tateishi T (2003) Redifferentiation of dedifferentiated bovine chondrocytes when cultured in vitro in a PLGA-collagen hybrid mesh. *FEBS Lett.* **542**, 95–9.

Cheng NC, Estes BT, Awad HA and Guilak F (2009) Chondrogenic differentiation of adipose-derived adult stem cells by a porous scaffold derived from native articular cartilage extracellular matrix. *Tissue Eng. Part A* **15**, 231–41.

Comisar WA, Hsiong SX, Kong HJ, Mooney DJ and Linderman JJ (2006) Multi-scale modeling to predict ligand presentation within RGD nanopatterned hydrogels. *Biomaterials* **27**, 2322–9.

Cool SM and Nurcombe V (2005) Substrate induction of osteogenesis from marrow-derived mesenchymal precursors. *Stem Cells Dev.* **14**, 632–42.

Cornwell KG, Downing BR and Pins GD (2004) Characterizing fibroblast migration on discrete collagen threads for applications in tissue regeneration. *J. Biomed. Mat. Res. A* **71**, 55–62.

Cornwell KG and Pins GD (2007) Discrete crosslinked fibrin microthread scaffolds for tissue regeneration. *J. Biomed. Mat. Res. A* **82**, 104–12.

Cornwell KG, Lei P, Andreadis ST and Pins GD (2007) Crosslinking of discrete self-assembled collagen threads: Effects on mechanical strength and cell-matrix interactions. *J. Biomed. Mat. Res. A* **80**, 362–71.

Cukierman E, Pankov R, Stevens DR and Yamada KM (2001) Taking cell-matrix adhesions to the third dimension. *Science* **294**, 1708–12.

Cukierman E, Pankov R and Yamada KM (2002) Cell interactions with three-dimensional matrices. *Curr. Opin. Cell Biol.* **14**, 633–9.

Currey JD (2002) *Bones: Structure and Mechanics*. Princeton, NJ: Princeton University Press.

Davies PF (1995) Flow-mediated endothelial mechanotransduction. *Physiol. Rev.* **75**, 519–60.

Dawson M, Germaine JT and Gibson LJ (2007) Permeability of open-cell foams under compressive strain. *Int. J. Solids Struct.* **44**, 5133–45.

Dembo M and Wang Y-L (1999) Stresses at the cell-to-substrate interface during locomotion of fibroblasts. *Biophys. J.* **76**, 2307–16.

Deshpande VS, McMeeking RM and Evans AG (2006) A bio-chemo-mechanical model for cell contractility. *Proc. Natl Acad. Sci. USA* **103**, 14015–20.

Dikovsky D, Bianco-Peled H and Seliktar D (2006) The effect of structural alterations of PEG-fibrinogen hydrogel scaffolds on 3-D cellular morphology and cellular migration. *Biomaterials* **27**, 1496–506.

Dikovsky D, Bianco-Peled H and Seliktar D (2008) Defining the role of matrix compliance and proteolysis in three-dimensional cell spreading and remodeling. *Biophys J.* **94**, 2914–25.

Discher DE, Janmey P and Wang YL (2005) Tissue cells feel and respond to the stiffness of their substrate. *Science* **310**, 1139–43.

Discher DE, Mooney DJ and Zandstra PW (2009) Growth factors, matrices, and forces combine and control stem cells. *Science* **324**, 1673–7.

Eastwood M, Mudera VC, McGrouther DA and Brown RA (1998) Effect of precise mechanical loading on fibroblast populated collagen lattices: morphological changes. *Cell Motility and the Cytoskeleton* **40**, 13–21.

Eggli S, Hunziker EB and Schenck RK (1988) Quantitation of structural features characterizing weight- and less-weight-bearing regions in articular cartilage: a stereological analysis of medial femoral condyles in young adult rabbits. *Anat. Rec.* **222**, 217–27.

Elsdale T and Bard J (1972) Collagen substrata for studies on cell behavior. *J. Cell Biol.* **54**, 626–37.

Engelmayr GC, Jr., Sales VL, Mayer JE, Jr. and Sacks MS (2006) Cyclic flexure and laminar flow synergistically accelerate mesenchymal stem cell-mediated engineered tissue formation: Implications for engineered heart valve tissues. *Biomaterials* **27**, 6083–95.

Engelmayr GC, Jr., Cheng M, Bettinger CJ, Borenstein JT, Langer R and Freed LE (2008) Accordion-like honeycombs for tissue engineering of cardiac anisotropy. *Natural Mat.* **7**, 1003–10.

Engler A, Bacakova L, Newman C, Hategan A, Griffin M and Discher D (2004a) Substrate compliance versus ligand density in cell on gel responses. *Biophys. J.* **86**, 617–28.

Engler AJ, Griffin MA, Sen S, Bonnemann CG, Sweeney HL and Discher DE (2004b) Myotubes differentiate optimally on substrates with tissue-like stiffness: pathological implications for soft or stiff microenvironments. *J. Cell Biol.* **166**, 877–87.

Engler AJ, Sen S, Sweeney HL and Discher DE (2006) Matrix elasticity directs stem cell lineage specification. *Cell* **126**, 677–89.

Engler AJ, Sweeney HL, Discher DE and Schwarzbauer JE (2007) Extracellular matrix elasticity directs stem cell differentiation. *J. Musculoskelet. Neuronal Interact.* **7**, 335.

Engler AJ, Carag-Krieger C, Johnson CP *et al.* (2008) Embryonic cardiomyocytes beat best on a matrix with heart-like elasticity: scar-like rigidity inhibits beating. *J. Cell Sci.* **121**, 3794–802.

Even-Ram S and Yamada KM (2005) Cell migration in 3D matrix. *Curr. Opin. Cell Biol.* **17**, 524–32.

Everaerts F, Torrianni M, Hendriks M and Feijen J (2007a) Biomechanical properties of carbodiimide crosslinked collagen: influence of the formation of ester crosslinks. *J. Biomed. Mat. Res. A* **85**, 547–55.

Everaerts F, Torrianni M, Hendriks M and Feijen J (2007b) Quantification of carboxyl groups in carbodiimide cross-linked collagen sponges. *J. Biomed. Mat. Res. A* **83**, 1176–83.

Farrell E, Byrne EM, Fischer J *et al.* (2007) A comparison of the osteogenic potential of adult rat mesenchymal stem cells cultured in 2-D and on 3-D collagen glycosaminoglycan scaffolds. *Technol. Health Care* **15**, 19–31.

Freed LE, Vunjak-Novakovic G, Biron RJ *et al.* (1994) Biodegradable polymer scaffolds for tissue engineering. *Biotechnology (NY)* **12**, 689–93.

Freyman TM, Yannas IV and Gibson LJ (2001a) Cellular materials as porous scaffolds for tissue engineering. *Prog. Mat. Sci.* **46**, 273–82.

Freyman TM, Yannas IV, Pek Y-S, Yokoo R and Gibson LJ (2001b) Micromechanics of fibroblast contraction of a collagen-GAG matrix. *Exp. Cell Res.* **269**, 140–53.

Freyman TM, Yannas IV, Yokoo R and Gibson LJ (2001c) Fibroblast contraction of a collagen-GAG matrix. *Biomaterials* **22**, 2883–91.

Freyman TM, Yannas IV, Yokoo R and Gibson LJ (2002) Fibroblast contractile force is independent of the stiffness which resists the contraction. *Exp. Cell Res.* **272**, 153–62.

Friedl P, Zanker KS and Brocker EB (1998) Cell migration strategies in 3-D extracellular matrix: differences in morphology, cell matrix interactions, and integrin function. *Microsc. Res. Tech.* **43**, 369–78.

Friedl P and Brocker EB (2000) The biology of cell locomotion within three-dimensional extracellular matrix. *Cell. Mol. Life Sci.* **57**, 41–64.

Fringer J and Grinnell F (2001) Fibroblast quiescence in floating or released collagen matrices: contribution of the ERK signaling pathway and actin cytoskeletal organization. *J. Biol. Chem.* **276**, 31047–52.

Fringer J and Grinnell F (2003) Fibroblast quiescence in floating collagen matrices: decrease in serum activation of MEK and Raf but not Ras. *J. Biol. Chem.* **278**, 20612–17.

Galbraith CG and Sheetz MP (1997) A micromachined device provides a new bend on fibroblast traction forces. *Proc. Natl Acad. Sci. USA* **94**, 9114–18.

Geiger B, Bershadsky A, Pankov R and Yamada KM (2001) Transmembrane extracellular matrix-cytoskeleton crosstalk. *Nature Rev. Molecul. Cell Biol.* **2**, 793–805.

Gentleman E, Lay AN, Dickerson DA, Nauman EA, Livesay GA and Dee KC (2003) Mechanical characterization of collagen fibers and scaffolds for tissue engineering. *Biomaterials* **24**, 3805–13.

Gentleman E, Dee KC, Livesay GA and Nauman EA (2007) Operating curves to characterize the contraction of fibroblast-seeded collagen gel/collagen fiber composite biomaterials: effect of fiber mass. *Plast. Reconstr. Surg.* **119**, 508–16.

Gibson LJ and Ashby MF (1997) *Cellular Solids: Structure and Properties.* Cambridge: Cambridge University Press.

Grimmer JF, Gunnlaugsson CB, Alsberg E *et al.* (2004) Tracheal reconstruction using tissue-engineered cartilage. *Arch. Otolaryngol. Head Neck Surg.* **130**, 1191–6.

Grinnell F, Ho CH, Lin YC and Skuta G (1999) Differences in the regulation of fibroblast contraction of floating versus stressed collagen matrices. *J. Biol. Chem.* **274**, 918–23.

Grinnell F, Ho CH, Tamariz E, Lee DJ and Skuta G (2003) Dendritic fibroblasts in three-dimensional collagen matrices. *Molec. Biol. Cell* **14**, 384–95.

Grzesiak JJ, Pierschbacher MD, Amodeo MF, Malaney TI and Glass JR (1997) Enhancement of cell interactions with collagen/glycosaminoglycan matrices by RGD derivatization. *Biomaterials* **18**, 1625–32.

Guan JL (1997a) Focal adhesion kinase in integrin signaling. *Matrix Biol.* **16**, 195–200.

Guan JL (1997b) Role of focal adhesion kinase in integrin signaling. *Int J. Biochem. Cell Biol.* **29**, 1085–96.

Gunzer M, Friedl P, Niggemann B, Brocker EB, Kampgen E and Zanker KS (2000) Migration of dendritic cells within 3-D collagen lattices is dependent on tissue origin, state of maturation, and matrix structure and is maintained by proinflammatory cytokines. *J. Leukoc. Biol.* **67**, 622–9.

Harley BA, Spilker MH, Wu JW *et al.* (2004) Optimal degradation rate for collagen chambers used for regeneration of peripheral nerves over long gaps. *Cells Tissues Organs* **176**, 153–65.

Harley BA and Yannas IV (2006) Induced peripheral nerve regeneration using scaffolds. *Minerva Biotecnologica* **18**, 97–120.

Harley BA, Freyman TM, Wong MQ and Gibson LJ (2007a) A new technique for calculating individual dermal fibroblast contractile forces generated within collagen-GAG scaffolds. *Biophys. J.* **93**, 2911–22.

Harley BA, Leung JH, Silva EC and Gibson LJ (2007b) Mechanical characterization of collagen-glycosaminoglycan scaffolds. *Acta Biomater.* **3**, 463–74.

Harley BAC and Gibson LJ (2008) In vivo and in vitro applications of collagen-GAG scaffolds. *Chem. Eng. J.* **137**, 102–21.

Harley BA, Kim HD, Zaman MH, Yannas IV, Lauffenburger DA and Gibson LJ (2008) Microarchitecture of three-dimensional scaffolds influences cell migration behavior via junction interactions. *Biophys J.* **95**, 4013–24.

Harley BA, Lynn AK, Wissner-Gross Z, Bonfield W, Yannas IV and Gibson LJ (2010a) Design of a multiphase osteochondral scaffold II: fabrication of a mineralized collagen-GAG scaffold. *J. Biomed. Mat. Res. A* **92**, 1066–77.

Harley BA, Lynn AK, Wissner-Gross Z, Bonfield W, Yannas IV and Gibson LJ (2010b) Design of a multiphase osteochondral scaffold III: fabrication of layered scaffolds with continuous interfaces. *J. Biomed. Mat. Res. A* **92**, 1078–93.

Harris (1980) Silicone rubber substrata: a new wrinkle in the study of cell locomotion. *Science* **208**, 177–9.

Harris AK, Stopak D and Wild P (1981) Fibroblast traction as a mechanism for collagen morphogenesis. *Nature* **290**, 249–51.

Haugh MG, Jaasma MJ and O'Brien FJ (2009) The effect of dehydrothermal treatment on the mechanical and structural properties of collagen-GAG scaffolds. *J. Biomed. Mat. Res. A.* **89**, 363–9.

Holland TA, Tessmar JKV, Tabata Y and Mikos AG (2004) Transforming growth factor-beta 1 release from oligo(poly(ethylene glycol) fumarate) hydrogels in conditions that model the cartilage wound healing environment. *J. Controlled Release* **94**, 101–14.

Hsiong SX, Huebsch N, Fischbach C, Kong HJ and Mooney DJ (2008) Integrin-adhesion ligand bond formation of preosteoblasts and stem cells in three-dimensional RGD presenting matrices. *Biomacromolecules* **9**, 1843–51.

Hsu WC, Spilker MH, Yannas IV and Rubin PA (2000) Inhibition of conjunctival scarring and contraction by a porous collagen-glycosaminoglycan implant. *Invest. Ophthalmol. Vis. Sci.* **41**, 2404–11.

Hui EE and Bhatia SN (2007) Micromechanical control of cell-cell interactions. *Proc. Natl Acad. Sci. USA* **104**, 5722–6.

Hulbert SF, Young FA, Mathews RS, Klawitter JJ, Talbert CD and Stelling FH (1970) Potential of ceramic materials as permanently implantable skeletal prostheses. *J. Biomed. Mat. Res.* **4**, 433–46.

Hung-Jen Shao, Chiang Sang Chen, I-Chi Lee, Jyh-Horng Wang and Tai-Horng Young (2009) Designing a three-dimensional expanded polytetrafluoroethylene–poly(lactic-co-glycolic acid) scaffold for tissue engineering. *Artificial Organs* **33**, 309–17.

Hunter SA, Noyes FR, Haridas B, Levy MS and Butler DL (2005) Meniscal material properties are minimally affected by matrix stabilization using glutaraldehyde and glycation with ribose. *J. Orthop. Res.* **23**, 555–61.

Hutmacher DW, Goh JC and Teoh SH (2001) An introduction to biodegradable materials for tissue engineering applications. *Ann. Acad. Med. Singapore* **30**, 183–91.

Huttenlocher A, Ginsberg MH and Horwitz AF (1996) Modulation of cell migration by integrin-mediated cytoskeletal linkages and ligand-binding affinity. *J. Cell Biol.* **134**, 1551–62.

Hwang NS, Varghese S, Theprungsirikul P, Canver A and Elisseeff J (2006) Enhanced chondrogenic differentiation of murine embryonic stem cells in hydrogels with glucosamine. *Biomaterials* **27**, 6015–23.

Ifkovits JL and Burdick JA (2007) Review: Photopolymerizable and degradable biomaterials for tissue engineering applications. *Tissue Eng.* **13**, 2369–85.

Imberti B, Seliktar D, Nerem RM and Remuzzi A (2002) The response of endothelial cells to fluid shear stress using a co-culture model of the arterial wall. *Endothelium* **9**, 11–23.

Jaasma MJ and O'Brien FJ (2008) Mechanical stimulation of osteoblasts using steady and dynamic fluid flow. *Tissue Eng. Part A* **14**, 1213–23.

Jaworski J and Klapperich CM (2006) Fibroblast remodeling activity at two- and three-dimensional collagen-glycosaminoglycan interfaces. *Biomaterials* **27**, 4212–20.

Jeon O, Bouhadir KH, Mansour JM and Alsberg E (2009) Photocrosslinked alginate hydrogels with tunable biodegradation rates and mechanical properties. *Biomaterials* **30**, 2724–34.

Jiang H and Grinnell F (2005) Cell-matrix entanglements and mechanical anchorage of fibroblasts in three-dimensional collagen matrices. *Mol. Biol. Cell* **16**, 5070–6.

Jungreuthmayer C, Donahue SW, Jaasma MJ *et al.* (2009a) A comparative study of shear stresses in collagen-glycosaminoglycan and calcium phosphate scaffolds in bone tissue-engineering bioreactors. *Tissue Eng. Part A* **15**, 1141–9.

Jungreuthmayer C, Jaasma MJ, Al-Munajjed AA, Zanghellini J, Kelly DJ and O'Brien FJ (2009b) Deformation simulation of cells seeded on a collagen-GAG scaffold in a flow perfusion bioreactor using a sequential 3D CFD-elastostatics model. *Med. Eng. Phys.* **31**, 420–7.

Kanungo BP and Gibson LJ (2009) Density-property relationships in collagen-glycosaminoglycan scaffolds. *Acta Biomater* **6**, 344–53.

Karp JM, Shoichet MS and Davies JE (2003) Bone formation on two-dimensional poly(DL-lactide-co-glycolide) (PLGA) films and three-dimensional PLGA tissue engineering scaffolds in vitro. *J. Biomed. Mat. Res.* **64A**, 388–96.

Kikuchi M, Matsumoto HN, Yamada T, Koyama Y, Takakuda K and Tanaka J (2004) Glutaraldehyde cross-linked hydroxyapatite/collagen self-organized nanocomposites. *Biomaterials* **25**, 63–9.

Kim BS, Nikolovski J, Bonadio J, Smiley E and Mooney DJ (1999) Engineered smooth muscle tissues: Regulating cell phenotype with the scaffold. *Exper. Cell Res.* **251**, 318–28.

Kim HD and Valentini RF (2002) Retention and activity of BMP-2 in hyaluronic acid-based scaffolds in vitro. *J. Biomed. Mat. Res.* **59**, 573–84.

Kim YT, Haftel VK, Kumar S and Bellamkonda RV (2008) The role of aligned polymer fiber-based constructs in the bridging of long peripheral nerve gaps. *Biomaterials* **29**, 3117–27.

Kinner B, Capito RM and Spector M (2005) Regeneration of articular cartilage. *Adv. Biochem. Eng. Biotechnol.* **94**, 91–123.

Kisiday J, Jin M, Kurz B *et al.* (2002) Self-assembling peptide hydrogel fosters chondrocyte extracellular matrix production and cell division: implications for cartilage tissue repair. *Proc. Natl Acad. Sci. USA* **99**, 9996–10001.

Kisiday JD, Jin M, DiMicco MA, Kurz B and Grodzinsky AJ (2004) Effects of dynamic compressive loading on chondrocyte biosynthesis in self-assembling peptide scaffolds. *J. Biomech.* **37**, 595–604.

Kong HJ, Smith MK and Mooney DJ (2003) Designing alginate hydrogels to maintain viability of immobilized cells. *Biomaterials* **24**, 4023–9.

Kong HJ, Alsberg E, Kaigler D, Lee KY and Mooney DJ (2004a) Controlling degradation of hydrogels via the size of cross-linked junctions. *Adv. Mat.* **16**, 1917–21.

Kong HJ, Kaigler D, Kim K and Mooney DJ (2004b) Controlling rigidity and degradation of alginate hydrogels via molecular weight distribution. *Biomacromolecules* **5**, 1720–7.

Kong HJ, Liu J, Riddle K, Matsumoto T, Leach K and Mooney DJ (2005) Non-viral gene delivery regulated by stiffness of cell adhesion substrates. *Natural Mat.* **4**, 460–4.

Kong HJ, Boontheekul T and Mooney DJ (2006) Quantifying the relation between adhesion ligand-receptor bond formation and cell phenotype. *Proc. Natl Acad. Sci. USA* **103**, 18534–9.

Kong HJ, Hsiong S and Mooney DJ (2007a) Nanoscale cell adhesion ligand presentation regulates nonviral gene delivery and expression. *Nano. Lett.* **7**, 161–6.

Kong HJ, Kim CJ, Huebsch N, Weitz D and Mooney DJ (2007b) Noninvasive probing of the spatial organization of polymer chains in hydrogels using fluorescence resonance energy transfer (FRET). *J. Amer. Chem. Soc.* **129**, 4518–9.

Kong HJ, Kim ES, Huang YC and Mooney DJ (2008) Design of biodegradable hydrogel for the local and sustained delivery of angiogenic plasmid DNA. *Pharmaceutical Research Res.* **25**, 1230–8.

Kontulainen S, Sievanen H, Kannus P, Pasanen M and Vuori I (2003) Effect of long-term impact-loading on mass, size, and estimated strength of humerus and radius of female racquet-sports players: a peripheral quantitative computed tomography study between young and old starters and controls. *J. Bone Miner. Res.* **18**, 352–9.

Kuhne JH, Bartl R, Frisch B, Hammer C, Jansson V and Zimmer M (1994) Bone formation in coralline hydroxyapatite. Effects of pore size studied in rabbits. *Acta Orthop. Scand.* **65**, 246–52.

Kuntz RM and Saltzman WM (1997) Neutrophil motility in extracellular matrix gels: mesh size and adhesion affect speed of migration. *Biophys J.* **72**, 1472–80.

Larsson T, Aspden RM and Heinegard D (1991) Effects of mechanical load on cartilage matrix biosynthesis in vitro. *Matrix* **11**, 388–94.

Lauffenburger DA and Horwitz AF (1996) Cell migration: a physically integrated molecular process. *Cell* **84**, 359–69.

Lee J, Leonard M, Oliver T, Ishihara A and Jacobson K (1994) Traction forces generated by locomoting keratocytes. *J. Cell. Biol.* **127**, 1957–64.

Lee CR, Grodzinsky AJ and Spector M (2001a) The effects of crosslinking of collagen-glycosaminoglycan scaffolds on compressive stiffness, chondrocyte-mediated contraction, proliferation, and biosynthesis. *Biomaterials* **22**, 3145–54.

Lee JE, Park JC, Hwang YS, Kim JK, Kim JG and Sub H (2001b) Characterization of UV-irradiated dense/porous collagen membranes: morphology, enzymatic degradation, and mechanical properties. *Yonsei Med. J.* **42**, 172–9.

Lee CR, Grodzinsky AJ, Hsu H-P and Spector M (2003) Effects of a cultured autologous chondrocyte-seeded type II collagen scaffold on the healing of a chondral defect in a canine model. *J. Orthop. Res.* **21**, 272–81.

Li S, Kim M, Hu YL, Jalali S, Schlaepfer DD, Hunter T, Chien S and Shyy JY (1997) Fluid shear stress activation of focal adhesion kinase. Linking to mitogen-activated protein kinases. *J. Biol. Chem.* **272**, 30455–62.

Liu H and Roy K (2005) Biomimetic three-dimensional cultures significantly increase hematopoietic differentiation efficacy of embryonic stem cells. *Tissue Eng.* **11**, 319–30.

Liu H, Collins SF and Suggs LJ (2006a) Three-dimensional culture for expansion and differentiation of mouse embryonic stem cells. *Biomaterials* **27**, 6004–14.

Liu H, Lin J and Roy K (2006b) Effect of 3D scaffold and dynamic culture condition on the global gene expression profile of mouse embryonic stem cells. *Biomaterials* **27**, 5978–89.

Liu H, Slamovich EB and Webster TJ (2006c) Less harmful acidic degradation of poly(lacticcoglycolic acid) bone tissue engineering scaffolds through titania nanoparticle addition. *Int J. Nanomedicine* **1**, 541–5.

Liu Y, Ramanath HS and Wang DA (2008) Tendon tissue engineering using scaffold enhancing strategies. *Trends Biotechnol.* **26**, 201–9.

Lo C-M, Wang H-B, Dembo M and Wang Y-L (2000) Cell movement is guided by the rigidity of the substrate. *Biophys. J.* **79**, 144–52.

Lo CM, Buxton DB, Chua GC, Dembo M, Adelstein RS and Wang YL (2004) Nonmuscle myosin IIb is involved in the guidance of fibroblast migration. *Mol Biol. Cell* **15**, 982–9.

Logsdon CD, Simeone DM, Binkley C *et al.* (2003) Molecular profiling of pancreatic adenocarcinoma and chronic pancreatitis identifies multiple genes differentially regulated in pancreatic cancer. *Cancer Res.* **63**, 2649–57.

Louie LK, Yannas IV, Hsu HP and Spector M (1997) Healing of tendon defects implanted with a porous collagen-GAG matrix: histological evaluation. *Tissue Eng.* **3**, 187–95.

Lu LC, Zhu X, Valenzuela RG, Currier BL and Yaszemski MJ (2001) Biodegradable polymer scaffolds for cartilage tissue engineering. *Clin. Orthopaed. Related Res.* **391**, S251–70.

Lutolf MP, Lauer-Fields JL, Schmoekel HG *et al.* (2003) Synthetic matrix metalloproteinase-sensitive hydrogels for the conduction of tissue regeneration: engineering cell-invasion characteristics. *Proc. Natl Acad. Sci. USA* **100**, 5413–18.

Maaser K, Wolf K, Klein CE *et al.* (1999) Functional hierarchy of simultaneously expressed adhesion receptors: integrin alpha2beta1 but not CD44 mediates MV3 melanoma cell migration and matrix reorganization within three-dimensional hyaluronan-containing collagen matrices. *Mol Biol. Cell* **10**, 3067–79.

Mahoney MJ and Anseth KS (2006) Three-dimensional growth and function of neural tissue in degradable polyethylene glycol hydrogels. *Biomaterials* **27**, 2265–74.

Martens PJ, Bryant SJ and Anseth KS (2003) Tailoring the degradation of hydrogels formed from multivinyl poly(ethylene glycol) and poly(vinyl alcohol) macromers for cartilage tissue engineering. *Biomacromolecules* **4**, 283–92.

Masters KS, Shah DN, Leinwand LA and Anseth KS (2005) Crosslinked hyaluronan scaffolds as a biologically active carrier for valvular interstitial cells. *Biomaterials* **26**, 2517–25.

McGarry JG, Klein-Nulend J, Mullender MG and Prendergast PJ (2005) A comparison of strain and fluid shear stress in stimulating bone cell responses – a computational and experimental study. *FASEB J.* **19**, 482–4.

Mochitate K, Pawelek P and Grinnell F (1991) Stress relaxation of contracted collagen gels: disruption of actin filament bundles, release of cell surface fibronectin, and down-regulation of DNA and protein synthesis. *Exp. Cell Res.* **193**, 198–207.

Moffat KL, Kwei AS, Spalazzi JP, Doty SB, Levine WN and Lu HH (2009a) Novel nanofiber-based scaffold for rotator cuff repair and augmentation. *Tissue Eng. Part A* **15**, 115–26.

Moffat KL, Wang IN, Rodeo SA and Lu HH (2009b) Orthopedic interface tissue engineering for the biological fixation of soft tissue grafts. *Clin. Sports Med.* **28**, 157–76.

Mueller SM, Shortkroff S, Schneider TO, Breinan HA, Yannas IV and Spector M (1999) Meniscus cells seeded in type I and type II collagen-GAG matrices in vitro. *Biomaterials* **20**, 701–9.

Munevar S, Wang Y and Dembo M (2001a) Traction force microscopy of migrating normal and H-ras transformed 3T3 fibroblasts. *Biophys. J.* **80**, 1744–57.

Munevar S, Wang YL and Dembo M (2001b) Distinct roles of frontal and rear cell-substrate adhesions in fibroblast migration. *Mol. Biol. Cell* **12**, 3947–54.

Murphy CM, Haugh MG and O'Brien FJ (2009) The effect of mean pore size on cell attachment, proliferation and migration in collagen-glycosaminoglycan scaffolds for bone tissue engineering. *Biomaterials* **31**, 461–6.

Nakanishi Y, Chen G, Komuro H *et al.* (2003) Tissue-engineered urinary bladder wall using PLGA mesh-collagen hybrid scaffolds: a comparison study of collagen sponge and gel as a scaffold. *J. Pediatr. Surg.* **38**, 1781–4.

Nehrer S, Breinan HA, Ramappa A *et al.* (1997) Matrix collagen type and pore size influence behavior of seeded canine chondrocytes. *Biomaterials* **18**, 769–76.

Nuttelman CR, Mortisen DJ, Henry SM and Anseth KS (2001) Attachment of fibronectin to poly(vinyl alcohol) hydrogels promotes NIH3T3 cell adhesion, proliferation, and migration. *J. Biomed. Mat. Res.* **57**, 217–23.

Nuttelman CR, Henry SM and Anseth KS (2002) Synthesis and characterization of photo-crosslinkable, degradable poly(vinyl alcohol)-based tissue engineering scaffolds. *Biomaterials* **23**, 3617–26.

Nuttelman CR, Tripodi MC and Anseth KS (2004) In vitro osteogenic differentiation of human mesenchymal stem cells photoencapsulated in PEG hydrogels. *J. Biomed. Mat. Res. A* **68**, 773–82.

O'Brien FJ, Harley BA, Yannas IV and Gibson LJ (2005) The effect of pore size on cell adhesion in collagen-GAG scaffolds. *Biomaterials* **26**, 433–41.

O'Brien FJ, Harley BA, Waller MA, Yannas IV, Gibson LJ and Prendergast PJ (2007) The effect of pore size on permeability and cell attachment in collagen scaffolds for tissue engineering. *Technol. Health Care* **15**, 3–17.

Ohan MP, Weadock KS and Dunn MG (2002) Synergistic effects of glucose and ultraviolet irradiation on the physical properties of collagen. *J. Biomed. Mat. Res.* **60**, 384–91.

Olde Damink LHH, Dijkstra PJ, van Luyn MJA, Van Wachem PB, Nieuwenhuis P and Feijen J (1996a) Cross-linking of dermal sheep collagen using a water soluble carbodiimide. *Biomaterials* **17**, 765–73.

Olde Damink LHH, Dijkstra PJ, vanLuyn MJA, vanWachem PB, Nieuwenhuis P and Feijen J (1996b) In vitro degradation of dermal sheep collagen cross-linked using a water-soluble carbodiimide. *Biomaterials* **17**, 679–84.

Oliver T, Dembo M and Jacobson K (1995) Traction forces in locomoting cells. *Cell Motil. Cytoskeleton* **31**, 225–40.

Panorchan P, Lee JS, Kole TP, Tseng Y and Wirtz D (2006) Microrheology and ROCK signaling of human endothelial cells embedded in a 3D matrix. *Biophys. J.* **91**, 3499–507.

Pedersen JA and Swartz MA (2005) Mechanobiology in the third dimension. *Ann. Biomed. Eng* **33**, 1469–90.

Pedersen JA, Boschetti F and Swartz MA (2007) Effects of extracellular fiber architecture on cell membrane shear stress in a 3D fibrous matrix. *J. Biomech.* **40**, 1484–92.

Pek YS, Spector M, Yannas IV and Gibson LJ (2004) Degradation of a collagen-chondroitin-6-sulfate matrix by collagenase and by chondroitinase. *Biomaterials* **25**, 473–82.

Pelham RJ and Wang Y-L (1997) Cell locomotion and focal adhesions are regulated by substrate flexibility. *Proc. Natl Acad. Sci. USA* **9**, 13661–5.

Pelham RJ and Wang Y-L (1999) High resolution detection of mechanical forces exerted by locomoting fibroblasts on the substrate. *Mol. Biol. Cell* **10**, 935–45.

Petrie TA, Capadona JR, Reyes CD and Garcia AJ (2006) Integrin specificity and enhanced cellular activities associated with surfaces presenting a recombinant fibronectin fragment compared to RGD supports. *Biomaterials* **27**, 5459–70.

Peyton SR, Kim PD, Ghajar CM, Seliktar D and Putnam AJ (2008) The effects of matrix stiffness and RhoA on the phenotypic plasticity of smooth muscle cells in a 3-D biosynthetic hydrogel system. *Biomaterials* **29**, 2597–607.

Peyton SR and Putnam AJ (2005) Extracellular matrix rigidity governs smooth muscle cell motility in a biphasic fashion. *J. Cell Physiol.* **204**, 198–209.

Pieper JS, Oosterhof A, Dijkstra PJ, Veerkamp JH and van Kuppevelt TH (1999) Preparation and characterization of porous crosslinked collagneous matrices containing bioavailable chondroitin sulfate. *Biomaterials* **20**, 847–58.

Pieper JS, Hafmans T, Veerkamp JH and van Kuppevelt TH (2000) Development of tailor-made collagen-glycosaminoglycan matrices: EDC/NHS crosslinking, and ultrastructural aspects. *Biomaterials* **21**, 581–93.

Pirone DM, Liu WF, Ruiz SA *et al.* (2006) An inhibitory role for FAK in regulating proliferation: a link between limited adhesion and RhoA-ROCK signaling. *J. Cell Biol.* **174**, 277–88.

Pizzo AM, Kokini K, Vaughn LC, Waisner BZ and Voytik-Harbin SL (2005) Extracellular matrix (ECM) microstructural composition regulates local cell-ECM biomechanics and fundamental fibroblast behavior: a multidimensional perspective. *J. Appl. Physiol.* **98**, 1909–21.

Prajapati RT, Chavally-Mis B, Herbage D, Eastwood M and Brown RA (2000a) Mechanical loading regulates protease production by fibroblasts in three-dimensional collagen substrates. *Wound Repair Regen* **8**, 226–37.

Prajapati RT, Eastwood M and Brown RA (2000b) Duration and orientation of mechanical loads determine fibroblast cyto-mechanical activation: monitored by protease release. *Wound Repair Regen* **8**, 238–46.

Prendergast PJ, Huiskes R and Soballe K (1997) Biophysical stimuli on cells during tissue differentiation at implant interfaces. *J. Biomech.* **30**, 539–48.

Raeber GP, Lutolf MP and Hubbell JA (2005) Molecularly engineered PEG hydrogels: a novel model system for proteolytically mediated cell migration. *Biophys. J.* **89**, 1374–88.

Rehfeldt F, Engler AJ, Eckhardt A, Ahmed F and Discher DE (2007) Cell responses to the mechanochemical microenvironment – implications for regenerative medicine and drug delivery. *Adv. Drug Deliv. Rev.* **59**, 1329–39.

Rhee S and Grinnell F (2007) Fibroblast mechanics in 3D collagen matrices. *Adv. Drug Deliv. Rev.* **59**, 1299–305.

Rhee S, Jiang H, Ho CH and Grinnell F (2007) Microtubule function in fibroblast spreading is modulated according to the tension state of cell-matrix interactions. *Proc. Natl Acad. Sci. USA* **104**, 5425–30.

Rowley JA and Mooney DJ (2002) Alginate type and RGD density control myoblast phenotype. *J. Biomed. Mat. Res.* **60**, 217–23.

Roy P, Petroll WM, Cavanagh HD, Chuong CJ and Jster JV (1997) An in vitro force measurement assay to study the early mechanical interaction between corneal fibroblasts and collagen matrix. *Exper. Cell Res.* **232**, 106–17.

Roy P, Petroll WM, Chuong CJ, Cavanagh HD and Jester JV (1999) Effect of cell migration on the maintenance of tension on a collagen matrix. *Ann. Biomed. Eng.* **27**, 721–30.

Rydholm AE, Held NL, Benoit DS, Bowman CN and Anseth KS (2008) Modifying network chemistry in thiol-acrylate photopolymers through postpolymerization functionalization to control cell-material interactions. *J. Biomed. Mat. Res. A* **86**, 23–30.

Ryu W, Min SW, Hammerick KE *et al.* (2007) The construction of three-dimensional microfluidic scaffolds of biodegradable polymers by solvent vapor based bonding of micro-molded layers. *Biomaterials* **28**, 1174–84.

Saad L and Spector M (2004) Effects of collagen type on the behavior of adult canine annulus fibrosus cells in collagen-glycosaminoglycan scaffolds. *J. Biomed. Mat. Res. A* **71**, 233–41.

Sachlos E and Czernuszka JT (2003) Making tissue engineering scaffolds work. Review: the application of solid freeform fabrication technology to the production of tissue engineering scaffolds. *Eur Cell Mater* **5**, 29–39; discussion 39–40.

Samuel RE, Lee CR, Ghivizzani SC *et al.* (2002) Delivery of plasmid DNA to articular chondrocytes via novel collagen-glycosaminoglycan matrices. *Human Gene Therapy* **13**, 791–802.

Sannino A, Harley BA, Hastings AZ and Yannas IV (2005) A novel technique to fabricate cylindrical and tubular structures with a patterned porosity. International (PCT) Patent Application.

Satcher R, Dewey CF, Jr. and Hartwig JH (1997) Mechanical remodeling of the endothelial surface and actin cytoskeleton induced by fluid flow. *Microcirculation* **4**, 439–53.

Schlaepfer DD and Hunter T (1998) Integrin signalling and tyrosine phosphorylation: just the FAKs? *Trends Cell Biol.* **8**, 151–7.

Schmidt O, Mizrahi J, Elisseeff J and Seliktar D (2006) Immobilized fibrinogen in PEG hydrogels does not improve chondrocyte-mediated matrix deposition in response to mechanical stimulation. *Biotechnol. Bioeng.* **95**, 1061–9.

Schulz-Torres D, Freyman TM, Yannas IV and Spector M (2000) Tendon cell contraction of collagen-GAG matrices in vitro: effect of crosslinking. *Biomaterials* **21**, 1607–19.

Seliktar D, Zisch AH, Lutolf MP, Wrana JL and Hubbell JA (2004) MMP-2 sensitive, VEGF-bearing bioactive hydrogels for promotion of vascular healing. *J. Biomed. Mat. Res. A* **68**, 704–16.

Sethi KK, Yannas IV, Mudera V, Eastwood M, McFarland C and Brown RA (2002) Evidence for sequential utilization of fibronectin, vitronectin, and collagen during fibroblast-mediated collagen contraction. *Wound Repair Regen.* **10**, 397–408.

Shafritz TA, Rosenberg LC and Yannas IV (1994) Specific effects of glycosaminoglycans in an analog of extracellular matrix that delays wound contraction and induces regeneration. *Wound Repair Regen.* **2**, 270–6.

Shah DN, Recktenwall-Work SM and Anseth KS (2008) The effect of bioactive hydrogels on the secretion of extracellular matrix molecules by valvular interstitial cells. *Biomaterials* **29**, 2060–72.

Shapira-Schweitzer K and Seliktar D (2007) Matrix stiffness affects spontaneous contraction of cardiomyocytes cultured within a PEGylated fibrinogen biomaterial. *Acta Biomater.* **3**, 33–41.

Silva EA, Kim ES, Kong HJ and Mooney DJ (2008) Material-based deployment enhances efficacy of endothelial progenitor cells. *Proc. Natl Acad. Sci. USA* **105**, 14347–52.

Silver FH, Yannas IV and Salzman EW (1974) Glycosaminoglycan inhibition of collagen-induced platelet aggregation. *Thrombic Res.* **13**, 267–77.

Spalazzi JP, Doty SB, Moffat KL, Levine WN and Lu HH (2006) Development of controlled matrix heterogeneity on a triphasic scaffold for orthopedic interface tissue engineering. *Tissue Eng.* **12**, 3497–508.

Steinert AF, Palmer GD, Capito R *et al.* (2007) Genetically enhanced engineering of meniscus tissue using ex vivo delivery of transforming growth factor-beta1 complementary deoxyribonucleic acid. *Tissue Eng* 13, 2227–37.

Stops AJ, McMahon LA, O'Mahoney D, Prendergast PJ and McHugh PE (2008) A finite element prediction of strain on cells in a highly porous collagen-glycosaminoglycan scaffold. *J. Biomech. Eng.* **130**, 061001 (11 pages).

Swann AC and Seedhom BB (1993) The stiffness of normal articular cartilage and the predominant acting stress levels: Implications for the aetiology of osteoarthrosis. *Brit. J. Rheumatol.* **32**, 16–25.

Tabe Y, Jin L, Tsutsumi-Ishii Y *et al.* (2007) Activation of integrin-linked kinase is a critical prosurvival pathway induced in leukemic cells by bone marrow-derived stromal cells. *Cancer Res.* **67**, 684–94.

Takai E, Landesberg R, Katz RW, Hung CT and Guo XE (2006) Substrate modulation of osteoblast adhesion strength, focal adhesion kinase activation, and responsiveness to mechanical stimuli. *Mol. Cell Biomech.* **3**, 1–12.

Tamariz E and Grinnell F (2002) Modulation of fibroblast morphology and adhesion during collagen matrix remodeling. *Mol. Biol. Cell* **13**, 3915–29.

Tanzer ML (2006) Current concepts of extracellular matrix. *J. Orthop. Sci.* **11**, 326–31.

Taqvi S and Roy K (2006) Influence of scaffold physical properties and stromal cell coculture on hematopoietic differentiation of mouse embryonic stem cells. *Biomaterials* **27**, 6024–31.

Tessmar JK and Gopferich AM (2007) Customized PEG-derived copolymers for tissue-engineering applications. *Macromol. Biosci.* **7**, 23–39.

Thery M, Racine V, Pepin A *et al.* (2005) The extracellular matrix guides the orientation of the cell division axis. *Nat. Cell Biol.* **7**, 947–53.

Thompson MT, Berg MC, Tobias IS, Rubner MF and Van Vliet KJ (2005) Tuning compliance of nanoscale polyelectrolyte multilayers to modulate cell adhesion. *Biomaterials* **26**, 6836–45.

Torres DS, Freyman TM, Yannas IV and Spector M (2000) Tendon cell contraction of collagen-GAG matrices in vitro: effect of cross-linking. *Biomaterials* **21**, 1607–19.

van Tienen TG, Heijkants RGJC, Buma P, de Groot JH, Pennings AJ and Veth RPH (2002) Tissue ingrowth and degradation of two biodegradable porous polymers with different porosities and pore sizes. *Biomaterials* **23**, 1731–8.

Vance J, Galley S, Liu DF and Donahue SW (2005) Mechanical stimulation of MC3T3 osteoblastic cells in a bone tissue-engineering bioreactor enhances prostaglandin E2 release. *Tissue Eng* **11**, 1832–9.

Vanwanseele B, Lucchinetti E and Stussi E (2002) The effects of immobilization on the characteristics of articular cartilage: current concepts and future directions. *Osteoarth. Cartil.* **10**, 408–19.

Vasita R, Shanmugam K and Katti DS (2008) Improved biomaterials for tissue engineering applications: surface modification of polymers. *Curr. Topics Med. Chem.* **8**, 341–53.

Veilleux NH, Yannas IV and Spector M (2004) Effect of passage number and collagen type on the proliferative, biosynthetic, and contractile activity of adult canine articular chondrocytes in type I and II collagen-glycosaminoglycan matrices in vitro. *Tissue Eng* **10**, 119–27.

Vickers SM, Squitieri LS and Spector M (2006) The effects of cross-linking type II collagen-GAG scaffolds on chondrogenesis in vitro: dynamic pore reduction promotes cartilage formation. *Tissue Eng.* **12**, 1345–55.

Vogel V and Sheetz M (2006) Local force and geometry sensing regulate cell functions. *Nat. Rev. Mol. Cell Biol.* **7**, 265–75.

Wang F, Li Z, Tamama K, Sen CK and Guan J (2009) Fabrication and characterization of pro-survival growth factor releasing, anisotropic scaffolds for enhanced mesenchymal stem cell survival/growth and orientation. *Biomacromolecules* **10**, 2609–18.

Wang H-B, Dembo M and Wang Y-L (2000) Substrate flexibility regulates growth and apoptosis of normal but not transformed cells. *Amer. J. Physiol. Cell Physiol.* **279**, C1345–50.

Wang H-B, Dembo M, Hanks SK and Wang Y-L (2001) Focal adhesion kinase is involved in mechanosensing during fibroblast migration. *Proc. Natl. Acad. Sci. USA* **98**, 11295–300.

Wang N, Butler JP and Ingber D (1993) Mechanotransduction across the cell surface and through the cytoskeleton. *Science* **260**, 1124–7.

Wang N, Tolic-Norrelykke IM, Chen JX *et al.* (2002) Cell prestress. I. Stiffness and prestress are closely associated in adherent contractile cells. *Amer. J. Phys. Cell Phys.* **282**, C606–16.

Wang S, Cui W and Bei J (2005) Bulk and surface modifications of polylactide. *Anal. Bioanal. Chem.* **381**, 547–56.

Wang Y, McNamara LM, Schaffler MB and Weinbaum S (2007) A model for the role of integrins in flow induced mechanotransduction in osteocytes. *Proc. Natl Acad. Sci. USA* **104**, 15941–6.

Webb AR, Yang J and Ameer GA (2004) Biodegradable polyester elastomers in tissue engineering. *Expert Opin. Biol. Ther.* **4**, 801–12.

Weber LM, Hayda KN and Anseth KS (2008) Cell-matrix interactions improve beta-cell survival and insulin secretion in three-dimensional culture. *Tissue Eng. Part A* **14**, 1959–68.

Wei Z, Deshpande VS, McMeeking RM and Evans AG (2008) Analysis and interpretation of stress fiber organization in cells subject to cyclic stretch. *J. Biomech. Eng.* **130**, 031009.

Willerth SM, Arendas KJ, Gottlieb DI and Sakiyama-Elbert SE (2006) Optimization of fibrin scaffolds for differentiation of murine embryonic stem cells into neural lineage cells. *Biomaterials* **27**, 5990–6003.

Williams LR, Longo FM, Powell HC, Lundborg G and Varon S (1983) Spatial-temporal progress of peripheral nerve regeneration within a silicone chamber: parameters for a bioassay. *J. Comp. Neurol.* **218**, 460–70.

Winer JP, Oake S and Janmey PA (2009) Non-linear elasticity of extracellular matrices enables contractile cells to communicate local position and orientation. *PLoS ONE* **4**, e6382.

Wolf K, Alexander S, Schacht V *et al.* (2009) Collagen-based cell migration models in vitro and in vivo. *Semin. Cell Dev. Biol.* **20**, 931–41.

Wrobel LK, Fray TR, Molloy JE, Adams JJ, Armitage MP and Sparrow JC (2002) Contractility of single human dermal myofibroblasts and fibroblasts. *Cell Motil. Cytoskeleton* **52**, 82–90.

Yannas IV and Tobolsky AV (1967) Cross-linking of gelatine by dehydration. *Nature* **215**, 509–10.

Yannas IV, Burke JF, Huang C and Gordon PL (1975a) Correlation of in vivo collagen degradation rate with in vitro measurements. *J. Biomed. Mat. Res.* **9**, 623–8.

Yannas IV, Burke JF, Huang C and Gordon PL (1975b) Suppression of in vivo degradability and of immunogenicity of collagen by reaction with glycosaminoglycans. *Polym. Prepr. Amer. Chem. Soc.* **16**, 209–14.

Yannas IV, Burke JF, Gordon PL, Huang C and Rubenstein RH (1980) Design of an artificial skin II: control of chemical-composition. *J. Biomed. Mat. Res.* **14**, 107–32.

Yannas IV, Lee E, Orgill DP, Skrabut EM and Murphy GF (1989) Synthesis and characterization of a model extracellular matrix that induces partial regeneration of adult mammalian skin. *Proc. Natl. Acad. Sci. USA* **86**, 933–7.

Yannas IV (1992) Tissue regeneration by use of collagen–glycosaminoglycan copolymers. *Clinic. Mater.* **9**, 179–87.

Yannas IV (2000) In vivo synthesis of tissues and organs. In *Principles of Tissue Engineering*, 2nd edn., ed. Lanza RP, Langer R and Vacanti J. San Diego, CA: Academic Press.

Yannas IV (2001) *Tissue and Organ Regeneration in Adults*. New York: Springer-Verlag.

Yannas IV (2005) Facts and theories of induced organ regeneration. *Adv. Biochem. Eng. Biotechnol.* **93**, 1–31.

Zaleskas JM, Kinner B, Freyman TM, Yannas IV, Gibson LJ and Spector M (2004) Contractile forces generated by articular chondrocytes in collagen–glycosaminoglycan matrices. *Biomaterials* **25**, 1299–308.

Zaman MH, Kamm RD, Matsudaira P and Lauffenburger DA (2005) Computational model for cell migration in three-dimensional matrices. *Biophys. J.* **89**, 1389–97.

Zaman MH, Trapani LM, Sieminski AL *et al.* (2006) Migration of tumor cells in 3D matrices is governed by matrix stiffness along with cell-matrix adhesion and proteolysis. *Proc. Natl Acad. Sci. USA* **103**, 10889–94.

Zaman MH (2007) Understanding the molecular basis for differential binding of integrins to collagen and gelatin. *Biophys. J.* **92**, L17–19.

Zaman MH, Matsudaira P and Lauffenburger DA (2007) Understanding effects of matrix protease and matrix organization on directional persistence and translational speed in three-dimensional cell migration. *Ann. Biomed. Eng.* **35**, 91–100.

Index

Printed in the United States
by Baker & Taylor Publisher Services